Research Notes in Mathematics

Submission of proposals for consideration
Suggestions for publication, in the form of outlines and representative samples, are invited by the editorial board for assessment. Intending authors should contact either the main editor or another member of the editorial board, citing the relevant AMS subject classifications. Refereeing is by members of the board and other mathematical authorities in the topic concerned, located throughout the world.

Preparation of accepted manuscripts
On acceptance of a proposal, the publisher will supply full instructions for the preparation of manuscripts in a form suitable for direct photo-lithographic reproduction. Specially printed grid sheets are provided and a contribution is offered by the publisher towards the cost of typing.

Illustrations should be prepared by the authors, ready for direct reproduction without further improvement. The use of hand-drawn symbols should be avoided wherever possible, in order to maintain maximum clarity of the text.

The publisher will be pleased to give any guidance necessary during the preparation of a typescript, and will be happy to answer any queries.

Important note
In order to avoid later retyping, intending authors are strongly urged not to begin final preparation of a typescript before receiving the publisher's guidelines and special paper. In this way it is hoped to preserve the uniform appearance of the series.

Titles in this series

1. Improperly posed boundary value problems
A Carasso and A P Stone
2. Lie algebras generated by finite dimensional ideals
I N Stewart
3. Bifurcation problems in nonlinear elasticity
R W Dickey
4. Partial differential equations in the complex domain
D L Colton
5. Quasilinear hyperbolic systems and waves
A Jeffrey
6. Solution of boundary value problems by the method of integral operators
D L Colton
7. Taylor expansions and catastrophes
T Poston and I N Stewart
8. Function theoretic methods in differential equations
R P Gilbert and R J Weinacht
9. Differential topology with a view to applications
D R J Chillingworth
10. Characteristic classes of foliations
H V Pittie
11. Stochastic integration and generalized martingales
A U Kussmaul
12. Zeta-functions: An introduction to algebraic geometry
A D Thomas
13. Explicit *a priori* inequalities with applications to boundary value problems
V G Sigillito
14. Nonlinear diffusion
W E Fitzgibbon III and H F Walker
15. Unsolved problems concerning lattice points
J Hammer
16. Edge-colourings of graphs
S Fiorini and R J Wilson
17. Nonlinear analysis and mechanics: Heriot-Watt Symposium Volume I
R J Knops
18. Actions of fine abelian groups
C Kosniowski
19. Closed graph theorems and webbed spaces
M De Wilde
20. Singular perturbation techniques applied to integro-differential equations
H Grabmüller
21. Retarded functional differential equations: A global point of view
S E A Mohammed
22. Multiparameter spectral theory in Hilbert space
B D Sleeman
24. Mathematical modelling techniques
R Aris
25. Singular points of smooth mappings
C G Gibson
26. Nonlinear evolution equations solvable by the spectral transform
F Calogero
27. Nonlinear analysis and mechanics: Heriot-Watt Symposium Volume II
R J Knops
28. Constructive functional analysis
D S Bridges
29. Elongational flows: Aspects of the behaviour of model elasticoviscous fluids
C J S Petrie
30. Nonlinear analysis and mechanics: Heriot-Watt Symposium Volume III
R J Knops
31. Fractional calculus and integral transforms of generalized functions
A C McBride
32. Complex manifold techniques in theoretical physics
D E Lerner and P D Sommers
33. Hilbert's third problem: scissors congruence
C-H Sah
34. Graph theory and combinatorics
R J Wilson
35. The Tricomi equation with applications to the theory of plane transonic flow
A R Manwell
36. Abstract differential equations
S D Zaidman
37. Advances in twistor theory
L P Hughston and R S Ward
38. Operator theory and functional analysis
I Erdelyi
39. Nonlinear analysis and mechanics: Heriot-Watt Symposium Volume IV
R J Knops
40. Singular systems of differential equations
S L Campbell
41. N-dimensional crystallography
R L E Schwarzenberger
42. Nonlinear partial differential equations in physical problems
D Graffi
43. Shifts and periodicity for right invertible operators
D Przeworska-Rolewicz
44. Rings with chain conditions
A W Chatters and C R Hajarnavis
45. Moduli, deformations and classifications of compact complex manifolds
D Sundararaman
46. Nonlinear problems of analysis in geometry and mechanics
M Atteia, D Bancel and I Gumowski
47. Algorithmic methods in optimal control
W A Gruver and E Sachs
48. Abstract Cauchy problems and functional differential equations
F Kappel and W Schappacher
49. Sequence spaces
W H Ruckle
50. Recent contributions to nonlinear partial differential equations
H Berestycki and H Brezis
51. Subnormal operators
J B Conway
52. Wave propagation in viscoelastic media
F Mainardi
53. Nonlinear partial differential equations and their applications: Collège de France Seminar. Volume I
H Brezis and J L Lions
54. Geometry of Coxeter groups
H Hiller
55. Cusps of Gauss mappings
T Banchoff, T Gaffney and C McCrory

56 An approach to algebraic K-theory
A J Berrick
57 Convex analysis and optimization
J-P Aubin and R B Vintner
58 Convex analysis with applications in the differentiation of convex functions
J R Giles
59 Weak and variational methods for moving boundary problems
C M Elliott and J R Ockendon
60 Nonlinear partial differential equations and their applications: Collège de France Seminar. Volume II
H Brezis and J L Lions
61 Singular systems of differential equations II
S L Campbell
62 Rates of convergence in the central limit theorem
Peter Hall
63 Solution of differential equations by means of one-parameter groups
J M Hill
64 Hankel operators on Hilbert space
S C Power
65 Schrödinger-type operators with continuous spectra
M S P Eastham and H Kalf
66 Recent applications of generalized inverses
S L Campbell
67 Riesz and Fredholm theory in Banach algebra
B A Barnes, G J Murphy, M R F Smyth and T T West
68 Evolution equations and their applications
F Kappel and W Schappacher
69 Generalized solutions of Hamilton-Jacobi equations
P L Lions
70 Nonlinear partial differential equations and their applications: Collège de France Seminar. Volume III
H Brezis and J L Lions
71 Spectral theory and wave operators for the Schrödinger equation
A M Berthier
72 Approximation of Hilbert space operators I
D A Herrero
73 Vector valued Nevanlinna Theory
H J W Ziegler
74 Instability, nonexistence and weighted energy methods in fluid dynamics and related theories
B Straughan
75 Local bifurcation and symmetry
A Vanderbauwhede
76 Clifford analysis
F Brackx, R Delanghe and F Sommen
77 Nonlinear equivalence, reduction of PDEs to ODEs and fast convergent numerical methods
E E Rosinger
78 Free boundary problems, theory and applications. Volume I
A Fasano and M Primicerio
79 Free boundary problems, theory and applications. Volume II
A Fasano and M Primicerio
80 Symplectic geometry
A Crumeyrolle and J Grifone
81 An algorithmic analysis of a communication model with retransmission of flawed messages
D M Lucantoni
82 Geometric games and their applications
W H Ruckle
83 Additive groups of rings
S Feigelstock
84 Nonlinear partial differential equations and their applications: Collège de France Seminar. Volume IV
H Brezis and J L Lions
85 Multiplicative functionals on topological algebras
T Husain
86 Hamilton-Jacobi equations in Hilbert spaces
V Barbu and G Da Prato
87 Harmonic maps with symmetry, harmonic morphisms and deformations of metrics
P Baird
88 Similarity solutions of nonlinear partial differential equations
L Dresner
89 Contributions to nonlinear partial differential equations
C Bardos, A Damlamian, J I Díaz and J Hernández
90 Banach and Hilbert spaces of vector-valued functions
J Burbea and P Masani
91 Control and observation of neutral systems
D Salamon
92 Banach bundles, Banach modules and automorphisms of C*-algebras
M J Dupré and R M Gillette
93 Nonlinear partial differential equations and their applications: Collège de France Seminar. Volume V
H Brezis and J L Lions

Contributions to nonlinear partial differential equations

C Bardos
Université de Paris-Nord
A Damlamian
Centre de Mathématiques, Ecole Polytechnique Palaiseau
J I Díaz
Universidad Complutense de Madrid
&
J Hernández
Universidad Autónoma, Madrid (Editors)

Contributions to nonlinear partial differential equations

Pitman Advanced Publishing Program
BOSTON · LONDON · MELBOURNE

PITMAN BOOKS LIMITED
128 Long Acre, London WC2E 9AN

PITMAN PUBLISHING INC
1020 Plain Street, Marshfield, Massachusetts 02050

Associated Companies
Pitman Publishing Pty Ltd, Melbourne
Pitman Publishing New Zealand Ltd, Wellington
Copp Clark Pitman, Toronto

First published 1983

AMS Subject Classifications: (main) 35J65, 35K60, 35L70
(subsidiary) 47H20, 49A29

Library of Congress Cataloging in Publication Data
Main entry under title:

Contributions to nonlinear partial differential equations.

(Research notes in mathematics; 89)
"Pitman advanced publishing program."
Proceedings of an international meeting held in Madrid on Dec. 14–17, 1981.
Includes bibliographical references.
1. Differential equations, Partial—Addresses, essays, lectures. 2. Nonlinear operators—Addresses, essays, lectures. I. Bardos, C. (Claude), 1940–
II. Series.
QA377.C763 1983 515.3′53 83-8112
ISBN 0-273-08595-6

British Library Cataloguing in Publication Data

Contributions to nonlinear partial differential equations.
—(Research notes in mathematics; 89)
1. Differential equations, Nonlinear—
Congresses
I. Bardos, C. II. Series
515.3′55 QA370

ISBN 0-273-08595-6

Reproduced and printed by photolithography
in Great Britain by Biddles Ltd, Guildford

Contents

Preface

This volume contains the proceedings of an International meeting devoted to non-linear partial differential equations. It was organized by the Universidad Autónoma and the Universidad Complutense of Madrid and held in Madrid on 14-17 December 1981.

Most of the communications presented here deal with genuine non-linear problems and they give a comprehensive survey of recent results and applications to realistic problems in physics or engineering (e.g. porous media, combustion, plasma physics, non-linear waves,...). Some communications also deal with linear problems in so far as they are actually motivated by non-linear theory (e.g. Schrödinger equation in L^p-spaces or product of distributions).

The Spanish organizers would like to point out that such a meeting would have been impossible only ten years ago. The progress in Spanish mathematics to the present level is in no small part due to the active collaboration of the French School. Among many they would like to express their gratitude to Professors J. L. Lions and H. Brézis.

The organization of the meeting was made possible through the cooperation programme between the French and Spanish Governments which supported the 'actions intégrées' involving both Madrid Universities and the Universities of Paris 6 and Paris 13. Within this programme the Cultural and Scientific offices of the French Embassy in Madrid have been extremely helpful and we would like to thank them here.

We would also like to thank the Asociación Matématica Española and the Project no 4285-79 of the Universidad Complutense for their support.

J. Carrillo and G. Díaz took a very important part in the day-to-day organization of the meeting, for which we express our appreciation.

Finally we express our gratitude to Pitman Books Ltd for their technical support in the publication and distribution of this volume.

Madrid and Paris, September 1982

List of participants

Alvarez, J. (Madrid)
Attouch, H. (Paris)
Bachelot, A. (Bordeaux)
Balabane, M. (Paris)
Bailet-Intissar, J. (Rabat)
Baillon, J.B. (Villeurbanne)
Bardos, C. (Paris)
Barthelemy, L. (Besançon)
Benachour, S. (Argel)
Bénilan, Ph. (Besançon)
Bernis, F. (Barcelona)
Bidaut-Véron, M.F. (Tours)
Brézis, H. (Paris)
Carrillo, J. (Madrid)
Casal, A. (Madrid)
Casas, A. (Madrid)
Catte, F. (Besançon)
Cazenave, T. (Paris)
Coron, J.M. (Paris)
Damlamian, A. (Paris)
Díaz, G. (Madrid)
Díaz, J.I. (Madrid)
Dou, A. (Madrid)
Esteban, J.R. (Madrid)
Esteban, M.J. (Paris)
Fleckinger, J. (Toulouse)
Ghidouche, A. (Paris)
Gutiérrez, A. (Madrid)
Hanouzet, B. (Bordeaux)
Hernández, J. (Madrid)
Herrero, M.A. (Madrid)
Lasry, J.M. (Paris)
Lions, P.L. (Paris)
Louro, B. (Lisboa)
Peral, I. (Madrid)
Pierre, M. (Grenoble)
Point, N. (Paris)
Rodrigues, J.F. (Lisboa)
Sanches, L. (Lisboa)
Schatzman, M. (Paris)
Vázquez, J.L. (Madrid)
Vegas, J.M. (Madrid)
Véron, L. (Tours)
Yarur, C. (Madrid)

J AGUIRRE & I PERAL

Existence of periodic solutions for a class of nonlinear equations

1. INTRODUCTION

In this note we prove the existence of nontrivial periodic solutions in L^p for the equation

$$u_{tt} + \Delta_n^{2m} u + |u|^{p-2} u = 0 \tag{1}$$

where $u = u(x,t)$, $x \in \mathbb{R}^n$, Δ_n is the n-dimensional Laplacian, m is an integer greater than or equal to n/2 and $p > 1$. For this we show that the linear part of the equation has an inverse that is compact from L^p into $L^{p'}$, $1 < p < \infty$, which allows us to apply the variational methods developed in [1].

We also consider the corresponding problem for the regularized two-dimensional wave operator

$$u_{tt} - \Delta_2 u + \varepsilon^2 \Delta_2^2 u. \tag{2}$$

We find that when $\varepsilon = 1/n$, n an odd integer, the same method can be applied, but if $\varepsilon = 1/n$, n even, the corresponding inverse operator is not compact, and the method fails.

2. EXISTENCE OF NONTRIVIAL SOLUTION FOR PROBLEM (1)

Let $n > 1$ and m be integers, $m \geqslant n/2$, $p > 1$, $\Omega = (0,\pi)^n$ and $Q = \Omega \times (0,2\pi)$. Consider the problem

$$\left.\begin{aligned} & u_{tt} + \Delta_n^{2m} u + |u|^{p-2} u = 0 \text{ on } Q \\ & \Delta_n^k u(x,t)\Big|_{x\in\partial\Omega} = 0, \quad 0 < t < 2\pi, \quad 0 \leqslant k < 2m \\ & u(x,0) = u(x,2\pi). \end{aligned}\right\} \tag{3}$$

THEOREM 1. There exist nontrivial (weak) solutions $u \in L^p$ of (3).

The proof of the theorem starts with the study of the linear part of the equation. Let $A_0 = \partial^2/\partial t^2 + \Delta_n^{2m}$ be defined on $L^2(Q)$ with domain

$$D(A_0) = \{u \in C^\infty(\bar{Q}) : \Delta_n^k u\big|_{\partial\Omega} = 0, \quad 0 \leq k < 2m,\ u(x,0) = u(x,\pi)\}.$$

It admits a self-adjoint extension A with domain

$$D(A) = \{u \in L^2(Q) : Au \in L^2\}$$

and whose spectral decomposition is given by

$$Au = \sum_{(j,k)\in\mathbb{N}^n\times\mathbf{Z}} (|j|^{4m} - k^2)u_{j,k}\sin j_1x_1 \ldots \sin j_nx_n e^{ikt}$$

if

$$u = \sum_{(j,k)\in\mathbb{N}^n\times\mathbf{Z}} u_{j,k} \sin j_1x_1 \ldots \sin j_nx_n\, e^{ikt},$$

where $j = (j_1,\ldots,j_n)$ and $|j|$ denotes the Euclidean length of j.

The kernel and range of A are given by

$$N(A) = \{u \in L^2(Q) : u_{j,k} = 0 \text{ if } |j|^{4m} \neq k^2\}$$

$$R(A) = \{u \in L^2(Q) : u_{j,k} = 0 \text{ if } |j|^{4m} = k^2\}.$$

For $f \in R$ we define

$$\tilde{K}f = \sum_{|j|^{4m}\neq k^2} (|j|^{4m} - k^2)^{-1} f_{j,k}\sin j_1x_1 \ldots \sin j_nx_n\, e^{ikt}.$$

If P is the projection onto R, let $K = \tilde{K}P$. Given $f \in L^2(Q)$, it is easily seen that Kf is the restriction to Q of $k_*\tilde{f}$, where $\tilde{f}$ is the extension to $(0,2\pi)^n \times (0,2\pi) = \tilde{Q}$ of f with same sine series, and

$$k(x,t) = \sum_{\substack{(j,k)\in\mathbf{Z}^n\times\mathbf{Z} \\ |j|^{4m}\neq k^2}} (|j|^{4m} - k^2)^{-1} e^{ij.x}e^{ikt}.$$

<u>LEMMA 2</u>. $k \in L^p(\tilde{Q})$, $1 \leq p < \infty$.

<u>Proof</u>. Let $q > 1$; then

$$\sum_{|j|^{4m}\neq k^2} \| j|^{4m} - k^2|^{-q} \leq \sum_j |j|^{-2mq} \sum_{|k|\neq |j|^{2m}} \||j|^{2m} - |k||^{-q}$$

$$\leq C \sum_j |j|^{-2mq} < \infty$$

whenever $2mq > n$.

The lemma follows now from the Hausdorff-Young theorem.

LEMMA 3. K is compact from $L^p(Q)$ into $L^{p'}(Q)$, $1 < p < \infty$, $\frac{1}{p} + \frac{1}{p'} = 1$.

Proof. It is a consequence of Lemma 2 and the following result:

If (X,μ) is a σ-finite measure space and $k(x,y) \in L^{p'}(d\mu \otimes d\mu)$, then the integral operator $TF(x) = \int_X k(x,y)f(y)d\mu(y)$ is compact from L^p into $L^{p'}$.

Once we have the compactness of K, all the machinery developed in [1] comes into play, proving Theorem 1.

3. THE REGULARIZED WAVE OPERATOR IN TWO SPATIAL DIMENSIONS

Fix the spatial dimension 2 and consider the following problem

$$\left.\begin{aligned} &u_{tt} - \Delta u + \frac{1}{n^2}\Delta^2 u + |u|^{p-2}u = 0, \text{ on } Q,\ p > 1,\ n \in \mathbb{N} \\ &u|_{\partial\Omega} = \Delta u|_{\partial\Omega} = 0 \\ &u(x,y,0) = u(x,y,2\pi). \end{aligned}\right\} \quad (4)$$

We prove the existence of nontrivial periodic solutions of (4) for odd n. When n is even, we show that the methods of the second paragraph do not apply.

THEOREM 4. Let $q > 1$, $n \in \mathbb{N}$ and

$$A_n = \{(j,k,m) \in \mathbb{N}^2 \times \mathbb{Z} : j^2 + k^2 + \left(\frac{j^2 + k^2}{n}\right)^2 \neq m^2\}.$$

Define

$$S = \sum_{A_n} |j^2 + k^2 + \left(\frac{j^2 + k^2}{n}\right)^2 - m^2|^{-q}.$$

Then

(a) If n is even, $S = \infty$

(b) If n is odd, $S \leqslant Cn^{2q+4}$.

Proof. (a) It is enough to sum over the indices of the form

$$(j,k,m) = (n\ell,\ 0,\ n\ell^2 + \tfrac{n}{2}),\quad \ell = 1,2,3,\ldots .$$

(b) We give first some results needed for the proof.

PROPOSITION 5. Let $r(\ell)$ be the number of solutions in the integers of the equation $j^2 + k^2 = \ell$. Then for all $\delta > 0$, $r(\ell) = O(\ell^{\delta})$.

This is just a classical result of number theory (see [2]).

PROPOSITION 6. Let $q > 1$, a^2 not an integer and $d(a) = \inf\limits_{n\in\mathbb{N}} |a-n|$. Then

(a) $$\sum_{m=0}^{\infty} |m^2 - a^2|^{-q} \leqslant \left(C + \frac{2}{d(a)^q}\right)a^{-q}$$

(b) $$\sum_{m=0}^{\infty} |m^2 - a^2|^{-q} \leqslant Ca^{-q} + \frac{2}{d(a^2)^q} .$$

The proof is just an easy computation.

PROPOSITION 7. Let ℓ and s be positive integers such that

$\ell + \left(\frac{\ell}{2s+1}\right)^2$ is not an integer. Then

(a) $$d\left(\ell + \left(\frac{\ell}{2s+1}\right)^2\right) \geqslant \frac{1}{(2s+1)^2}$$

(b) $$d\left(\left\{\ell + \left(\frac{\ell}{2s+1}\right)^2\right\}^{1/2}\right) \geqslant \frac{1}{6(2s+1)} \text{ if } \ell \geqslant 8s^2(s+1)^2.$$

Proof. (a) is clear. To see (b) we let $N = 2s^2 + 2s$ so that $(2s+1)^2 = 2N + 1$ and $\ell \geqslant 2N^2$. Then we have

$$\left.\begin{aligned} &(2N+1)\ell + \ell^2 - (\ell + N)^2 \geqslant \frac{\ell}{2} \\ &(\ell + N + 1)^2 - (2N+1)\ell - \ell^2 \geqslant \ell \end{aligned}\right\}$$

and

$$\left.\begin{aligned}\{(2N+1)\ell + \ell^2\}^{1/2} - (\ell+N) &\geqslant \frac{\ell}{2(\ell+N+\{(2N+1)\ell+\ell^2\}^{1/2})} \geqslant \frac{1}{6}\\ \ell+N+1 - \{(2N+1)\ell + \ell^2\}^{1/2} &\geqslant \frac{\ell}{\ell+N+1+\{(2N+1)\ell+\ell^2\}^{1/2}} \geqslant \frac{1}{4}\,.\end{aligned}\right\}$$

Thus $\ell + N < \{(2N+1)\ell + \ell^2\}^{1/2} < \ell+N+1$. Dividing by 2s+1 we get

$$\frac{\ell+N}{2s+1} < \left\{\ell + \left(\frac{\ell}{2s+1}\right)^2\right\}^{1/2} < \frac{\ell+N+1}{2s+1}$$

and since between $\frac{\ell+N}{2s+1}$ and $\frac{\ell+N+1}{2s+1}$ there is no integer, we finally obtain

$$d(\ell + \left(\frac{\ell}{2s+1}\right)^2)^{1/2} \geqslant \frac{1}{6}\cdot\frac{1}{2s+1}\,.$$

We can now finish the proof of (b).

$$\begin{aligned}S &= C + \sum_{\substack{j^2+k^2\neq 0\\ A_n}} \left|j^2 + k^2 + \left(\frac{j^2+k^2}{2s+1}\right)^2 - m^2\right|^{-q}\\ &= C + \sum_{\ell=1}^{\infty} r(\ell) \sum_{m} \left|\ell + \left(\frac{\ell}{2s+1}\right)^2 - m^2\right|^{-q}\\ &\leqslant C + C\sum_{\ell=1}^{\infty} \ell^{\delta} \sum_{m=0}^{\infty} \left|\ell + \left(\frac{\ell}{2s+1}\right)^2 - m^2\right|^{-q}, \text{ for any } \delta > 0.\end{aligned}$$

Now if $\ell \geqslant 8s^2(s+1)^2$

$$\begin{aligned}\sum_{m=0}^{\infty} \left|\ell + \left(\frac{\ell}{2s+1}\right)^2 - m^2\right|^{-q} &\leqslant (C + 2.6^q.(2s+1)^q)\left\{\ell + \left(\frac{\ell}{2s+1}\right)^2\right\}^{-q/2}\\ &\leqslant Cs^{2q}\ell^{-q}\end{aligned}$$

If $\ell < 8s^2(s+1)^2$

$$\sum_{m=0}^{\infty} \left|\ell + \left(\frac{\ell}{2s+1}\right)^2 - m^2\right|^{-q} \leqslant C\left\{\ell + \left(\frac{\ell}{2s+1}\right)^2\right\}^{-q/2} + 2(2s+1)^{2q} \leqslant Cs^{2q}.$$

Thus, choosing $\delta > 0$ so that $q - \delta > 1$ we get the desired result.

An immediate consequence is that the trigonometric series

$$\Sigma \left(j^2 + k^2 + \left(\frac{j^2 + k^2}{2s+1}\right)^2 - m^2\right) e^{i(jx+ky+mt)}$$

is the Fourier series of a function in L^p for $1 \leq p < \infty$. Thus the methods of [1] prove the existence of nontrivial periodic solutions of (4) when n is odd. On the other hand, if n is even, the trigonometric series is not in any L^p, and the corresponding operator (though bounded on L^2) is not compact. So we see that when n is even, (4) behaves much like the corresponding problem for the wave operator.

The authors are grateful to A. Córdoba, who pointed out some number theoretical facts which resulted in a great simplification of some of the proofs.

BIBLIOGRAPHY

[1] H. Brézis, J.M. Coron and L. Nirenberg, Comm. Pure Appl. Math. 33, 667-684 (1980).

[2] Hardy and Wright, An Introduction to the Theory of Numbers (4th ed.), Oxford (1971).

J. Aguirre[1] and I. Peral[2]
División de Matemáticas
Universidad Autónoma
Madrid-34
Spain

1. J. Aguirre is grateful to I. Peral for introducing him to this problem.
2. I. Peral was partially supported by a research contract with M.E.C. of Spain.

A BACHELOT

Inverse scattering problem for the nonlinear Klein–Gordon equation

1. INTRODUCTION

In this paper we shall study the inverse scattering problem for the nonlinear Klein-Gordon equation

$$u_{tt} - \Delta_x u + m^2 u = f(x,u), \quad x \in \mathbb{R}^3.$$

In [4] Morawetz and Strauss showed that the scattering operator uniquely determines any interaction function f(u) which is odd and analytic. We extend this result to

$$f(x,u) = \sum_{k=1}^{\infty} a_k(x) \; |u|^{2k} \; u.$$

The fundamental argument is that the solution u_m of the Klein-Gordon equation with mass $m > 0$

$$u_{tt} - \Delta_x u + m^2 u = 0 \tag{KG_m}$$

converges to the solution of the wave equation in $L^4(\mathbb{R}^4)$ as $m \to 0$. In Section 2, we treat the case $f(x,u) = q(x) \; |u|^2 u$ (see [1], [2]); in Section 3, we establish a theorem of convergence in $L^q(\mathbb{R}^{n+1})$ of solutions of the inhomogeneous Klein-Gordon equation as $m \to m_0$; in Section 4, we derive our general result of inverse scattering.

2. THE CUBIC INTERACTION

THEOREM 1. Let q be a function in $W^{1,\infty}(\mathbb{R}^3)$. Then q is determined by the scattering operator S for the nonlinear Klein-Gordon equation

$$u_{tt} - \Delta_x u + m^2 u = q(x) \; |u|^2 u, \quad x \in \mathbb{R}^3. \tag{NLKG}$$

More precisely, given $x_0 \in \mathbb{R}^3$, $\lambda > 0$, let $g \neq 0$ in $S(\mathbb{R}^3)$ be so that

$$|\xi|^{-\frac{1}{2}} \; \hat{g} \in L^2(\mathbb{R}^3_\xi).$$

Let ϕ_λ be the solution of (KG_m) so that $\phi_\lambda(0,x) = 0$, $\phi_{\lambda_t}(0,x) = \lambda g(\lambda(x-x_0))$ and let u be the solution of $u_{tt} - \Delta_x u = 0$, $u(0,x) = 0$, $u_t(0,x) = g(x)$; then $q(x_0)$ is given by

$$q(x_0) = [\iint |u(t,x)|^4 dt\, dx]^{-1} \lim_{\lambda\to+\infty} \lambda^4 \lim_{\varepsilon\to 0} \varepsilon^{-5}[W(S(\varepsilon\phi_\lambda),$$

$$S(\varepsilon^2\phi_\lambda)) - \varepsilon^3 W(\phi_\lambda,\phi_\lambda)]$$

where $W(f,f')$ is defined for f, f' in $C^1(\mathbb{R}_t, L^2(\mathbb{R}^3_x))$ by

$$W(f,f') = \int f\, \bar{f}'_t - f_t\, \bar{f}'\, dx.$$

<u>Proof</u>: The existence of the scattering operator for small regular data is proved in [6]. We choose u_-, v_- solutions of (KG_m) so that $u_+ = Su_-$, $v_+ = Sv_-$ exist.

Let u, v be solutions of (NLKG) which verify

$$\|u(t) - u_\pm(t)\|_e \to 0, \quad \|v(t) - v_\pm(t)\|_e \to 0, \text{ as } t \to \pm\infty,$$

where $\| \quad \|_e$ denotes the energy norm. Following [4] consider $W(u,v)$

$$W(u,v) = \int (u(t)\, \bar{v}_t(t) - u_t(t)\bar{v}(t))\, dx.$$

By differentiation and integration we obtain

$$W(u(T), v(T)) - W(u(-T),v(-T)) = \int_{-T}^{+T} \int q\, u\, \bar{v}(|u|^2 - |v|^2) dt\, dx$$

and by a passage to the limit $T \to +\infty$

$$W(Su_-, Sv_-) - W(u_-, v_-) = \int\int q\, u\, \bar{v}(|u|^2 - |v|^2) dt\, dx.$$

Now, choose $u_- = \varepsilon\phi_\lambda$, $v_- = \varepsilon^2\phi_\lambda$, ε small enough; in order to estimate the integral we represent u as

$$u(t) = u_-(t) + \int_{-\infty}^{t} R(t-s) \underset{x}{*} q|u|^2 u\, ds,$$

where R is the Riemann function of (KG_m).

We solve this equation by the standard Picard method in

$$\{u \in L^4(\mathbb{R}^4) \ / \ \|u\|_{L^4(\mathbb{R}^4)} \leqslant 2\,\|u_-\|_{L^4(\mathbb{R}^4)}\}$$

for u_- small enough. By the L^p-L^q estimate for R (see [3]) and the standard singular integral inequality, we obtain

$$|u(t) - u_-(t)|_{L^4(\mathbb{R}^3)} \leqslant C \int_{-\infty}^{t} |t-s|^{-\frac{1}{2}} |u(s)|^3_{L^4(\mathbb{R}^3)} \, ds,$$

$$\|u-u_-\|_{L^4(\mathbb{R}^4)} \leqslant C' \,\|u\|^3_{L^4(\mathbb{R}^4)} \leqslant 8\, C' \,\|u_-\|^3_{L^4(\mathbb{R}^4)}.$$

Hence

$$\|u-\varepsilon\,\phi_\lambda\|_{L^4(\mathbb{R}^4)} = 0(\varepsilon^3), \quad \|v-\varepsilon^2\phi_\lambda\|_{L^4(\mathbb{R}^4)} = 0(\varepsilon^6).$$

Applying these estimates, we obtain

$$W(S(\varepsilon\,\phi_\lambda),\, S(\varepsilon^2\phi_\lambda)) - \varepsilon^3 W(\phi_\lambda,\phi_\lambda) = \varepsilon^5 \iint q|\phi_\lambda|^4 \, dt\, dx + 0(\varepsilon^7).$$

From change variable $x' = \lambda(x-x_0)$, $t' = \lambda t$ we have

$$\iint q(x_0 + \frac{x'}{\lambda})|u_\lambda(t',x')|^4 \, dt'\, dx'$$

$$= \lambda^4 \lim_{\varepsilon\to 0} \varepsilon^{-5}[W(S(\varepsilon\phi_\lambda),\, S(\varepsilon^2\phi_\lambda)) - \varepsilon^3 W(\phi_\lambda,\phi_\lambda)],$$

where $u_\lambda(t', x') = \phi_\lambda(t,x)$ is a solution of

$$u_{t't'} - \Delta_{x'}u + \left(\frac{m}{\lambda}\right)^2 u = 0; \; u_\lambda(0,x') = 0; \; u_{\lambda_{t'}}(0,x') = g(x'). \qquad \text{(E)}$$

Therefore, in order to determine $q(x_0)$ it suffices to prove that solutions of equation $(KG)_m$ converge in $L^4(\mathbb{R}^4)$ to the solution of the wave equation as $m \to 0$.

3. CONVERGENCE OF SOLUTIONS OF (KG_m) AS $m \to m_0$

THEOREM 2. Let $m \geqslant 0$; we denote by $u_m = T_m(f,g,h)$ the solution of

$$u_{tt} - \Delta_x u + m^2 u = h(t,x), \qquad x \in \mathbb{R}^n,$$

$$u(0,x) = f(x) \colon u_t(0,x) = g(x).$$

Let H, H_0 be Hilbert spaces defined by

$$H = \{(f,g)/(1 + |\xi|^2)^{1/4}\, \hat{f} \text{ and } (1 + |\xi|^2)^{-1/4}\, \hat{g} \in L^2(\mathbb{R}^n_\xi)\},$$

$$H_0 = \{(f,g)/(1 + |\xi|^2)^{1/4}\, \hat{f} \text{ and } |\xi|^{-1/2}\, \hat{g} \in L^2(\mathbb{R}^n_\xi)\}.$$

For $n \geqslant 2$ we have the following results:

(a) if $m, m_0 > 0$, then $T_m \to T_{m_0}$ in $\mathcal{L}(H \times L^p(\mathbb{R}^{n+1}), L^q(\mathbb{R}^{n+1})$ under the topology of simple convergence as $m \to m_0$, with p, q so that

$$\frac{1}{p} + \frac{1}{q} = 1;\ 2\,\frac{n+2}{n} \leqslant q \leqslant 2\,\frac{n+1}{n-1}.$$

(b) $T_m \to T_0$ in $\mathcal{L}(H_0 \times L^p(\mathbb{R}^{n+1}), L^q(\mathbb{R}^{n+1}))$ under the topology of simple convergence as $m \to 0$ with p, q so that

$$\frac{1}{p} + \frac{1}{q} = 1;\ q = 2\,\frac{n+1}{n-1}.$$

<u>Proof</u>: Strichartz [8] proved the equicontinuity of T_m, hence it suffices to show the convergence on a dense subset of data (f,g,h). First we study solutions of the homogeneous equation $u_m = T_m(f,g,0)$; u_m may be written:

$$u_m(t,x) = F^{-1}_{\tau,\xi}\,[\tfrac{1}{2}(\sqrt{m^2 + |\xi|^2}\hat{f}(\xi) + i\ \mathrm{sgn}(\tau)\ \hat{g}(\tau))\ d\mu_m],$$

where $d\mu_m$ is a measure given by

$$\langle d\mu_m, \phi(\tau,\xi)\rangle = \int (\phi(\sqrt{m^2 + |\xi|^2},\xi) + \phi(-\sqrt{m^2 + |\xi|^2},\xi))(m^2 + |\xi|^2)^{-\frac{1}{2}}\, d\xi.$$

Suppose that: $\hat{f}, \hat{g} \in \mathcal{D}(\mathbb{R}^n)$, $0 \notin \mathrm{supp}\, \hat{f} \cup \mathrm{supp}\, \hat{g}$.

We choose $\chi \in \mathcal{D}(\mathbb{R}^{n+1} \setminus \{0\})$ so that

$$\xi \in \mathrm{supp}\, \hat{f} \cup \mathrm{supp}\, \hat{g},\ 0 \leqslant \tau^2 - |\xi|^2 \leqslant (m_0 + 1)^2 \Rightarrow \chi(\tau,\xi) = 1.$$

We have for $0 \leqslant m \leqslant m_0 + 1$

$$u_m(t,x) = F^{-1}_{\tau,\xi}[\tfrac{1}{2}(\sqrt{m^2+|\xi|^2}\ \hat{f}(\xi) + i\ \mathrm{sgn}(\tau)\hat{g}(\xi))\chi] * F^{-1}_{\tau,\xi}\ (\chi . d\mu_m).$$

Hence $u_m - u_{m_o}$ may be written

$$u_m - u_{m_0} = F_m * F^{-1}_{\tau,\xi}[\chi(d\mu_m - d\mu_{m_0})] + F^{-1}_{\tau,\xi}[\tfrac{1}{2}(\sqrt{m^2+|\xi|^2} - \sqrt{m_0^2+|\xi|^2})\hat{f}(\xi).d\mu_{m_0}].$$

Let

$$S_{m_0} = \{(\tau,\xi)/-\tau^2 + |\xi|^2 + m_0^2 = 0\}.$$

Then $\frac{1}{2}\hat{f}(\xi)(\sqrt{m^2 + |\xi|^2} - \sqrt{m_0 + |\xi|^2})$ converges to 0 in $L^2(S_{m_0}, d\mu_{m_0})$ as $m \to m_0$; hence it follows from [8] that for the desired q

$$F^{-1}_{\tau,\xi}[\tfrac{1}{2}(\sqrt{m^2 + |\xi|^2} - \sqrt{m_0^2 + |\xi|^2})\hat{f}(\xi)\, d\mu_{m_0}] \to 0 \text{ in } L^q(\mathbb{R}^{n+1}) \text{ as } m \to m_0.$$

On the other hand, we notice that $(F_m)_{0 \leqslant m \leqslant m_0+1}$ is bounded in $L^1(\mathbb{R}^{n+1})$. Now consider the analytic family of operators T^z_m

$$T^z_m\,\phi = \phi * F^{-1}_{\tau,\xi}[\gamma(z)((-\tau^2 + |\xi|^2 + m^2)^z_+ - (-\tau^2 + |\xi|^2 + m_0^2)^z_+)\chi],$$

where $\gamma(z)$ is a suitable holomorphic function which cancels the poles of the distribution valued meromorphic function $(-\tau^2 + |\xi|^2 + m^2)^z_+$ and with a simple zero at $z = -1$ so that

$$T^{-1}_m\,\phi = \phi * F^{-1}_{\tau,\xi}[\chi(d\mu_m - d\mu_{m_0})].$$

More precisely we choose γ

$$\gamma(z) = [\Gamma(z+1)]^{-1}(z + \tfrac{n}{2})\sin[\pi(z + \tfrac{n}{2})], \text{ n odd}$$

$$\gamma(z) = [\Gamma(z+1)]^{-1}(z + \tfrac{n}{2})\sin[\pi(z + \tfrac{n}{2})]\,(z+1)^{-1}, \text{ n even.}$$

So that $(T^z_m)_{\mu \leqslant m \leqslant m_0+1}$ satisfies the hypothesis of Stein's interpolation theorem for $-(n+2)/2 \leqslant \mathrm{Re}\, z \leqslant 0$ if $\mu > 0$ and for $-(n+1)/2 \leqslant \mathrm{Re}\, z \leqslant 0$ if $\mu = 0$.

From the dominated convergence theorem it follows that

$$T^{iy}_m \to 0 \text{ in } \mathcal{L}(L^1(\mathbb{R}^{n+1}), L^2(\mathbb{R}^{n+1})) \text{ as } m \to m_0$$

and by an explicit computation we obtain that $(T^z_m)_{\mu \leqslant m \leqslant m_0+1}$ is bounded in $\mathcal{L}(L^1(\mathbb{R}^{n+1}), L^\infty(\mathbb{R}^{n+1}))$ for $-(n+2)/2 \leqslant \mathrm{Re}\, z \leqslant -(n+1)/2$ if $0 < \mu, m_0$ and for $\mathrm{Re}\, z = -(n+1)/2$ if $\mu = 0$. Stein's interpolation theorem shows that $T^{-1}_m \to 0$

in $\mathcal{L}(L^1(\mathbb{R}^{n+1}), L^q(\mathbb{R}^{n+1}))$ for suitable q as $m \to m_0$. This implies that $u_m \to u_{m_0}$ in $L^q(\mathbb{R}^{n+1})$ as $m \to m_0$.

Now, study the convergence of $u_m = T_m(0,0,h)$. Let E_m^+, (E_m^-) be the fundamental solution of (KG_m) in the forward (respectively backward) light cone. The problem is reduced to show that $(E_m^\pm - E_{m_0}^\pm) * \to 0$ simply in $\mathcal{L}(L^p(\mathbb{R}^{n+1}), L^q(\mathbb{R}^{n+1}))$ as $m \to m_0$. Following [7] we define an analytic family of operators

$$P_{m_\pm}^z h = \lim_{\sigma \to 0\pm} F_{\tau,\xi}^{-1}[((-\tau^2 + |\xi|^2 + m^2 + \sigma^2 - 2i\sigma\tau)^z - (-\tau^2 + |\xi|^2 + m_0^2 + \sigma^2 - 2i\sigma\tau)^z)\hat{h}(\tau,\xi)].$$

We have

$$P_{m_\pm}^{-1} h = (E_m^\pm - E_{m_0}^\pm) * h.$$

On the one hand it is clear that $P_{m_\pm}^{iy} \to 0$ simply in $\mathcal{L}(L^2(\mathbb{R}^{n+1}), L^2(\mathbb{R}^{n+1}))$ as $m \to m_0$, on the other hand, it follows from [7] that $(P_{m_\pm}^z)_{\mu \leqslant m \leqslant m_0+1}$ is bounded in $\mathcal{L}(L^1(\mathbb{R}^{n+1}), L^\infty(\mathbb{R}^{n+1}))$ for $-(n+2)/2 \leqslant \mathrm{Re}\, z \leqslant -(n+1)/2$ if $0 < \mu, m_0$ and for $\mathrm{Re}\, z = -(n+1)/2$ if $\mu = 0$. A slight modification of Stein's interpolation theorem proves that $P_{m_\pm}^{-1} \to 0$ simply in $\mathcal{L}(L^p(\mathbb{R}^{n+1}), L^q(\mathbb{R}^{n+1}))$. The proof of Theorem 2 is therefore complete.

4. ANALYTIC INTERACTION

THEOREM 3. Let $(a_{2n+1})_{n \geqslant 1}$ be a bounded sequence in $W^{2,\infty}(\mathbb{R}^3)$. Then this sequence is determined by the scattering operator S for the equation

$$u_{tt} - \Delta_x u + m^2 u = \sum_{n=1}^{\infty} a_{2n+1}(x)\, |u|^{2n} u; \; 0 < m, \; x \in \mathbb{R}^3.$$

Remark: Notice that this result holds whatever the sign of a_{2n+1}; indeed, the existence of S is useful only for small data (in fact only for the small regular wave packet). We sketch the proof: the scattering operator is defined for free solutions u_- so that $\|u_-\|_{scat}$ is small enough where

$$\|u_-\|_{scat}^2 = \sup_{t \in \mathbb{R}} \|u(t)\|_e^2 + \sup_{t \in \mathbb{R}} (1 + |t|)^3 \|u(t)\|_{L^\infty(\mathbb{R}_x^3)}^2$$

(see for instance [5]).

Notice that

$$\sup_{q \geq 4} \|\cdot\|_{L^q(\mathbb{R}^4)} \leq c \|\cdot\|_{scat}.$$

Now with the notation of Section 2 we have

$$u(t) = u_-(t) + \int_{-\infty}^{t} R(t-s) \underset{x}{*} f(x,u(s))ds.$$

This equation is solved by the Picard method for the $\| \ \|_{scat}$ norm, hence

$$u(t) = \sum_{k=0}^{\infty} \mathcal{R}^k(u_-) \tag{*}$$

convergence taking place in the $\| \ \|_{scat}$ norm and where

$$\mathcal{R}(f) = \int_{-\infty}^{t} R(t-s) \underset{x}{*} f(x,s)\ ds.$$

Following [4] we choose $u_- = \varepsilon\ \phi_\lambda$, $v_- = 2\varepsilon\phi_-$, and we have

$$W(S(2\varepsilon\ \phi_\lambda),(\varepsilon\ \phi_\lambda)) - 2\varepsilon^2 W(\phi_\lambda,\phi_\lambda) = \iint (u\overline{f(x,v)} - f(x,u)\bar{v})\ dt\ dx.$$

Applying (*) to u and v we see that

$$W(S(2\varepsilon\ \phi_\lambda),\ S(\varepsilon\ \phi_\lambda)) - 2\ \varepsilon^2 W(\phi_\lambda,\phi_\lambda) = \sum_{n \geq 4} \varepsilon^h\ Q_n(\lambda)$$

where

$$Q_n(\lambda) = (2^{n-1}-2) \iint a_{n-1}\ |\phi_\lambda|^n\ dx\ dt + Q_n'(\lambda),$$

where Q_n' involves only $a_{n-2}, a_{n-3}, \ldots, a_3$ ($a_{2k} = 0$ by definition). We assume by induction that $Q_n'(\lambda)$ is already determined so that S determines

$$\iint a_{n-1}(x_0 + \frac{x'}{\lambda})\ |u_\lambda(t',x')|^n\ dt'\ dx',\ n \text{ even},$$

u_λ being solution of (E). Theorem 2 ensures that u_λ converges in $L^4(\mathbb{R}^4)$ as $\lambda \to +\infty$, hence the problem is reduced showing that

$$\sup_{\lambda} \|u_\lambda\|_{L^\infty(\mathbb{R}^4)} < \infty .$$

By the Sobolev inequality

$$|u_\lambda(t)|_{L^\infty(\mathbb{R}^3)} \leq c \; |\nabla u_\lambda(t)|^{3/4}_{L^4(\mathbb{R}^3)} \; |u_\lambda(t)|^{1/4}_{L^4(\mathbb{R}^3)}.$$

We have

$$|u_\lambda(t)|_{L^4(\mathbb{R}^3)} \leq |u_\lambda(t)|^{\frac{1}{2}}_{L^2(\mathbb{R}^3)} \; |u_\lambda(t)|^{\frac{1}{2}}_{L^6(\mathbb{R}^3)}$$

$$|u_\lambda(t)|_{L^6(\mathbb{R}^3)} \leq c \; |\nabla u_\lambda(t)|_{L^2(\mathbb{R}^3)}$$

and by the conservation of energy

$$|u_\lambda(t)|_{L^2(\mathbb{R}^3)} \leq | \; |\xi|^{-1}\hat{g}|_{L^2(\mathbb{R}^3_\xi)} \; ; \; |\nabla u_\lambda(t)|_{L^2(\mathbb{R}^3)} \leq |g|_{L^2(\mathbb{R}^3)}.$$

Hence

$$|u_\lambda(t)|_{L^4(\mathbb{R}^3)} \leq c \; |g|^{\frac{1}{2}}_{L^2(\mathbb{R}^3)} \; | \; |\xi|^{-1}\hat{g}|^{\frac{1}{2}}_{L^2(\mathbb{R}^3_\xi)}.$$

Even

$$|\nabla u_\lambda(t)|_{L^4(\mathbb{R}^3)} \leq c \; | \; |\xi| \; \hat{g}|^{\frac{1}{2}}_{L^2(\mathbb{R}^3_\xi)} \; |g|^{\frac{1}{2}}_{L^2(\mathbb{R}^3)}.$$

Finally

$$\sup_\lambda \; |u_\lambda|_{L^\infty(\mathbb{R}^4)} \leq c \; |g|^{\frac{1}{2}}_{L^2(\mathbb{R}^3)} \; | \; |\xi| \; \hat{g}|^{3/8}_{L^2(\mathbb{R}^3_\xi)} \; | \; |\xi|^{-1} \; \hat{g}|^{1/8}_{L^2(\mathbb{R}^3_\xi)}.$$

We choose g so that $\hat{g} \in \mathcal{D}(\mathbb{R}^3)$ $0 \notin \text{supp}\,\hat{g}$.

This completes the proof of Theorem 3.

REFERENCES

[1] A. Bachelot, C.R. Acad. Sci. Paris, serie A, 293, 121-124 (1981).

[2] A. Bachelot, Thèse, Université de Bordeaux I (1981).

[3] B. Marshall, W. Strauss, and S. Wainger, J. Math. Pures Appl. 59, 417-440 (1980).

[4] C. Morawetz and W. Strauss, Comm. Pure Appl. Math. 25, 1-31 (1972).

[5] M. Reed and B. Simon, Scattering Theory , Academic Press, New York, (1979).

[6] J. Segal, Ann. Sci. Ecole Norm. Sup. (4), 1, 459-497 (1968).

[7] R.S. Strichartz, Indiana Univ. Math. J. 24, 499-525 (1974).

[8] R.S. Strichartz, Duke Math. J. 44, 705-714, (1977).

A. Bachelot
Université de Bordeaux I
U.E.R. de Mathématiques et Informatique
Laboratoire associé no 226
351, Cours de la Libération
33405 Talence Cedex
France

M BALABANE & H EMAMI-RAD

Pseudo differential parabolic systems in $L^p(\mathbb{R}^n)$

1. INTRODUCTION

Let us consider the Cauchy problem:

$$\frac{\partial U}{\partial t} = A(D)U; \quad U(0) = U_0 \qquad (*)$$

$A(\xi)$ is an $N \times N$ matrix-valued function on $\mathbb{R}^n$, infinitely differentiable on $\mathbb{R}^n \setminus \{0\}$, positively homogeneous of degree $\alpha > 0$;

$$D = \left(\frac{1}{i} \frac{\partial}{\partial x_1}, \ldots, \frac{1}{i} \frac{\partial}{\partial x_n}\right);$$

$A(D)$ is the operator from the Schwartz space $[S(\mathbb{R}^n)]^N$ to $S'[(\mathbb{R}^n)]^N$ defined by

$$F(A(D)\phi)(\xi) = A(\xi) \cdot F(\phi)(\xi) \text{ for any } \phi \in [S(\mathbb{R}^n)]^N$$

(here F is the usual Fourier transform on $[S(\mathbb{R}^n)]^N$).

It is well known that given $U_0 \in [L^p(\mathbb{R}^n)]^N$, the Cauchy problem is not well posed in $[L^p(\mathbb{R}^n)]^N$ for $p \neq 2$ (Littman [6]). The special case of the wave equation is studied by Peral [7]. The case of the first-order hyperbolic system (i.e. $A(\xi) = \sum_{j=1}^{j=n} \xi_j A_j$ is proved by Brenner [2] to be well posed in $L^p(\mathbb{R}^n)$, $p \neq 2$, if and only if the A_j are Hermitian, mutually commuting constant matrices. In [1], the authors studied the Schrödinger equation, and proved that this problem is well posed in $L^p(\mathbb{R}^n)$ as a distribution in the t variable. Because of this, smooth distribution groups were defined. Strengthening this abstract tool turns out to be a good approach for studying systems.

In this paper, we define smooth distribution semi-groups on Banach spaces, a special case of regular distribution semi-groups (Lions [5]). We define their order and give their essential properties. We characterize them by the properties of the resolvent of their generators (Section 2). An approximation theorem of Hille-Yosida type is then given. In Section 3, we establish a technique enabling us to study the regularizing effects, for the distribution semi-groups, of acting on the 'regular vectors' of the Banach space. This leads to an abstract asymptotic estimate. In Section 4 we study the Cauchy problem quoted above. Viewed as an ordinary differential equation in $[L^p(\mathbb{R}^n)]^N$, it turns out that its solution defines a smooth distribution semi-group under the hypothesis: (H1) $A(\xi)$ is diagonalizable, and (H2) the eigenvalues of $A(\xi)$ lie in the upper half plane of $\mathbb{C}$. This distribution is of order $n|\frac{1}{p} - \frac{1}{2}|$. Then applying the results of Section 2 gives an estimate of the complex powers of the resolvant of A(D), viewed as a closed (unbounded) operator in $[L^p(\mathbb{R}^n)]^N$. In Section 5 we examine the case where the Cauchy data belongs to $[W^{m,p}(\mathbb{R}^n)]^N$, with $m > n\alpha|\frac{1}{p} - \frac{1}{2}|$, and applying Section 3 we prove that the solution of the Cauchy problem is actually a function, recovering results proved in special cases by Brenner [2] and Peral [7]. Furthermore the results of Section 3 enable us to assert that the solution grows no faster than $t^{n\alpha|(1/p)-(1/2)|}$, in the L^p norm, when t goes to infinity. In Section 6 we give a converse result asserting that the only case where the Cauchy problem is well posed in $[L^p(\mathbb{R}^n)]^N$, $p \neq 2$, is the case where A(D) is a diagonal differential system of the first order, up to linear canonical transformation.

2. SMOOTH DISTRIBUTION SEMI-GROUPS

$\mathcal{D}(\mathbb{R}_+)$ is the Schwartz space of infinitely differentiable, compactly supported functions on the open set $\mathbb{R}_+$, and $L^1(\mathbb{R}_+)$ is the usual Lebesgue space on $\mathbb{R}_+$. We define on $\mathcal{D}(\mathbb{R}_+)$, the family of semi-norms:

$$p_k(\phi) = \sum_{\ell=1}^{\ell=k} \|t^\ell \frac{d^\ell}{dt^\ell} \phi\|_{L^1} \quad \text{for any } \phi \in \mathcal{D}(\mathbb{R}_+) \text{ and any integer } k.$$

Let T be the completion of $\mathcal{D}(\mathbb{R}_+)$ for this family of semi-norms. T is a Fréchet space, and a Fréchet algebra for the usual (additive) convolution on $\mathbb{R}_+$.

For any integer k, let T_k be the completion of $\mathcal{D}(\mathbb{R}_+)$ for the norm p_k. Then T_k is a Banach algebra for the usual (additive) convolution on $\mathbb{R}_+$.

Remark 1. For any $\phi \in T$, any integer k, and any real $s > 0$, $p_k(\frac{1}{s}\phi(\frac{\cdot}{s}))$ is independent of s.

DEFINITION 1. Let X be a Banach space. A smooth distribution semi-group over X (σDSG in short) is a continuous linear mapping G from T to $\mathcal{L}(X)$ such that:

(a) $G(\phi*\psi) = G(\phi) \circ G(\psi)$ for any ϕ and ψ in T

(b) There exists an everywhere dense subspace D in X such that, for any $x \in D$, $G \otimes x$ is a continuous function, and its value at $t = 0$ is x.

DEFINITION 2. Let G be a σDSG over a Banach space X.

G is of order k (where K is an integer) if G extends to a continuous mapping from T_k to $\mathcal{L}(X)$.

For $\phi \in \mathcal{D}(\mathbb{R})$, let us define $\check{\phi}$ by: $\check{\phi}(t) = \phi(-t)$ for $t \in \mathbb{R}$. If X is a Banach space and G a mapping from $\mathcal{D}(\mathbb{R})$ to $\mathcal{L}(X)$, let us define $\check{G}$ by: $\check{G}(\phi) = G(\check{\phi})$ for any $\phi \in \mathcal{D}(\mathbb{R})$.

DEFINITION 3. Let X be a Banach space. A smooth distribution group over X(σDG in short) is a mapping G from $\mathcal{D}(\mathbb{R})$ to $\mathcal{L}(X)$ such that:

(a) $G(\phi*\psi) = G(\phi) \circ G(\psi)$ for any ϕ and ψ in $\mathcal{D}(\mathbb{R})$

(b) $G = G_+ + G_-$ where G_+ and $\check{G}_-$ are σDSG.

PROPOSITION 1. Let G be a σDSG over X. Then

(a) For any $\phi \in \mathcal{D}(\mathbb{R}_+)$ such that $\int \phi = 1$, $G\left(\frac{1}{s}\ \phi(\frac{\cdot}{s})\right)$ converges strongly to the identity of X, when s goes to zero;

(b) $N = \bigcap_{\phi \in \mathcal{D}(\mathbb{R}_+)} \operatorname{Ker} G(\phi) = \{0\}$;

(c) $R = \bigcup_{\phi \in \mathcal{D}(\mathbb{R}_+)} \operatorname{Im} G(\phi)$ is everywhere dense in X.

Proof. (a) For $x \in D$, $G\left(\frac{1}{s}\ \phi(\frac{\cdot}{s})\right)x$ goes to x by applying Lebesgue's theorem. The strong convergence of these operators to I follows then from the equicontinuity quoted in Remark 1.

(b) and (c) are straight applications of (a).

Remark 2. A σDSG is a special case of a regular distribution semi-group defined by Lions [5]. The properties (b) and (c) of Proposition 1, which have to be taken as axioms when defining regular DSG, appear here as consequences of the choice of T as the space of test functions. Moreover this space of test functions enables us to characterize the order of the σDSG, the weights t^k appearing in the semi-norms being the natural ones to introduce for handling differential equations in $L^p(\mathbb{R}^n)$. Finally the fact that for Re $\lambda > 0$, $Y(t)e^{-\lambda t}$ belongs to T_k for any k enables us to prove easily estimates on the resolvent of the infinitesimal generator of a σDSG, which we introduce now.

Following Lions [5], to define $G(\theta)$ for any $\theta \in E'(\mathbb{R})$ we have:

DEFINITION 4. Let G be a σDSG over a Banach space X. The infinitesimal generator of G is the unbounded operator on X, closure of the closable operator $G(-\delta')$ defined on R by:

$$G(-\delta')\ G(\phi)x = G(-\phi')x \text{ for any } \phi \in \mathcal{D}(\mathbb{R}_+) \text{ and } x \in X.$$

Proof of the consistency of this definition and of the closability of $G(-\delta')$ follows directly from (a) and (c) of Proposition 1.

PROPOSITION 2. Let X be a Banach space and A a closed operator on X. A generates a σDG if and only if A and -A generate σDSG.

THEOREM 1. Let X be a Banach space, G a σDSG on X, of order k, A its infinitesimal generator. If $\lambda \in \mathbb{C}$, Re $\lambda > 0$, then λ belongs to the resolvent set of A. Moreover, for any $\mu \in \mathbb{C}$, Re $\mu > 0$, we have:

$$||(\lambda I-A)^{-\mu}||_{\mathcal{L}(X)} \leq C|\mu|^k \frac{\Gamma(\text{Re }\mu)}{|\Gamma(\mu)|} \frac{|\lambda|^k}{(\text{Re }\lambda)^{k+\text{Re }\mu}},$$

where C is a constant independent of λ and μ.

Proof. We first consider positive integer values of μ. In this case, a slight modification of the proof given in Balabane - Emami-Rad [1] shows that

$$(\mu - 1)!\ (\lambda I - A)^{-\mu} = G(Y(t)t^{\mu-1}\ e^{-\lambda t}).$$

By the fact that $Y(t)t^{\mu-1}e^{-\lambda t} \in T_k$, and by the definition of the order of a σDSG, the estimate follows from the computation of $p_k(Y(t)t^{\mu-1}e^{-\lambda t})$.

We now consider the case $\mu \notin \mathbb{N}$, Re $\mu > k+1$. Defining complex powers by a Dunford integral, we have:

$$(\lambda I-A)^{-\mu} = \frac{1}{2i\pi} \int_\Gamma (\lambda-z)^{-\mu}(z-A)^{-1}\, dz,$$

Γ being the line Re $z = \beta$ with $0 < \beta < \text{Re}\ \lambda$. The estimate proved on $\|(z-A)^{-1}\|$ and the hypothesis Re $\mu > k+1$, making this integral convergent. The formula given for the resolvent gives

$$(\lambda I-A)^{-\mu} = \frac{1}{2i\pi} \int_\Gamma (\lambda-z)^{-\mu} G(Y(t)e^{-\lambda t})dt = G\left(\frac{Y(t)}{2i\pi} \int_\Gamma (\lambda-z)^{-\mu} e^{-\lambda t}\, dt\right),$$

this last equality being valid because the integral converges for the p_k norm in the t variable.

Changing the integration path to a path asymptotic to the half line $\lambda + \mathbb{R}_+$ at infinity, and using the Hankel expression for the Γ-function (Whittaker and Watson [10]) we have:

$$(\lambda I-A)^{-\mu} = G\left(\frac{(\sin \pi\mu)\Gamma(1-\mu)}{\pi} Y(t)t^{\mu-1}e^{-\lambda t}\right) = G\left(Y(t)\, \frac{t^{\mu-1}}{\Gamma(\mu)}\, e^{-\lambda t}\right).$$

By analytic continuation in the μ variable, this formula remains valid for Re $\mu > 0$. (The function $Y(t)t^{\mu-1}e^{-\lambda t}$ being in T_k for Re $\mu > 0$.)

The estimate then follows from:

$$\|(\lambda I-A)^{-\mu}\| \leq C\, p_k(Y(t)\, \frac{t^{\mu-1}}{\Gamma(\mu)}\, e^{-\lambda t})$$

and the computation of this last semi norm.

Remark 3. If μ takes positive integer values, Theorem 1 gives an estimate of the iterated resolvent: for $\ell \in \mathbb{N}^*$

$$\|(\lambda I-A)^{-\ell}\| \leq C\, \ell^k\ \ |\lambda|^k\ (\text{Re}\ \lambda)^{-k-\ell}.$$

It is interesting to note that for $k = 0$, this is exactly the Hille-Yosida estimate for the generator of a strongly continuous semi-group. Moreover the coefficient $C\ell^k$ grows polynomially with ℓ, whereas it is a constant in the strongly continuous semi-group case, and grows exponentially in the regular distribution semi-group case (Chazarain [3]).

We now come to converse results:

THEOREM 2. Let X be a Banach space and A a closed, densely defined linear operator in X. If for any $\lambda \in \mathbb{C}$ with Re $\lambda > 0$, λ belongs to the resolvent set of A, and if we have the estimate

$$\|(\lambda I-A)^{-1}\|_{\mathcal{L}(X)} \leqslant C|\lambda|^{k} \ |\text{Re } \lambda|^{-k-1},$$

where C is independent of λ, then A generates a σDSG of order k + 2.

Proof. This is slight modification of the proof given in Balabane - Emami-Rad [1].

We end this section by an approximation theorem of Hille-Yosida type. With that aim, and under the hypothesis of Theorem 2, we consider the bounded operators $A_n = A(I-n^{-1}A)^{-1}$, where n is a positive integer. The spectrum of A_n lies in the disc $\{\lambda \in \mathbb{C};\ |\lambda + \frac{n}{2}| < \frac{n}{2}\}$. So, for any fixed n, A_n generates an exponential e^{tA_n}, bounded for $t \in \mathbb{R}_+$. Let G_n denote the σDSG defined by:

$$G_n(\phi) = \int_{\mathbb{R}_+} \phi(t)e^{tA_n}dt \text{ for any } \phi \in \mathcal{D}(\mathbb{R}_+).$$

THEOREM 3. Under the hypothesis of Theorem 2, A_n converges strongly to A on D(A), and G_n converges to G, strongly on T_{k+2}.

Proof. The first assertion follows from the fact that the family (A_n) is equicontinuous when viewed as a family of operators acting from the Banach space D(A) equipped with the graph norm to X. The strong convergence then follows from the strong convergence on $D(A^{k+1})$.

In the same way, and for any $\phi \in T_{k+2}$, the second assertion is proved by proving that the family $G_n(\phi)$ is equicontinuous, with norm less than $p_{k+2}(\phi)$.

3. TRANSFERRING REGULARITY

THEOREM 4. Let X be a reflexive Banach space, and G a σDSG on X of order k. Let A be its infinitesimal generator. Then for any $x \in D(A^k)$, the distribution $G \otimes x$, denoted by $e^{tA}x$, reduces to a function and we have:

$$\|e^{tA}x\|_X \leqslant M(\|x\| + \|A^k x\|)(1+t^k).$$

To prove this theorem, we need the following:

LEMMA 1. There exist k functions $(\alpha_i)_{i=0,\ldots k-1}$ in $\mathcal{D}(\mathbb{R}_+)$ such that for any $\phi \in \mathcal{D}(\mathbb{R}_+)$, there exists $\psi \in \mathcal{D}(\mathbb{R}_+)$ with

$$\phi = \sum_{i=0}^{k-1} \left(\int t^i \phi(t)dt\right)\alpha_i + \frac{d^k}{dt^k}\psi.$$

Proof. Denoting by t^i the distributions associated with t^i, and by D the usual differentiation in $\mathcal{D}(\mathbb{R}_+)$, we know that

$$\operatorname{Im} D^k = \bigcap_{i=0}^{k-1} \operatorname{Ker} t^i.$$

For any i = 0,...,k-1, take

$$\alpha_i \in \operatorname{Ker} t^j \text{ for } j = 0,\ldots,k-1;\ j \neq i; \text{ and } \int t^i \alpha_i = 1.$$

Then $(\alpha_i)_{i=0,\ldots,k-1}$ is a basis of a supplement of $\operatorname{Im} D^k$ in $\mathcal{D}(\mathbb{R}_+)$ which fits the property.

LEMMA 2. Let $x \in D(A^k)$, and (ϕ_n) a sequence of positive functions of $\mathcal{D}(\mathbb{R}_+)$, approximating δ_s. Then $G(\phi_n)x$ is bounded and we have

$$\|G(\phi_n)x\| \leqslant C(1+s^k)(\|A^kx\| + \|x\|).$$

Proof. Let ψ_n be the function associated with ϕ_n by Lemma 1; then

$$\|G(\phi_n)x\| \leqslant \sum_{i=0}^{k-1} |\lambda_n^i| \; \|G(\alpha_i)\| \; \|x\| + \|G(\psi_n^{(k)})x\|,$$

where $\lambda_n^i = \int t^i \phi_n$. By definition of (ϕ_n), λ_n^i converges to s^i, and

$$\|G(\psi_n^{(k)})x\| = \|G(\psi_n)A^kx\| \leqslant \|A^kx\| \, p_k(\psi_n)$$

$$\leqslant \|A^kx\| \left(\|t^k \phi_n\|_{L^1} + \sum_{i=0}^{k-1} |\lambda_n^i| \int t^k |\alpha_i(t)| dt\right).$$

ϕ_n being positive, this expression converges to

$$\|A^kx\| \left(s^k + \sum_{i=0}^{k-1} \left(\int t^k |\alpha_i|\right) s^i\right).$$

Remark 4. Other terms in p_k are handled in a similar way.

LEMMA 3. For $x \in D(A^{k+2})$, the distribution $G \otimes x$ reduces to a continuous function $e^{tA}x$ with

$$\|e^{tA}x\|_X \leqslant C(1 + t^{k+2})(\|x\| + \|A^{k+2}x\|).$$

Proof. Writing the resolvent identity and using the Dunford integral representation of G (Balabane - Emami-Rad [1]), we have

$$G(\phi)x = \sum_{i=0}^{k+1} (\int t^i \phi)A^i x + \int_0^\infty \phi(t) [\frac{1}{2i\pi}\int_\Gamma e^{\lambda t} \lambda^{-k-2}(\lambda-A)^{-1}d\lambda]A^{k+2}x \, dt,$$

where Γ is the line $\{\mathrm{Re}\ \lambda = \alpha t\}$ with $\alpha > 0$. Using the resolvent inequality after a change of variable $\lambda t = \mu$ gives the result.

Proof of the theorem. For $x \in D(A^k)$, let (x_ℓ) be the sequence in $D(A^{k+2})$, which converges to x for the $D(A^k)$ norm (see Lions [5]). Then $e^{sA}x_\ell$ is a Cauchy sequence, by

$$\|e^{sA}x_\ell - e^{sA}x_{\ell'}\| \leqslant \|e^{sA}x_\ell - G(\phi_n)x_\ell\|$$

$$+ \|G(\phi_n)x_\ell - G(\phi_n)x_{\ell'}\| + \|e^{sA}x_{\ell'} - G(\phi_n)x_{\ell'}\|.$$

The second term is bounded by $C(\|x_\ell - x_{\ell'}\|_{D(A^k)})(1+s^k)$. A classical $\varepsilon/3$ argument proves the sequence is a Cauchy one.

To prove that its limit is the distribution $G \otimes x$, we note that the local boundedness of $e^{sA}x_\ell$ in the s-variable and its pointwise convergence imply its convergence in the distribution sense, and the result follows by the convergence of the abstract differential equation it satisfies: $\partial U/\partial t = AU + \delta \otimes x_n$. To prove the estimate, note that $G \otimes x$ is the weak limit of $G(\phi_n)x$, the norm of which is bounded by $C\|x\|_{D(A^k)}(1+s^k)$.

4. SYSTEMS IN $L^p(\mathbb{R}^n)$

(a) Fourier multipliers

DEFINITION 5. Let $\omega \in C^\infty(\mathbb{R}^n \setminus \{0\})$. ω is called a Fourier multiplier in L^p if for any $\phi \in S(\mathbb{R}^n)$ the following inequality holds:

$$\| \bar{F}(\omega . F\phi) \|_{L^p} \leqslant A_p \| \phi \|_{L^p},$$

where A_p is independent of ϕ. The least such constant A_p is called the norm of ω in the space of p-Fourier multipliers. It is denoted by $\| \omega \|_{M_p}$.

LEMMA 4. For any $\omega \in M_p$, we have: $\|\omega(t\xi)\|_{M_p} = \|\omega(\xi)\|_{M_p}$. If $\omega_1 \in M_p$ and $\omega_2 \in M_p$, then $\omega_1\omega_2 \in M_p$ and we have

$$\|\omega_1\omega_2\|_{M_p} \leqslant \|\omega_1\|_{M_p} \cdot \|\omega_2\|_{M_p}.$$

THEOREM 5. (Stein [9]; Sjöstrand [8].) Let $\omega \in C^\infty(\mathbb{R}^n \setminus \{0\})$ and $1 < p < \infty$. Then:

(a) if ω is positively homogeneous of negative (or zero) degree, outside of some finite ball, and bounded, then $\omega \in M_p$.

(b) if ω is positively homogeneous of degree $\alpha > 0$, with $\mathrm{Im}(\omega(\xi)) \geqslant 0$, then for $k > n|\frac{1}{p} - \frac{1}{2}|$:

$$\int_0^1 (1-s)^{k-1} e^{is\omega(\xi)} ds \in M_p.$$

Proof. Part (a) is a straightforward application of a theorem of Stein [9], p. 96.

Part (b) is a slight modification of a theorem proved by Sjöstrand [8].

(b) Functional spaces; pseudo differential operators

DEFINITION 6. Let $1 < p < \infty$; $m > 0$, $L^{m,p}$ is the completion of $S(\mathbb{R}^n)$ for the norm

$$\| \phi \|_{L^{m,p}} = \| \phi \|_{L^p} + \| \bar{F}(|\xi|^m . F\phi) \|_{L^p}.$$

PROPOSITION 3. $L^{m,p}$ is a Banach space, and for integer m, $L^{m,p}$ is the Sobolev space $W^{m,p}$.

Now we define pseudo differential operators in $L^p(\mathbb{R}^n)$. Let $a(\xi)$ be a positively homogeneous function of degree $\alpha > 0$, $a(\xi) \in C^{\infty}(\mathbb{R}^n \setminus \{0\})$. For any $\phi \in S(\mathbb{R}^n)$, we define $a_p(D)\phi$ by:

$$a_p(D)\phi = \bar{F}(a(\xi)F\phi).$$

This defines a bounded operator from $L^{\alpha,p}$ to L^p by virtue of the inequality

$$\|a_p(D)\phi\|_{L^p} \leqslant \|a(\xi)|\xi|^{-\alpha}\|_{M_p} \; \|\phi\|_{L^{\alpha,p}}.$$

Viewed as an unbounded operator in L^p with domain $L^{\alpha,p}$, this operator is closable by virtue of the continuity of multiplication and Fourier transform in $S'(\mathbb{R}^n)$. We also denote by $a_p(D)$ the closure of $a_p(D)$ in L^p.

(c) Cauchy systems in $L^p(\mathbb{R}^n)$; $1 < p < \infty$:

Let $A(\xi)$ be an $N \times N$ matrix-valued function defined on $\mathbb{R}^n$, infinitely differentiable on $\mathbb{R}^n \setminus \{0\}$, positively homogeneous of order $\alpha > 0$. Let us consider the Cauchy problem

$$\frac{\partial U}{\partial t} = i\, A_p(D)U; \quad U(0) = U_0 \in [S(\mathbb{R}^n)]^N. \tag{*}$$

Its solution is

$$U(t,x) = \bar{F}(e^{itA(\xi)}FU_0). \tag{**}$$

We are going to make the following assumptions:

(H1) $A(\xi)$ is diagonalizable: i.e. denoting by S_{n-1} the unit sphere of $\mathbb{R}^n$, there exists a C^{∞} mapping $P(\xi)$ from S_{n-1} to $GL_n(\mathbb{C})$, and N C^{∞}-mappings from S_{n-1} to $\mathbb{C}$: $(\lambda_i)_{i=1,\ldots,N}$, such that for any $\xi \in S_{n-1}$,

$$A(\xi) = P^{-1}(\xi) \begin{pmatrix} \lambda_1(\xi) & & 0 \\ & \ddots & \\ 0 & & \lambda_n(\xi) \end{pmatrix} P(\xi).$$

(H2) the system is parabolic: i.e. for any $i = 1,\ldots,N$, and for any $\xi \in S_{n-1}$, $\mathrm{Im}(\lambda_i(\xi)) \geqslant 0$.

Under (H1) and (H2), the Cauchy problem (*) has a solution for any $U_0 \in [L^p(\mathbb{R}^n)]^N$ which is a distribution in the t-variable, with values in $[L^p(\mathbb{R}^n)]^N$. We are going to study this solution.

First extend $P(\xi)$ to $\mathbb{R}^n \setminus \{0\}$ by homogeneity of degree zero, and $\lambda_i(\xi)$ by homogeneity of degree α.

THEOREM 6. Under the hypothesis (H1), (H2), there exists a σDSG G on $[L^p(\mathbb{R}^n)]^N$, whose generator is $iA_p(D)$. Its order k is any integer, $k > n|\frac{1}{p} - \frac{1}{2}|$. G solves the Cauchy problem (*) whenever $U_0 \in [L^p(\mathbb{R}^n)]^N$.

Proof. For any $U_0 \in [S(\mathbb{R}^n)]^N$ and any $\phi \in \mathcal{D}(\mathbb{R}_+)$, put

$$G(\phi)U_0 = \int_0^\infty \phi(t)(\bar{F}\,(e^{itA(\xi)}FU_0))\,dt.$$

We first prove the estimate

$$\|G(\phi)U_0\|_{[L^p(\mathbb{R}^n)]^N} \leqslant C\, p_k(\phi)\; \|U_0\|_{[L^p(\mathbb{R}^n)]^N}.$$

With this aim, and denoting by $\Delta(\xi)$ the diagonal form of $A(\xi)$, we have

$$G(\phi)U_0 = \bar{F}(P^{-1}(\xi)(\int_0^\infty \phi(t)e^{it\Delta(\xi)}dt)P(\xi)FU_0(\xi)).$$

So we have, with a slight abuse of notation,

$$\|G(\phi)U_0\|_{|L^p(\mathbb{R}^n)|^N} \leqslant \|P^{-1}(\xi)\|_{M_p}\, \|P(\xi)\|_{M_p}$$

$$\|\int_0^\infty \phi(t)e^{it\Delta(\xi)}dt\,\|_{M_p}\, \|U_0\|_{[L^p(\mathbb{R}^n)]^N},$$

the first two norms being finite by Theorem 5. For the third one we note that the matrix involved is diagonal, so that the only thing to compute is, for $j = 1,\dots,N$,

$$\|\int_0^\infty \phi(t)e^{it\lambda_j(\xi)}\,dt\,\|_{M_p}.$$

We have:

$$\int_0^\infty \phi(t)e^{it\lambda_j(\xi)} dt = \int_0^\infty \phi^{(k)}(t)[\int_0^t \frac{(t-s)^{k-1}}{(k-1)!} e^{is\lambda_j(\xi)} ds] dt$$

$$= \int_0^\infty t^k\phi^k(t)[\int_0^1 \frac{(1-s)^{k-1}}{(k-1)!} e^{is\lambda_j(t^{1/\alpha}\xi)} ds] dt$$

$$= \int_0^\infty t^k\phi^{(k)}(t)I(t,\xi) dt.$$

But $I(t,\xi)$ is a Fourier multiplier in $L^p(\mathbb{R}^n)$ (Theorem 5) with M_p norm independent of t (Lemma 4), for $k > n|\frac{1}{p} - \frac{1}{2}|$. So

$$\|G(\phi)U_0\|_{[L^p(\mathbb{R}^n)]^N} \leqslant \|P^{-1}(\xi)\|_{M_p} \|P(\xi)\|_{M_p} \| I(t,\xi)\|_{M_p} \|U_0\|_{L^p} p_k(\phi).$$

This inequality shows that G extends to a linear continuous mapping from T_k to $\mathcal{L}([L^p(\mathbb{R}^n)]^N)$. Furthermore G satisfies condition (a) of Definition 1, because the Fourier transform exchanges convolution and product. And G satisfies trivially condition (b) with $D = [S(\mathbb{R}^n)]^N$.

By construction, the infinitesimal generator of G is $iA_p(D)$, so G solves the Cauchy Problem (*).

COROLLARY 1. If the N functions $(\lambda_i(\xi))_{i=1,\ldots,N}$ are real valued, then $iA_p(D)$ generates a σDG.

G being a σDSG on $[L^p(\mathbb{R}^n)]^N$, the abstract machinery developed in 2 and 3 can apply, and we have:

COROLLARY 2. Under the hypothesis (H1) and (H2), for any $\lambda \in \mathbb{C}$, Re $\lambda > 0$, λ belongs to the resolvent set of $iA_p(D)$, and we have, for Re $\mu > 0$ and $k > n|\frac{1}{p} - \frac{1}{2}|$:

$$\|(\lambda - iA_p(D))^{-\mu}\|_{\mathcal{L}([L^p(\mathbb{R}^n)]^N)} \leqslant C\, |\mu|^k \frac{\Gamma(\mathrm{Re}\,\mu)}{|\Gamma(\mu)|} \frac{|\lambda|^k}{(\mathrm{Re}\,\lambda)^{k+\mathrm{Re}\,\mu}}.$$

Proof. This is a direct translation of the abstract Theorem 1.

COROLLARY 3. Under the hypothesis (H1) and (H2), and for any $U_0 \in [L^{\beta,p}]^N$ with $\beta > \alpha n|\frac{1}{p} - \frac{1}{2}|$, the Cauchy problem admits as a weak solution a function

U(t,x), and we have

$$\|U(t,x)\|_{[L^p(\mathbb{R}^n)]^N} \leq C(1 + t^k) \|U_0\|_{[L^{\beta,p}]^N},$$

where k is an integer greater than $n|p^{-1}-2^{-1}|$.

Proof. This is a direct translation of the abstract Theorem 4 and a use of the fact that $D(A_p(D))$ includes $L^{\alpha,p}$ by construction. This improves a result of Peral [7].

5. APPLICATION TO FIRST-ORDER DIFFERENTIAL SYSTEMS

PROPOSITION 4. The Cauchy problem (*) is well posed in $[L^p(\mathbb{R}^n)]^N$ in the usual sense, and so generates a σDSG of order zero, only if $A(\xi)$ is of order 1 and the real parts of its eigenvalues are linear functions on ξ.

Proof. $P(\xi)$, $P^{-1}(\xi)$ and $\exp(-t \text{ Im } \lambda_j(\xi))$ being Fourier multipliers, the fact that (*) is well posed in $[L^p(\mathbb{R}^n)]^N$ in the usual sense implies that for $j = 1,\ldots,N$,

$$\exp (it \text{ Re } (\lambda_j(\xi))) \in M_p.$$

Then by a classical result (Hörmander [4]; Brenner [2]) Re $\lambda_j(\xi)$ must be a linear function of ξ. And so the order of $A(\xi)$ must be equal to 1.

PROPOSITION 5. Let $A(\xi) = \sum_{i=1}^{n} \xi_i A_i$ where A_i are matrices independent of ξ. If $A(\xi)$ satisfies (H1) and (H2), then $iA_p(D)$ generates a σDSG of order $k > m|\frac{1}{p} - \frac{1}{2}|$ (m being the dimension of the subspace spanned by $A_1,\ldots,A_n$ in $M_N(\mathbb{C})$). Moreover, the Cauchy problem (*) for $A(\xi)$ is well posed in the usual sense in $[L^p(\mathbb{R}^n)]^N$ only if the real parts of the eigenvalues of $A(\xi)$ are linear functions of ξ.

Proof. Theorem 6 asserts that $iA_p(D)$ generates a σDSG of order $n|\frac{1}{p} - \frac{1}{2}|$. This order can be improved to $m|\frac{1}{p} - \frac{1}{2}|$ by a change of variables in the x-variables: let $Q = (q_i^j) \in GL_N(\mathbb{C})$ define the linear change of variables: $y = Qx$. Then the Cauchy problem

$$\frac{\partial U}{\partial t} = \sum_{i=1}^{n} A_i \frac{\partial U}{\partial x_i}$$

becomes

$$\frac{\partial V}{\partial t} = \sum_{j=1}^{n} B_j \frac{\partial V}{\partial y_j},$$

where $V = U \circ Q^{-1}$ and $B_j = \sum_{i=1}^{n} q_j^i A_i$.

So one can find $Q \in GL_N(\mathbb{C})$ such that $B_{m+1} = \ldots = B_n = 0$. Since the true number of y-variables of this problem is m, Theorem 6 gives the desired result.

REFERENCES

[1] M. Balabane and H. Emami-Rad, Smooth distribution group and Schrödinger Equation in $L^p(\mathbb{R}^N)$, J. Math. Anal. Appl. 70 (1), 61-71 (1979).

[2] P. Brenner, V. Thomée and L. Wahlbin, Besov Spaces and Applications to Difference Methods for Initial Value Problems, Lecture Notes in Math. 434. Springer Verlag, Berlin, Heidelberg, New York (1975).

[3] J.C. Chazarain, Problèmes de Cauchy abstraits et applications à quelques problèmes mixtes, J. Funct. Anal. 7, 386-446 (1971).

[4] L. Hörmander, Estimates for translation invariant operators in L^p spaces, Acta Math. 104, 93-140 (1960).

[5] J.L. Lions, Semi-Groupes distributions. Portugalae Math. 19, 141-164 (1960).

[6] W. Littman, The wave operator and L^p norms, J. Math. Mech. 12, 55-68 (1963).

[7] J. Peral, L^p estimates for the wave equation, J. Funct. Anal. 36, 114-145 (1980).

[8] S. Sjöstrand, On the Riesz means of solutions to Schrödinger equation, Ann. Scuola Norm. Sup. Pisa, 24, 331-348 (1970).

[9] E. Stein, Singular Integrals and Differentiability Properties of Functions, Princeton University Press, Princeton N.J. (1970).

[10] E.T. Whittaker and J.N. Watson, A Course of Modern Analysis , Cambridge University Press (1969).

M. Balabane and H. Emami-Rad
Université Paris Nord
93430 Villetaneuse
France

C BARDOS

A priori estimates and existence results for the Vlasov equation

1. INTRODUCTION

The main tool needed for the study of nonlinear evolution equations is a set of a priori estimates. These estimates are very often related to the description of the invariants of the system as given by the physicists.

They are used to show that the equations do have at least one solution (eventually a weak solution), which implies that the system makes sense. The next step is the study of smooth solutions; in general it is easy to prove that with smooth initial data, the solutions remain smooth for a finite time. The fact that the solutions may become singular after a finite time is related to the eventual appearance of turbulence.

In the first section we describe the Vlasov equations and the classical invariant, in which form there is no result guaranteeing the existence of a solution (even a weak solution) for large time.

Next we assume the magnetic field B is zero, which is equivalent to assuming that the electric field E is a gradient. In this situation we demonstrate the existence of a weak solution in any dimension, and of a smooth solution for all time in dimension two. The method can also be applied in dimensions greater than two, giving a lower bound for the time during which the solution remains regular. The method follows on one hand the ideas of Ukai and Okabe [2] or Wollman [4] to obtain a uniform bound of the total amount of charge; on the other hand the most precise method to obtain the differentiability of the solution seems to be the study of the distance between two particles of the fluid as has been done by Wolibner [3] and Kato [1] for the Euler equation.

2. DESCRIPTION OF THE EQUATIONS AND OF THE INVARIANTS

We deal with a plasma which is composed of N particles. We denote by m_α, q_α and $f_\alpha(x,v,t)$ the corresponding mass, charge and probability density for a particle α $(1 \leq \alpha \leq N)$ of the species at time t, position x and velocity v. Neglecting shocks, we assume that the variation of the function

$f_\alpha(x,v,t)$ is due only to the motion of the particle and to the action of the electromagnetic power. This leads to the equations:

$$\frac{\partial f_\alpha}{\partial t} + v\cdot\nabla_x f_\alpha + \frac{q_\alpha}{m_\alpha}\left(E + \frac{vB}{c}\right)\cdot\nabla_v f_\alpha = 0 \tag{1}$$

$$j = \sum_\alpha \int q_\alpha\, v\, f_\alpha(x,v,t)\, dv \tag{2}$$

$$\frac{1}{c}\frac{\partial E}{\partial t} + \frac{4\pi}{c} j = \nabla\wedge B \tag{3}$$

$$\frac{1}{c}\frac{\partial B}{\partial t} + \nabla\wedge E = 0. \tag{4}$$

From this system of equations one deduces the classical relation:

$$\frac{\partial\rho}{\partial t} + \nabla\cdot j = 0,\ \rho = \sum_\alpha \int_v q_\alpha\, f_\alpha(x,v,t)\, dv. \tag{5}$$

Therefore if for $t = 0$ one has $\nabla\cdot E = 4\pi\rho$ this relation remains true for all times.

Multiplication of the equation (1) by $\frac{1}{2}m_\alpha|v|^2$, and some further manipulation lead to the classical invariant which involves the electromagnetic and kinetic energy:

$$\iint \sum_\alpha \frac{m_\alpha}{2}|v|^2 f_\alpha(x,v,t)\, dx\, dv + \frac{1}{8\pi}\int (|E|^2 + |B|^2)\, dx = \text{constant}. \tag{6}$$

The equations (1) are transport equations and therefore they are related to the solutions of the Hamiltonian systems

$$\dot{x}(t) = v_\alpha(t);\quad \dot{v}_\alpha(t) = \frac{-q_\alpha}{m_\alpha}\left(E + \frac{v\wedge B}{c}\right)(x_\alpha(t), v_\alpha(t)) \tag{7}$$

by the relation

$$\frac{d}{dt} f_\alpha\, (x_\alpha(t), v_\alpha(t), t) = 0. \tag{8}$$

Therefore for smooth fields E and B the system (7) defines a group of volume-preserving isomorphisms of $\mathbb{R}^n_x \times \mathbb{R}^n_v$ into itself:

$$(x,v) \to (\text{expt } H_\alpha)(x,v) = (x_\alpha(t), v_\alpha(t)). \tag{9}$$

From (8) and (9) we deduce for $f_\alpha(x,v,t)$ the following invariants:

(i) the positivity of the function $f_\alpha(x,v,t)$ is preserved, and so is its sup norm.

(ii) more generally any L^p norm $(1 \leqslant p \leqslant \infty)$

$$\left(\iint |f_\alpha(x,v,t)|^p \, dx \, dv\right)^{1/p}$$

remains constant.

3. THE VLASOV-POISSON EQUATION

A classical approximation of the Vlasov equation is obtained by neglecting the magnetic field in equations (1) and (4) and keeping the relation $\nabla \cdot E = 4\pi\rho$, which leads to the so-called Vlasov-Poisson equation

$$\frac{\partial f_\alpha}{\partial t} + v \cdot \nabla_x f_\alpha - \frac{q_\alpha}{m_\alpha} \nabla_x \Phi \cdot \nabla_v f_\alpha = 0 \tag{10}$$

$$- \Delta\Phi = 4\pi \sum_2 q_\alpha \int f_\alpha(x,v,t) \, dx \, dv. \tag{11}$$

Since the relation $\nabla \wedge j = 0$ is in general not true, the equation (3) does not make sense, but the positivity of $f_\alpha(x,v,t)$ and the fact that the norm

$$\left(\iint (f_\alpha(x,v,t))^p \, dx \, dv\right)^{1/p} \quad (1 \leqslant p \leqslant \infty)$$

are constant remains true. This will be enough to study the existence and the regularity of the solution of (10), (11). In particular we have

THEOREM 1. For any family of initial data $f^0_\alpha(x,v)$ belonging to the space $L^1(\mathbb{R}^n_x \times \mathbb{R}^n_v) \cap L(\mathbb{R}^n_x \times \mathbb{R}^n_v)$ the Vlasov-Poisson equation admits at least one weak solution.

Remark 1. By a weak solution we mean a set of functions f_α, Φ which constitutes a solution in the sense of distribution of the system

$$\frac{\partial}{\partial t} f_\alpha + \sum_{i=1}^n \frac{\partial}{\partial x_i} (v_i f_\alpha) - \frac{\partial}{\partial v_i}\left(\frac{q_\alpha}{m_\alpha} f_\alpha \frac{\partial \Phi}{\partial x_i}\right) = 0 \tag{12}$$

$$\Delta\Phi = -4\pi \sum_\alpha q_\alpha \int f_\alpha(x,v,t) \, dv. \tag{13}$$

Sketch of the proof: We will omit the details and notice only the following essential facts.

(i) If the functions f_α belong to the space

$$L^\infty(\mathbb{R}_t^+; L^1(\mathbb{R}_x^n \times \mathbb{R}_v^n) \cap L^\infty(\mathbb{R}_x^n \times \mathbb{R}_v^n)),$$

then the functions $v_i \frac{\partial f_\alpha}{\partial x_i}$ and $\frac{\partial \Phi}{\partial x_i} f_\alpha$ belong to the space $L^\infty(\mathbb{R}_t^+; L^1(K))$ for any compact K of $\mathbb{R}_x^n \times \mathbb{R}_v^n$; therefore the relations (12) are well defined in the sense of distributions.

(ii) The a priori estimates which have been observed on the exact solution will remain valid for any convenient approximation $(f_\alpha^\varepsilon, \Phi^\varepsilon)$ of this solution. Then the only difficulty is to prove that we have

$$\lim_{\varepsilon \to 0} \frac{\partial \Phi^\varepsilon}{\partial x_i} f_\alpha^\varepsilon = \frac{\partial \Phi}{\partial x_i} f_\alpha \tag{14}$$

However f_α^ε is a sequence of uniformly bounded functions in L^∞; one can therefore define a subsequence with weak star convergence in L^∞. On the other hand, using the fact that $\Delta\Phi^\varepsilon$ is bounded in $L^1(\mathbb{R}^n)$ one can prove, by a compactness argument, the existence of a subsequence converging strongly in L^1_{loc}; this is enough for the relation (14).

Remark 2. One of two main ingredients for the proof of Theorem 1 is the estimate

$$\iint f_\alpha(x,v,t)\, dx\, dv = \text{cte} \tag{15}$$

on the other hand, there is no information on the local charge

$$\int f_\alpha(x,v,t)\, dv = \rho_\alpha(x,t),$$

and this in turn will be the main difficulty in proving the existence of a smooth solution.

Intuitively the fact that the ρ_α may become arbitrarily large would be related to the fact that some particle may reach arbitrarily large velocity. It is not clear if this may happen or if it has some physical interpretation in connection with turbulence.

As far as the mathematical results are concerned, there is an important difference between the case $n = 2$ and the case $n \geq 3$. This is given in

THEOREM 2. Assume that the initial data $f^o_\alpha(x,v)$ satisfy the relation:

$$\left.\begin{aligned} &\iint f^o_\alpha(x,v)\,dx\,dv \leq M_\alpha < \infty \\ &\sup_{x,v} f^o_\alpha(x,v) = N_\alpha < \infty \end{aligned}\right\} \tag{16}$$

and

$$\sup_{(x,v)} (1 + |v|)^\gamma f_\alpha(x,v) \leq k^\alpha < \infty \quad (\gamma > n) \tag{17}$$

then the Vlasov-Poisson equation has a weak solution which satisfies the additional a priori estimate:

$$\left.\begin{aligned} &\sup_x \int f_\alpha(x,v,t)dv \leq C\, e^{K|t|} \qquad \text{for } n = 2 \\ &\sup_x \int f_\alpha(x,v,t)dv \leq C(1-|t|\; K\, D^{n-2})^{-1/(n-2)} \quad \text{for } n \geq 3. \end{aligned}\right\} \tag{18}$$

Sketch of the proof of Theorem 2: Once again we will only give the a priori estimate. We denote by $G_n(x)$ the Green function of the operator in $\mathbb{R}^n$.

$$G_n(x) = \frac{C_n}{|x|^{n-2}} \text{ for } n \geq 3,\ G_n(x) = \frac{1}{2\pi} \log \frac{1}{|x|} \text{ for } n = 2.$$

We denote by $\rho(x,t)$ the total amount of charge:

$$\rho(x,t) = \sum_\alpha \int q_\alpha\, f_\alpha(x,v,t)\, dv.$$

From the relation:

$$\Phi(x,t) = \int G_n(x-x')\, \rho(x',t)\, dx' \tag{19}$$

we deduce the estimate.

$$\begin{aligned} |\nabla_x(G * \rho)|_\infty &\leq C\{\int_{|x-x'|<r} \frac{\rho(x')}{|x-x'|^{n-1}} dx' + \int_{|x-x'|>r} \frac{\rho(x')dx'}{|x-x'|^{n-1}}\} \\ &\leq C\{|\rho(\cdot,t)|_\infty r + \frac{1}{r^{n-1}} |\rho(\cdot,t)|_1\}, \end{aligned} \tag{20}$$

where C denotes a constant independent of ρ, and $r > 0$. Taking the minimum with respect to r of the right-hand side of (20) we obtain the relation:

$$|\nabla_x \Phi|_\infty \leqslant C|\rho(\cdot,t)|_1^{1/n}\ |\rho(\cdot,t)|_\infty^{(n-1)/n}. \tag{21}$$

Now we introduce the following partition of the space $\mathbb{R}_v^n$ into two regions I and II, defined by the relations:

$$(I) = \{v \mid |v| \geqslant 2\,\frac{q_\alpha}{m_\alpha} \int_0^t |\nabla\Phi(\cdot,s)|_\infty\ ds\},$$

$$(II) = \{v \mid |v| \leqslant 2\,\frac{q_\alpha}{m_\alpha} \int_0^t |\nabla\Phi(\cdot,s)|_\infty\ ds\}.$$

Using the relation (10) and the Hamiltonian flow e^{tH_α} defined by (7), we have

$$\int f_\alpha(x,v,t)dv = \int f_\alpha^o\,(e^{-tH_\alpha}(x,v))dv$$

$$\leqslant N_\alpha\,(\int_{II} dv) + \int_I f_\alpha^o(e^{-tH_\alpha}(x,v))dv. \tag{22}$$

By the mean value theorem we have

$$|v - (e^{-tH_\alpha}(x,v))_2| \leqslant \frac{q_\alpha}{m_\alpha}\int_0^t |\nabla\Phi(\cdot,s)|_\infty\ ds. \tag{23}$$

Therefore, using (17), we obtain for the second integral of the right-hand side of (22), the estimate

$$\int_I f_\alpha^o(e^{-tH_\alpha}(x,v))dv \leqslant K^\alpha \int_J \left(1 + \frac{|v|}{2}\right)^{-\gamma} dv \leqslant C. \tag{24}$$

With (20) and (24) we obtain for $\rho(\cdot,t)$ the relation

$$|\rho(\cdot,t)| \leqslant C + D(\int_0^t |\nabla\Phi(\cdot,s)|_\infty\ ds)^n. \tag{25}$$

Now the relation (21) gives the inequality

$$|\rho(\cdot,t)|_\infty \leqslant C + D\,(\int_0^t |\rho(\cdot,s)|_\infty^{(n-1)/n}\ ds)^n, \tag{26}$$

from which one deduces, by a comparison argument, the relation (18).

Once the relation (18) is proven we know that the Laplacian of the potential Φ is bounded in $L^\infty(\mathbb{R}_x^n)$. This is not enough, however, to obtain directly a proof of the regularity of the solution of Vlasov-Poisson equation, because the fact that $\Delta\Phi \in L^\infty$ does not imply that the vector field $\nabla\Phi$ is Lipschitzian. The same difficulty has been overcome by Wolibner [3] and Kato [1] for the Euler equation and following their ideas we show that, for smooth initial data, the charge $\rho(x,t)$ satisfies a Hölder regularity condition.

This will be enough to imply that $\nabla\Phi$ is Lipschitzian and therefore that the solution in dimension two is for all time,but in dimensions greater than two it is for a finite time, and is as regular as the initial data.

THEOREM 3. We assume that the initial data $f_\alpha^0(x,v)$ of the Vlasov-Poisson equation have the following properties: they belong to the space $W^{1,\infty}(\mathbb{R}_x^n \times \mathbb{R}_v^n)$ and they satisfy the estimates:

$$f_\alpha^0(x,v) \leqslant k_\alpha(1 + |v|)^{-\gamma} \qquad (\gamma > n) \tag{27}$$

$$|f_\alpha^0(x,v) - f_\alpha^0(y,w)|$$

$$\leqslant C_\alpha|(x,v) - (y,w)|^\eta(1 + |v|)^{-\gamma} \text{ for } |w| > \frac{|v|}{2} \ (\eta > 0) \tag{28}$$

then the Poisson-Vlasov equation has a solution which satisfies the relation (18) and for which the charge $\rho(x,t)$ is an Hölderian function of x.

Remark 3. The condition (27) and (28) are readily satisfied for smooth initial data with compact support.

Sketch of the proof: Following [2] or [3] we denote by s(t) the distance between two particles which follow the field H_α:

$$s(t) = |(x(y),v(t)) - (y(t),w(t))|.$$

Since we have

$$\dot{x}(t) = v(t), \quad \dot{v}(t) = \frac{q_\alpha}{m_\alpha}\nabla\Phi$$

and $\Delta\Phi$ bounded in $L^\infty(\mathbb{R}_x^n)$, we obtain the following estimate for s(t):

$$|\dot{s}(t)| \leqslant Cs(t)\left(1 + \log\frac{D}{s(t)}\right). \tag{29}$$

By a comparison argument with the solution of the differential equation:

$$\dot{y} = - Cy\left(1 + \log \frac{D}{y}\right) \tag{30}$$

We deduce the estimate

$$|s(t)|^{\eta e^{-Ct}} \geqslant C(t)\ |s(0)|^{\eta}. \tag{31}$$

Finally we use the fact $f_\alpha(x,v,t)$ is constant on the trajectories of M_α to obtain with (31) and the relations (27), (31) an estimate of the form:

$$|\int (f_\alpha(x,v,t) - f_\alpha(y,v,t))dv| \leqslant C(t)\ |x-y|^{\eta e^{-Ct}}, \tag{32}$$

which completes the proof of Theorem 3.

REFERENCES

[1] T. Kato, On the classical solution of the two dimensional non stationary Euler Equation, Arch. Rat. Mech. Anal. 35, 302-324 (1967).

[2] S. Ukai and T. Okabe, On the classical solution in the large of the two dimensional Vlasov equation, Osaka J. of Math. No. 15, 245-261 (1978).

[3] W. Wolibner, Un théorème sur l'existence d'un mouvement plan d'un fluide parfait homogène et incompressible pendant un temps infiniment long, Math. Z. 37, 727-738 (1933).

[4] S. Wollman, Global in time solutions of the two dimensional Vlasov-Poisson System. Comm. Pure and Appl. Math. 33, 173-197 (1980).

C. Bardos
Université de Paris-Nord
Avenue J-B Clément
93430 Villetaneuse
France

P BÉNILAN

A strong regularity L^p for solution of the porous media equation

Consider the porous media equation

$$u_t = \Delta u^m \text{ on } Q =]0,T[\times \Omega$$

where Ω is any open set in $\mathbb{R}^N$ and $m > 1$. We will always write u^m for $|u|^{m-1}u$. If $\Omega \neq \mathbb{R}^N$, we consider some boundary conditions.

This equation has been extensively studied; in particular in the nonnegative case, it has been shown, for given initial data, existence (and uniqueness) of a strong solution, that is a solution such that u_t and Δu^m are functions in $L^1_{loc}(Q)$. We will extend this result to the general case (see Proposition 1 of Section 2).

This paper is devoted to showing that the nonnegative solutions of the porous media equation have not only strong regularity L^1 but also some strong regularity L^p for some $p > 1$. This is interesting for many reasons; in particular this shows that the derivatives $\partial^2 u^m/\partial x_i \partial x_j$ are functions.

Actually we will give two different results. In Section 1 we prove an abstract result whose application to the porous media equation gives strong regularity L^p for any $1 \leqslant p < 1 + \frac{1}{m}$. In Section 2, besides the application of the abstract result and the general statement about strong solution, we show in the case $N = 1$ and $\Omega = \mathbb{R}$, that the nonnegative solutions of the porous media equation are classical if $1 < m < 2$ and have strong regularity L^p for any $1 \leqslant p < 1 + \frac{1}{m-2}$ (resp. $p = \infty$) if $m > 2$ (resp. $m = 2$).

1. AN ABSTRACT RESULT

Let Ω be a measure space with σ-finite measure and $\phi: L^2(\Omega) \to [0,\infty]$ be convex l.s.c. with $\phi(0) = 0$. We assume

(H1) the T-accretiveness for the L^1-norm of the subdifferential $\partial\phi$ which is characterized by the property (see [9]):

$$\phi(w \wedge (\hat{w}-k)) + \phi((w-k) \vee \hat{w}) \leqslant \phi(w) + \phi(\hat{w}) \qquad (1)$$

for any $w, \hat{w} \in L^2(\Omega)$.

(H2) the positive homogeneity of ϕ of order $1 + \alpha$, for some given $\alpha > 0$:

$$\phi(\lambda w) = \lambda^{1+\alpha}\phi(w) \text{ for any } w \in L^2(\Omega),\ \lambda > 0. \tag{2}$$

(H3) the density of $D(\phi) = \{w \in L^2(\Omega);\ \phi(w) < \infty\}$ in $L^2(\Omega)^+ = \{w \in L^2(\Omega);$ $w \geqslant 0$ a.e. on $\Omega\}$.

We consider the operator A in $L^1(\Omega)$ defined by

$$\left.\begin{array}{l} v \in Au \Longleftrightarrow v \in L^1(\Omega),\ u \in L^1(\Omega) \cap L^\infty(\Omega) \text{ and} \\ \phi(w) \geqslant \phi(u^m) + \int v(w-u^m) \text{ for any } w \in L^2(\Omega) \cap L^\infty(\Omega) \end{array}\right\} \tag{3}$$

where m is given and satisfies

$$m > 1,\ m\alpha \neq 1 \text{ and } u^m = |u|^{m-1}u. \tag{4}$$

THEOREM 1. Under the assumptions and notations above, for any $u_0 \in L^1(\Omega) \cap L^\infty(\Omega)$ with $u_0 \geqslant 0$ a.e. on Ω, there exists a unique u such that

$$\left.\begin{array}{l} u \in C([0,\infty[;\ L^1(\Omega)) \cap W^{1,1}_{loc}(]0,\infty[;\ L^1(\Omega)) \\ \frac{du}{dt}(t) + Au(t) \ni 0 \text{ a.e. } t \in]0,\infty[,\ u(0) = u_0. \end{array}\right\} \tag{5}$$

Moreover

$$t\,\frac{du}{dt} \in L^\infty(0,\infty;L^1(\Omega)) \tag{6}$$

and for any $1 < p < 1 + \frac{1}{m}$, $\frac{du}{dt} \in L^p_{loc}(]0,\infty[\times\ \Omega)$ in the sense

$$\iint_{I\times K} \left|\frac{du}{dt}\right|^p < \infty \tag{7}$$

for any compact set I in $]0,\infty[$ and integrable set K in Ω.

The solution u of (5) has many other properties which follow by the general theory of nonlinear semi-groups using

LEMMA 1. Let $\bar{A}$ be the closure of A in $L^1(\Omega)$:

$$\left.\begin{array}{l} v \in \bar{A}u \Longleftrightarrow \text{there exists } v_n \in Au_n \text{ such that} \\ u_n \to u,\ v_n \to v \text{ in } L^1(\Omega). \end{array}\right\} \tag{8}$$

Then

(1) $\bar{A}$ is a m-T-accretive submarkovian operator in $L^1(\Omega)$, namely: for any $\lambda > 0$, $T_\lambda = (I + \lambda\bar{A})^{-1}$ is an application everywhere defined on $L^1(\Omega)$ and satisfying

$$\int (T_\lambda u - T_\lambda \hat{u})^+ \leqslant \int (u-\hat{u})^+,$$

$$\text{inf ess } u \wedge 0 \leqslant T_\lambda u \leqslant \text{sup ess } u \vee 0,$$

for any $u, \hat{u} \in L^1(\Omega)$.

(2) $D(\bar{A})$ is dense in $L^1(\Omega)^+$

(3) $\bar{A}$ is positively homogeneous of order $m\alpha$:

$$\bar{A}(\lambda u) = \lambda^{m\alpha}\bar{A}(u) \text{ for any } u \in D(\bar{A}),\ \lambda > 0.$$

(4) A is the restriction of $\bar{A}$ to $(L^1(\Omega) \cap L^\infty(\Omega)) \times L^1(\Omega)$:

$$v \in Au \iff v \in \bar{A}u \text{ and } u \in L^\infty(\Omega).$$

Using the Crandall-Liggett theorem, $\bar{A}$ 'generates' a semi-group of applications $S(t) : \bar{D} \mapsto \bar{D}$, where $\bar{D}$ is the closure of $D(\bar{A})$ in $L^1(\Omega)$:

$$S(t)u_0 = \lim_{n\to\infty} T^n_{t/n}\, u_0 \text{ for any } u_0 \in \bar{D},\ t > 0$$

the limit being in $L^1(\Omega)$ uniformly for t bounded. Clearly we have

$$\int (S(t)u_0 - S(t)\hat{u}_0)^+ \leqslant \int (u_0 - \hat{u}_0)^+ \tag{9}$$

$$\text{inf ess } u_0 \wedge 0 \leqslant S(t)u_0 \leqslant \text{sup ess } u_o \vee 0. \tag{10}$$

Moreover, by the general theory of nonlinear semi-groups (see [3], [10] or [14]), a function u is a solution of

$$u \in C([0,\infty[;\ L^1(\Omega)) \cap W^{1,1}_{loc}(]0,\infty[;\ L^1(\Omega))$$

$$\frac{du}{dt}(t) + \bar{A}u(t) \ni 0 \text{ a.e. } t \in\ ]0,\infty[,\ u(0) = u_0$$

iff $\quad u_0 \in \bar{D},\ u(t) = S(t)u_0$ and $u \in W^{1,1}_{loc}(]0,\infty[;\ L^1(\Omega))$.

Then, according to (10) and Lemmas 1, 2 and 4, Theorem 1 is reduced to the statement:

for $u_0 \in L^1(\Omega) \cap L^\infty(\Omega)^+$, the function $u(t) = S(t)u_0$

satisfies $u \in W^{1,1}_{loc}(]0,\infty[;L^1(\Omega))$,

$$t\,\frac{du}{dt} \in L^\infty(0,\infty;L^1(\Omega)) \text{ and } \frac{du}{dt} \in L^p_{loc}(]0,\infty[\times\Omega) \tag{11}$$

for any $1 < p < 1 + \frac{1}{m}$.

In order to prove this, we will use first the results of [8] about semi-groups generated by homogeneous operators:

LEMMA 2. (see [8]): For any $u_0 \in \bar{D}$,

$$\|S(t+h)u_0 - S(t)u_0\|_{L^1} \leq \frac{2\|u_0\|_{L^1}}{|m\alpha - 1|}\,\frac{h}{t} \quad \text{for any } t > 0,\ h > 0. \tag{12}$$

If moreover $u_0 \geq 0$, then

$$(m\alpha-1)(S(t+h)u_0 - S(t)u_0) \geq -\frac{h}{t}\,S(t)u_0 \text{ for any } t > 0,\ h > 0. \tag{13}$$

Secondly, we will use an energy estimate:

LEMMA 3. Let $u_0 \in \bar{D} \cap L^{m+1}(\Omega)$; then the function $U(t) = |S(t)u_0|^{\frac{m-1}{2}}(S(t)u_0)$ satisfies

$$U \in W^{1,2}_{loc}(]0,\infty[;\ L^2(\Omega)) \text{ and}$$
$$\iint_{]0,\infty[\times\Omega} t\left(\frac{dU}{dt}\right)^2 \leq \frac{m+1}{4m}\int |u_0|^{m+1}. \tag{14}$$

The last argument is the following:

LEMMA 4. Let K be a measure space with finite measure, $v \in C([t_0,t_1];\ L^1(K))$, being nonnegative, nondecreasing in t and satisfying

$$v^\gamma \in W^{1,r}(t_0,t_1;\ L^r(K))$$

for some $\gamma > 1$ and $r > 1$. Then

$$v \in W^{1,p}(t_0,t_1;\ L^p(K))$$

for any $1 \leq p < 1 + \frac{1}{\gamma r' - 1}$ where $r' = \frac{r}{r-1}$.

Proof of Theorem 1: Let $u_0 \in L^1(\Omega) \cap L^\infty(\Omega)^+$, $u(t) = S(t)u_0$. We have

$$u \in C([0,\infty[; L^1(\Omega)), u \geqslant 0 \text{ (see (10).)}$$

Set $v(\tau) = \tau u(\tau^{m\alpha-1})$; we have $v \in C(]0,\infty[; L^1(\Omega))$ nonnegative and nondecreasing in τ (see (13)). Also by Lemma 3,

$$v^{\frac{m+1}{2}} \in W^{1,2}_{loc}(]0,\infty[; L^2(\Omega)).$$

Then by Lemma 4, $v \in W^{1,p}_{loc}(]0,\infty[; L^p_{loc}(\Omega))$ for any $1 \leqslant p < 1 + \frac{1}{m}$. It is the same for

$$u(t) = t^{-1/m\alpha-1}\ v(t^{1/m\alpha-1}).$$

We end up with (12) to obtain

$$t\frac{du}{dt} \in L^\infty(0,\infty; L^1(\Omega)).$$

Proof of Lemma 4: Let $\varepsilon > 0$ and set $v_\varepsilon(t) = (v(t) - \varepsilon)^+$: it is nondecreasing in t and $v_\varepsilon \in W^{1,r}(t_0,t_1; L^r(K))$ with

$$\frac{dv_\varepsilon}{dt}(t) = \frac{X_\varepsilon(t)}{\gamma v(t)^{\gamma-1}}\ \frac{d}{dt}\ v^\gamma(t) \text{ a.e. } t \in]0,\infty[$$

where $X_\varepsilon(t)$ is the characteristic function of $\{v(t) > \varepsilon\}$. (This follows for instance by the Appendix in [11].)

Let $p > 1$ and $0 \leqslant \nu \leqslant p$; we have

$$\left(\frac{dv_\varepsilon}{dt}\right)^p = \left(\frac{1}{\gamma}\ \frac{dv^\gamma}{dt}\right)^\nu\ \left(\frac{X_\varepsilon}{v^{1-\sigma}}\ \frac{dv_\varepsilon}{dt}\right)^{p-\nu}$$

with

$$\sigma = 1 - \frac{\nu(\gamma-1)}{p-\nu}.$$

By assumption

$$\left(\frac{dv^\gamma}{dt}\right)^\gamma \in L^{r/\nu}(]t_0,t_1[\times K).$$

Choose ν such that $p-\nu = 1 - \frac{\nu}{r}$, that is

$$\nu = (p-1)r'.$$

With this value

$$\sigma = 1 - \frac{r(p-1)(\gamma-1)}{r-p}$$

and the assumption $p < 1 + \frac{1}{\gamma r'-1}$ means $\sigma > 0$. With this assumption we have also $\nu < r$; by the Hölder inequality, then

$$\iint \left(\frac{dv_\varepsilon}{dt}\right)^p \leq \frac{1}{\gamma^\nu} \left(\iint\left(\frac{dv^\gamma}{dt}\right)^r\right)^{\nu/r} \left(\iint \frac{X_\varepsilon}{v^{1-\sigma}} \frac{dv_\varepsilon}{dt}\right)^{p-\nu}.$$

Now $(v^\sigma-\varepsilon^\sigma)^+ \in W^{1,r}(t_0,t_1; L^r(K))$ and

$$\frac{d}{dt}(v^\sigma-\varepsilon^\sigma)^+(t) = \frac{\sigma X_\varepsilon(t)}{v(t)^{1-\sigma}} \frac{dv_\varepsilon}{dt}(t) \quad \text{a.e. } t \in]0,\infty[$$

then

$$\iint \frac{X_\varepsilon}{v^{1-\sigma}} \frac{dv_\varepsilon}{dt} = \frac{1}{\sigma} \int_K (v^\sigma-\varepsilon^\sigma)^+(t)\Big|_{t=t_0}^{t=t_1} \leq \frac{1}{\sigma} \int_K v(t_1)^\sigma$$

$$\leq \frac{1}{\sigma} \text{ meas } K^{1-\sigma}\left(\int_K v(t_1)\right)^\sigma.$$

In this way, we obtain an estimate of $\iint\left(\frac{dv_\varepsilon}{dt}\right)^p$ independent of $\varepsilon > 0$. Then at the limit, $v \in W^{1,p}(t_0,t_1; L^p(K))$ and

$$\iint\left(\frac{dv}{dt}\right)^p \leq \frac{1}{\gamma^\nu \sigma^{p-\nu}} (\text{meas } K)^{\nu(\gamma-1)} \left(\iint\left(\frac{dv^\gamma}{dt}\right)^r\right)^{\nu/r} \left(\int v(t_1)\right)^{p-\nu\gamma}. \tag{15}$$

Remark 1: This lemma appears like an interpolation result; we cannot, however, relate it to a classical one. We will find a similar result for second-order derivatives (see Lemma 7). Some extension to σ-finite measure space is open and will be interesting; in particular we conjecture that the solution u of (5) is $W^{1,p}_{loc}(]0,\infty[; L^p(\Omega))$ but are not able to prove it.

Proof of Lemma 3: This lemma is a particular case of an energy estimate for evolution equations of the type

$$\frac{du}{dt} + \partial\phi\partial\Psi u \ni 0.$$

We give a formal computation for (14); a justification could be found in more general cases in [5].

Set $u(t) = S(t)u_0$; we have

$$\left(\frac{du}{dt}(t)\right)^2 = \left(\frac{m+1}{2}\right)^2 (u(t))^{m-1}\left(\frac{du}{dt}(t)\right)^2$$

$$= \frac{(m+1)^2}{4m} \frac{du^m}{dt}(t) \frac{du}{dt}(t) = -\frac{(m+1)^2}{4m} \partial\phi(u^m(t)) \frac{du^m}{dt}(t)$$

$$\frac{d}{dt}\phi(u^m(t)) = \int \partial\phi(u^m(t)) \frac{du^m}{dt}(t).$$

Then

$$\iint t\left(\frac{du}{dt}\right)^2 = -\frac{(m+1)^2}{4m} \int t \frac{d}{dt}\phi(u^m(t))\, dt$$

$$\leqslant \frac{(m+1)^2}{4m} \int \phi(u^m(t))\, dt$$

but

$$\phi(u^m(t)) \leqslant \int u^m(t)\partial\phi(u^m(t)) = -\int u^m(t) \frac{du}{dt}(t)$$

$$= -\frac{1}{m+1}\frac{d}{dt}\int u^{m+1}(t),$$

then the result follows.

Proof of Lemma 1: As is proved in [9], the assumption (1) is equivalent to

$$\left.\begin{array}{l}\phi(w-p(w-\hat{w})) + \phi(\hat{w}+p(w-\hat{w})) \leqslant \phi(w) + \phi(\hat{w}) \\ \text{for any } w, \hat{w} \in L^2(\Omega) \text{ and } p:\mathbb{R} \to \mathbb{R} \text{ Lipschitz} \\ \text{continuous with } 0 \leqslant p' \leqslant 1 \text{ and } p(0) = 0.\end{array}\right\} \quad (16)$$

Let $\tilde{A}$ be the operator in $L^1(\Omega)$ defined by

$$\left.\begin{array}{l}v \in \tilde{A}u \iff v \in L^1(\Omega),\ u \in L^1(\Omega) \text{ and} \\ \phi(w) \geqslant \phi(w-p(w-u^m)) + \int vp(w-u^m) \\ \text{for any } w \in L^2(\Omega) \text{ and } p \in P_0,\end{array}\right\} \quad (17)$$

where P_0 is the set of functions $p:\mathbb{R} \to \mathbb{R}$ bounded Lipschitz continuous with $0 \leqslant p' \leqslant 1$ and $p \equiv 0$ on a neighbourhood of 0.

Note first that $p(w-u^m) \in L^1(\Omega) \cap L^\infty(\Omega)$ for any $w \in L^2(\Omega)$, $u \in L^1(\Omega)$:

indeed p is bounded and if $p \equiv 0$ on $[-\delta,\delta]$,

$$\{p(w-u^m) \neq 0\} \subset \{|w| > \tfrac{\delta}{2}\} \cup \{|u| > (\tfrac{\delta}{2})^{1/m}\},$$

which is an integrable set. This shows that the terms in (17) are well defined and also easily that $\tilde{A}$ is closed in $L^1(\Omega)$.

Note now that $\tilde{A}$ is an extension of A: indeed let $v \in Au$, $w \in L^2(\Omega)$ and $p \in P_0$. By definition of A,

$$\phi(u^m + p(w-u^m)) \geq \phi(u^m) + \int v\, p(w-u^m),$$

$$\phi(u^m) \leq \int vu^m < \infty.$$

Then using (16) with $\hat{w} = u^m$, we obtain (17). Also A is the restriction of $\tilde{A}$ to

$$(L^1(\Omega) \cap L^\infty(\Omega)) \times L^1(\Omega);$$

indeed let $v \in \tilde{A}u$ with $u \in L^\infty(\Omega)$ and apply (17) with $p(r) = (\text{sign } r)(|r|-\delta)^+ \wedge k$ and let $\delta \to 0$, $k \to \infty$ to obtain (3).

The operator A is T-accretive: indeed let $v \in Au$, $\hat{v} \in A\hat{u}$; let $p \in P_0$ and apply (16) with $w = \hat{u}^m \in L^1(\Omega) \cap L^\infty(\Omega)$

$$\phi(\hat{u}^m) \geq \phi(\hat{u}^m - p(\hat{u}^m-u^m)) + \int vp(\hat{u}^m-u^m).$$

Using $\hat{p}(r) = -p(r)$, we have also

$$\phi(u^m) \geq \phi(u^m+p(\hat{u}^m-u^m)) - \int \hat{v}p(\hat{u}^m-u^m).$$

Add and use (16) to obtain

$$\int (\hat{v}-v)p(\hat{u}^m-u^m) \geq 0.$$

Apply with $p(r) = (r-\varepsilon)^+ \wedge \varepsilon$, divide by $\varepsilon > 0$ and let $\varepsilon \to 0$ to obtain

$$\int_{[\hat{u}>u]} \hat{v}-v = \int (v-\hat{v}) \operatorname{sign}_0^+(\hat{u}^m-u^m) \geq 0.$$

Note also that

$$\int j(u) \leq \int j(u+\lambda v) \text{ for any } v \in \tilde{A}u,\ \lambda > 0 \tag{18}$$

and $j:\mathbb{R} \to [0,\infty]$ convex l.s.c. with $j(0) = 0$. Indeed it is enough to prove it for the convex functions $j(r) = (r \pm k)^{\pm}$ with any $k > 0$; then (18) follows (and actually is equivalent) by

$$\int_{[u>k]} v \geqslant 0 \geqslant \int_{[u<-k]} v \quad \text{for any } v \in \tilde{A}u,\ k > 0. \tag{19}$$

But, using (16) with $w = 0$,

$$\int vp(u^m) \geqslant \phi(p(u^m)) \geqslant 0 \text{ for any } v \in \tilde{A}u,\ p \in \mathcal{P}_0.$$

Apply with $p(r) = (r-k)^+ \wedge \varepsilon$ (resp. $-[(-r-k)^+ \wedge \varepsilon]$), divide by $\varepsilon > 0$ and let $\varepsilon \to 0$ to obtain (19).

We show now that

$$R(I+\lambda A) \supset L^1(\Omega) \cap L^\infty(\Omega) \text{ for any } \lambda > 0$$

using the Brézis-Strauss method (see [12]): for $f \in L^1(\Omega) \cap L^\infty(\Omega)$, the function

$$\beta(r) = (\text{sign } r)(|r|^{1/m} \wedge \|f\|_{L^\infty}) - \frac{1}{m} \|f\|_{L^\infty}^{1/m-1} r.$$

is continuous nondecreasing with $\beta(0) = 0$; by Proposition 2.17 in [11], using (16), the operator $\beta + \lambda\partial\phi$ is maximal monotone in $L^2(\Omega)$; then there exists $w \in L^2(\Omega)$ such that

$$u = \frac{1}{m} \|f\|_{L^\infty}^{1/m-1} w + \beta(w) \in L^2(\Omega)$$

$$u + \lambda\partial\phi(w) \ni f.$$

Using the proof of (18) above, one easily sees that

$$u \in L^1(\Omega) \cap L^\infty(\Omega) \text{ and } \|u\|_{L^\infty} \leqslant \|f\|_{L^\infty}.$$

Then $w = u^m$ and $u + \lambda Au \ni f$.

By the results proved above, we can prove (1) and (4); the statement (3) follows directly by the assumption (2) on ϕ. To end up we prove (2): first by the assumption (3) on ϕ and using also (16), one can show that the set

$$\{u \in L^1(\Omega) \cap L^\infty(\Omega);\ u^m \in D(\phi)\}$$

is dense in $L^1(\Omega)^+$; let u be in this set and $u_\lambda = (I+\lambda A)^{-1}u$. We have $\|u_\lambda\|_{L^p} \leq \|u\|_{L^p}$ for any $1 \leq p \leq \infty$ (use (18)); let $\lambda_n \to 0$ such that $u_{\lambda_n} \rightharpoonup \bar{u}$ (in $L^p(\Omega)$ for any $1 < p < \infty$). Passing to the limit in the inequality

$$\phi(u^m) \geq \phi(u_\lambda^m) + \int \frac{u-u_\lambda}{\lambda}(u^m-u_\lambda^m),$$

we obtain

$$\int (u-\bar{u})(u^m-\bar{u}^m) \leq 0$$

and then $\bar{u} = u$. It follows that $u_\lambda \to u$ in $L^p(\Omega)$ for any $1 < p < \infty$ and also for $p = 1$ (see Lemma in [18]).

Remark 2. Lemma 1 is a particular case of general extensions of the Brézis-Strauss theorem (see [10]).

2. THE POROUS MEDIA EQUATION

First we apply the abstract result to the porous media equation:

THEOREM 2. Let Ω be any open set in $\mathbb{R}^N$, $m > 1$ and V be a closed subspace in $H^1(\Omega)$ containing $H_0^1(\Omega)$ and satisfying

$$w \in V \Longrightarrow (w-1)^+ \in V. \tag{20}$$

Then for any $u_0 \in L^1(\Omega)$ with $u_0 \geq 0$, there exists a unique solution of

$$\left.\begin{array}{l} u \in C([0,\infty[; L^1(\Omega)) \cap W^{1,1}_{loc}(]0,\infty[; L^1(\Omega)) \\ u(0) = u_0 \text{ and a.e. } t \in]0,\infty[\\ u(t) \in L^\infty(\Omega),\ u(t)^m \in V \text{ and} \\ \int \frac{du}{dt}(t)w = \int \text{grad } u(t)^m.\text{grad } w \text{ for every } w \in V. \end{array}\right\} \tag{21}$$

Moreover $t\,\frac{du}{dt} \in L^\infty(0,\infty; L^1(\Omega))$ and for any $1 < p < 1 + \frac{1}{m}$, $0 < t_0 < t_1 < \infty$ and $R > 0$

$$u_t = \Delta u^m \in L^p(]t_0,t_1[\times \Omega_R), \tag{22}$$

where $\Omega_R = \{x \in \Omega;\ |x| < R\}$.

Proof: Set

$$\phi(w) = \begin{cases} \frac{1}{2}\int |\text{grad } w|^2 & \text{if } w \in V \\ +\infty & \text{if } w \in L^2(\Omega)\setminus V \end{cases} \tag{23}$$

It is clear that ϕ is a l.s.c. convex function on $L^2(\Omega)$ satisfying the assumption (H1) (use (20)), (H2) (with $\alpha = 1$) and (H3) of Section 1. The operator A corresponding to this ϕ is single-valued, defined by $Au = \Delta u^m$ on

$$D(A) = \{u \in L^1(\Omega)\cap L^\infty(\Omega);\ u^m \in V,$$

$$v = -\Delta u^m \in L^1(\Omega) \text{ and}$$

$$\int vw = \int \text{grad } u^m. \text{ grad } w \text{ for any } w \in V\}.$$

To complete the proof of Theorem 2 by application of Theorem 1, we use the 'regularizing effect' $L^1(\Omega)$ into $L^\infty(\Omega)$: the semi-group $S(t)$ 'generated' by the m-T-accretive operator $\bar{A}$ (see Lemma 1) satisfies

$$S(t)u_0 \in L^\infty(\Omega) \text{ for any } u_0 \in L^1(\Omega)\ (= \bar{D}) \text{ and } t > 0$$

(see [6], [20]).

The first part of Theorem 2 (existence and uniqueness of a solution u of (21) for any $u_0 \in L^1(\Omega)^+$ satisfying moreover

$$t\,\frac{du}{dt} \in L^\infty(0,\infty;\ L^1(\Omega))$$

had already been stated in [2] in the case $\Omega = \mathbb{R}^N$ (and then $V = H^1(\mathbb{R}^N)$)): besides the estimate

$$cu_t \geqslant -\frac{u}{t}$$

(within the case $\Omega = \mathbb{R}^N$ the best constant $c = m-1 + \frac{2}{N}$) we were using the continuity of a nonnegative bounded weak solution of $u_t = \Delta u^m$ proved for the first time by Caffarelli and Friedmann in [13]. Actually, now the continuity of any bounded weak solution has been proved (see [17], [21]); then we can give the following

PROPOSITION 1. Let Ω, m, V as in Theorem 2. For any $u_o \in L^1(\Omega)$ there exists a unique solution u of (21). Moreover u is continuous on $]0,\infty[\times\,\Omega$,

u^m is continuous from $]0,\infty[$ into V, $t\,\frac{du}{dt} \in L^\infty(0,\infty;\, L^1(\Omega))$ and for any $\delta > 0$, $u \in L^\infty(]\delta,\infty[\times\,\Omega)$, $u^m \in L^\infty(\delta,\infty;V)$.

Proof: Using the function ϕ defined by (23) and the corresponding notations of Section 1, we have to show that $u(t) = S(t)u_0$ is $W^{1,1}_{loc}(]0,\infty[;\, L^1(\Omega))$ and satisfies the other properties claimed. Using the regularizing effect $L^1(\Omega)$ into $L^\infty(\Omega)$, we may assume $u_0 \in L^\infty(\Omega)$ and then $u \in L^\infty(]0,\infty[\times\,\Omega)$. Using

$$t\,\frac{d}{dt}\,\phi(u(t)^m) = -\,\frac{4mt}{(m+1)^2}\int\left|\frac{d}{dt}\,u(t)^{\frac{m+1}{2}}\right|^2 \in L^2(0,\infty)$$

and

$$\int_0^\infty \phi(u(t)^m)dt \leq \frac{1}{m+1}\int u_0^{m+1}$$

(see proof of Lemma 3), we may assume $u_0^m \in V$ and we have u^m continuous and bounded from $[0,\infty]$ into V. It is not difficult to show also that u is a weak solution of $u_t = \Delta u^m$ with the boundary conditions defined by V:

$$\left.\begin{aligned}&\iint uw_t = \iint \operatorname{grad} u^m.\operatorname{grad} w + \int u_0 w(0)\\ &\qquad\text{for any } w \in W^{1,1}(0,\infty;\, L^1(\Omega)) \cap L^1(0,\infty;\, V).\end{aligned}\right\}$$

All the arguments above could be developed in much more general cases and are today quite classical in the theory of nonlinear semi-groups. We now claim the continuity of u on $]0,\infty[\times\,\Omega$: it is a nontrivial result (see [17], [21]); it follows that

$$Q_0 = \{(t,x) \in\,]0,\infty[\times\,\Omega;\ u(t,x) \neq 0\}$$

is open in $\mathbb{R}^{N+1}$; by a classical regularity result $u \in C^\infty(Q_0)$. To complete the proof we use (see Lemma 2).

$$\left\|\frac{u(t+h)-u(t)}{h}\right\|_{L^1} \leq \frac{2\|u_0\|_{L^1}}{(m-1)t} \text{ for any } t > 0,\ h > 0. \tag{24}$$

This shows in particular that $\partial u/\partial t$ (in the sense of distribution) is a Radon measure on $Q = \,]0,\infty[\times\,\Omega$; since u is continuous, $\partial u/\partial t$ does not charge $Q \setminus Q_0 = \{u = 0\}$ and then

$$\frac{\partial u}{\partial t} = \frac{\partial u}{\partial t}\,\chi_{Q_0}$$

is a function on $Q(\frac{\partial u}{\partial t} \in L^1_{loc}(Q))$. In other words

$$u \in W^{1,1}_{loc}(]0,\infty[;\ L^1_{loc}(\Omega));$$

returning to (24),

$$t\,\frac{du}{dt} \in L^{\infty}(0,\infty;\ L^1(\Omega)).$$

Let us emphasize the differences between Theorem 2 and Proposition 1:

(1) In Theorem 2, we assume $u_0 \geqslant 0$ while we do no assumption on $u_0 \in L^1(\Omega)$ in Proposition 1.

(2) In Theorem 2, we obtain the regularity

$u_t = \Delta u^m \in L^p(]t_0,t_1[\times\Omega_R)$ for any $1 < p < 1 + \frac{1}{m}$ but we do not state such a result in Proposition 1: I conjecture that this regularity is still true without the assumption $u_0 \geqslant 0$.

(3) For the proof of existence of solution of (21): in Proposition 1, one main argument is the continuity of u; the method is the same as in [2]; in Theorem 2, we avoid this very sharp continuity result, replacing it by an estimate in L^p_{loc} for some $p > 1$.

The strong regularity L^p for any $1 < p < 1 + 1/m$, is not the best possible: in [19], it is proved in the case $m = 2$, $\Omega = \mathbb{R}^N$ that

$$u_t = \Delta u^m \in L^3(]\delta,\infty[\times \mathbb{R}^N).$$

In the case $N = 1$ and $\Omega = \mathbb{R}$, we can prove much more:

<u>THEOREM 3</u>. Let $Q =]0,T[\times \mathbb{R}$ for some $T > 0$ and $u \in L^{\infty}(Q)$ satisfying

$$u \geqslant 0,\ u_t = (u^m)_{xx} \text{ in } \mathcal{D}'(Q). \tag{25}$$

Then u, $(u^m)_x$ are continuous on Q and

(1) if $1 < m < 2$, $u_t = (u^m)_{xx}$ is continuous on Q.

(2) if $1 < m \leqslant 2$, $tu_t = t(u^m)_{xx} \in L^{\infty}(Q)$.

(3) if $m > 2$, $u_t = (u^m)_{xx} \in L^{\infty}(\delta,T;L^p_{loc}(\mathbb{R}))$ for any $1 \leqslant p < 1 + \frac{1}{m-2}$ and $\delta > 0$.

Remark 3: The regularity stated in (2) and (3) is the best possible in that direction as one can see on the similarity solutions

$$u(t,x) = t^{-1/(m+1)}\left(1 - \frac{m-1}{2m(m+1)}\,(xt^{-1/(m+1)})^2\right)^{+1/(m-1)}.$$

Following [1] (see also [2]) we set

$$v = \frac{m}{m-1}\,u^{m-1}. \tag{26}$$

The theorem is based on two one-sided estimates:

LEMMA 5. For $u \in L^\infty(Q)$ a solution of (25), the function v defined by (26) satisfies:

$$v_{xx} \geq -\frac{1}{(m+1)t} \quad \text{in } \mathcal{D}'(Q), \tag{27}$$

$$v_t \leq \frac{8m\,\|v\|_{L^\infty}}{(m+1)t}. \tag{28}$$

The estimate (27) is proved in [2] under smoothness assumptions on the solution u of (25), namely

$$\left.\begin{array}{l} u,\ \frac{1}{u},\ u_x,\ u_t,\ u_{xx},\ u_{xt},\ u_{xxx},\ u_{xxt} \\ \text{are continuous and bounded on } Q. \end{array}\right\} \tag{29}$$

The general case follows by

LEMMA 6. A solution $u \in L^\infty(Q)$ of (25) may be approximated by a sequence of solutions u_n of (25) satisfying the smoothness assumptions (29) and

$$\|u_n\|_{L^\infty} \leq \|u\|_{L^\infty},\ u_n \to u \quad \text{a.e. on } Q.$$

We will not prove this technical lemma which can be shown as well by classical as by semi-group methods (see [1], [7]).

Proof of estimate (28): By Lemma 6, we may assume that u satisfies (29). It is clear that v satisfies the same smoothness. We have

$$v_t = (m-1)vv_{xx} + v_x^2$$

$$v_{tx} = (m+1)v_x v_{xx} + (m-1)vv_{xxx}.$$

Set $p = v_t + \alpha v_x^2 = (m-1)vv_{xx} + (\alpha+1)v_x^2$ for some $\alpha \geqslant 0$ we will choose later. We have

$$p_x = (m-1)vv_{xxx} + (m + 2\alpha + 1)v_x v_{xx}$$

$$p_{xx} = v_{txx} + 2\alpha(v_{xx}^2 + v_x v_{xxx})$$

$$p_t = (m-1)vv_{xxt} + (m-1)v_t v_{xx} + 2(\alpha+1)(m+1)v_x^2 v_{xx} + 2(\alpha+1)(m-1)vv_x v_{xxx}$$

$$p_t - (m-1)vp_{xx} - 2v_x p_x = (m-1)v_{xx}[v_t + 2\alpha v_x^2 - 2\alpha vv_{xx}]$$

$$= \frac{1}{v}[p - (\alpha+1)v_x^2][\alpha\left(1 + \frac{2(\alpha+1)}{m-1}\right)v_x^2 - \left(\frac{2\alpha}{m-1} - 1\right)p].$$

Using the parabolic maximum principle, we have $p \leqslant \frac{C}{t}$ according as

$$-\frac{C}{t^2} \geqslant \frac{1}{v}\left[\frac{C}{t} - (\alpha+1)v_x^2\right]\left[\alpha\left(1 + \frac{2(\alpha+1)}{m-1}\right)v_x^2 - \left(\frac{2\alpha}{m-1} - 1\right)\frac{C}{t}\right]. \tag{30}$$

Now using (27)

$$v_x^2 \leqslant \frac{2\|v\|_{L^\infty}}{(m-1)t} \tag{31}$$

(by Taylor formula, for any $(t,x) \in Q$,

$$\|v\|_{L^\infty} \geqslant v(t,x+h) = v(t,x) + hv_x(t,x) + \frac{h^2}{2}v_{xx}(t,y)$$

$$\geqslant hv_x(t,x) - \frac{h^2}{2(m+1)t} \quad \text{for any } h \in \mathbb{R}).$$

Then setting $C = \dfrac{2c\|v\|_{L^\infty}}{(m+1)}$, (30) is implied by

$$\frac{c(m+1)v}{2\|v\|_{L^\infty}} + (c - (\alpha+1)\tau)(\alpha\left(1 + \frac{2(\alpha+1)}{m-1}\right)\tau - \left(\frac{2\alpha}{m-1} - 1\right)c) \leqslant 0$$

for any $\tau \in [0,1]$.

One can check that it is satisfied with $\alpha = m-1$, $c = 4m$.

<u>Proof of Theorem 3</u>: Using (31) (implied by (27)), we see that v is continuous on Q (and then also u) and

$$|(u^m)_x| = |uv_x| \leqslant u\left(\frac{2\|v\|_{L^\infty}}{(m+1)t}\right)^{\frac{1}{2}}.$$

We know then that $u \in C^\infty(\{u > 0\})$ and $(u^m)_x$ is continuous on Q with $(u^m)_x \equiv 0$ on $\{u = 0\}$.

Now (27) implies

$$u_t = (u^m)_{xx} \geqslant uv_{xx} \geqslant -\frac{u}{(m+1)t} .$$

If $1 < m \leqslant 2$, (28) implies

$$(u^m)_{xx} = u_t = \frac{u^{2-m}}{m} v_t \leqslant \frac{8m\|v\|_{L^\infty}}{(m+1)t} u^{2-m} .$$

It follows clearly that $tu_t \in L^\infty(Q)$, and that if $1 < m < 2$ u_t is continuous on Q with $u_t \equiv 0$ on $\{u = 0\}$.

In the case $m > 2$, we have also

$$(u^m)_{xx} \leqslant \frac{8m\|v\|_{L^\infty}}{(m+1)t} u^{2-m} \text{ on } \{u > 0\}.$$

Then part (3) of the theorem follows by applying Lemma 7 below with $w(x) = u^m(t,x)$, $\beta = \frac{m-1}{m}$, $\gamma = \frac{m-2}{m}$.

<u>LEMMA 7</u>. Let w be nonnegative continuous and bounded on $\mathbb{R}$ satisfying

$$(w^\beta)_{xx} \geqslant -c \text{ in } \mathcal{D}'(\mathbb{R}) \text{ for some } 0 < \beta < 1,$$

$$w_{xx} \leqslant cw^{-\gamma} \text{ in } \mathcal{D}'(\{w > 0\}) \text{ for some } \gamma > 0.$$

Then for any $1 \leqslant p < 1 + \frac{\beta-1}{\gamma}$, $w_{xx} \in L^p_{loc}(\mathbb{R})$.

<u>Proof</u>: The first assumption implies (see (31))

$$(w^\beta)_x^2 \leqslant 2c \|w\|^\beta_{L^\infty}$$

and then $w \in C^1(\mathbb{R})$ with

$$|w_x| \leqslant \frac{w^{1-\beta}}{\beta}(2c\ \|w\|^{\beta}_{L^\infty})^{\frac{1}{2}}.$$

Also

$$w_{xx} \geqslant \frac{w^{1-\beta}}{\beta}\ (w^\beta)_{xx} \geqslant -\frac{c}{\beta}\, w^{1-\beta} \text{ in } \mathcal{D}'(\mathbb{R}).$$

Then w_x is locally of bounded variation on $\mathbb{R}$ since, using the second assumption, w_x is locally Lipschitz on

$$\{w > 0\} \supset \{w_x \neq 0\}, \quad w_{xx} \in L^1_{loc}(\mathbb{R})$$

(w_{xx} does not charge $\{w_x = 0\} \supset \{w = 0\}$).

We have for $1 \leqslant p < 1 + (1-\beta)/\gamma$,

$$\begin{aligned}
&|w_{xx} + \frac{c}{\beta}\, w^{1-\beta}|^p \\
&= (w^\gamma w_{xx} + \frac{c}{\beta}\, w^{1-\beta+\gamma})^{p-1}\left(\frac{w_{xx}}{w^{\gamma(p-1)}} + \frac{c}{\beta}\, w^{1-\beta-\gamma(p-1)}\right) \\
&\leqslant C(w^{\sigma+\beta-1} w_{xx} + \frac{c}{\beta}\, w^\sigma) \quad \text{a.e. on } \{w > 0\}
\end{aligned}$$

with

$$C = \left(c + \frac{c}{\beta}\, \|w\|^{1-\beta+\gamma}_{L^\infty}\right)^{p-1}, \quad \sigma = 1-\beta-\gamma(p-1) > 0.$$

Now

$$\begin{aligned}
w^{\sigma+\beta-1}\, w_{xx} &= \frac{1-\beta}{\beta\sigma}\, (w^\sigma(w^\beta)_x)_x - \frac{1-\beta-\sigma}{\beta\sigma}\, w^\sigma(w^\beta)_{xx} \\
&\leqslant \frac{1-\beta}{\beta\sigma}\, (w^\sigma(w^\beta)_x)_x + \frac{c\gamma(p-1)}{\beta\sigma}\, w^\sigma;
\end{aligned}$$

then

$$|w_{xx} + \frac{c}{\beta}\, w^{1-\beta}|^p \leqslant \frac{C(1-\beta)}{\beta\sigma}\, [(w^\sigma(w^\beta)_x)_x + cw^\sigma]$$

and

$$\int_{-R}^{+R} |w_{xx} + \frac{c}{\beta}\, w^{1-\beta}|^p < \infty \quad \text{for any } R \text{ finite.}$$

Remark 4: Lemma 7 is of the same type as Lemma 4 and we can repeat Remark 1: in particular we conjecture that, under the assumption $w^{1-\beta} \in L^1(\mathbb{R})$,

$w_{xx} \in L^p(\mathbb{R})$ for any $1 \leqslant p < 1 + \frac{1-\beta}{\gamma}$ (it is clear that $w_{xx} \in L^1(\mathbb{R})$). These Lemmas 4 and 7, which are the key for Theorems 1 and 3, can be extended to more general situations, but we do not yet see clearly how to state them in a general abstract framework.

Remark 5: The proof of Theorem 3 gives many other results. First the estimates of Lemma 5 show that $tv_t \in L^\infty(Q)$ and then for $m \geqslant 2$,

$$|u(t+h,x) - u(t,x)| \leqslant \frac{C}{t} h^{1/(m-1)} \quad \text{for any } t > 0,\ h > 0$$

(this result has been proved under some more assumptions by Di Benedetto [16]). Now using Lemma 7 with $w(x) = u^r(t,x)$, we see that

$$(u^r)_{xx} \in L^\infty(\delta,T;\ L^2_{loc}(\mathbb{R})) \text{ for } r > \frac{3}{2}(m-1)$$

(a similar result has been obtained by a completely different way by Benachour and Monlay [4]).

Remark 6: The same results as in Theorem 3 can be obtained, with some more computations, for the equation

$$u_t = (u^m)_{xx} + a(u^n)_x$$

in the case $n \geqslant m > 1$. Also they can be extended to the equation

$$u_t = \phi(u)_{xx}$$

under some restrictive condition on ϕ (see [15] for an extension of the estimate (27)).

REFERENCES

[1] D.G. Aronson, Regularity properties of flows through porous media, SIAM J. Appl. Math., 17 461-467 (1969).

[2] D.G. Aronson and Ph. Bénilan, Régularité des solutions de l'équation des milieux poreux dans $\mathbb{R}^N$, C. R. Acad. Sci. Paris, Série A-B 288, 103-105 (1979).

[3] V. Barbu, *Non-linear semi-groups and differential equations in Banach spaces.* Nordhoff Inf. Publ. Co., Leyden (1976).

[4] S. Benachour and M.S. Monlay, Régularité des solutions agissant la filtration d'un fluide dans un milieu poreux (to appear).

[5] Ph. Bénilan, Sur un problème d'évolution non monotone dans $L^2(\Omega)$. Publ. Math. Fac. Sc. Besançon, No. 2 (1976).

[6] Ph. Bénilan, Opérateurs et semi-groupes dans les espaces $L^p(1 \leqslant p \leqslant \infty)$, in Funct. An. and Num. An. Japan-France Seminar 1976, ed. H. Fujita. Japan Soc. Prom. Sc. 15-53 (1978).

[7] Ph. Bénilan and M.G. Crandall, The continuous dependence on ϕ of the solutions of $u_t - \Delta\phi(u) = 0$, Ind. Un. Math. J. 30(2), 161-171 (1981).

[8] Ph. Bénilan and M.G. Crandall, Regularizing effects of homogeneous evolution equations, Techn. Sum. Rep. 2076, M.R.C. Univ. of Wisc. Madison.

[9] Ph. Bénilan and C. Picard, Quelques aspects non linéaires du principe du maximum, Lect. Notes in Math. 713, Springer-Verlag, 1-37 (1979).

[10] Ph. Bénilan, M.G. Crandall and A. Pazy, Nonlinear evolution equations governed by accretive operators (to appear).

[11] H. Brézis, Opérateurs maximaux monotones et semi-groupes de contractions dans les espaces de Hilbert, North Holland, Amsterdam (1977).

[12] H. Brézis and W. Strauss. Semi-linear elliptic equations in L^1. J. Math. Soc. Japan. 25, 565-590 (1973).

[13] L.A. Caffarelli and A. Friedmann, Proc. Nat. Acad. Sc. USA, 75, 2084 (1978).

[14] M.G. Crandall, An introduction to evolution governed by accretive operators. An International Symp. Cesari, Hale and LaSalle (eds.), Academic Press, N.Y (1975).

[15] M.G. Crandall and M. Pierre, Regularizing effects for $u_t - \Delta\phi(u) = 0$. Techn. Sum. Rep. 2166, M.R.C. Univ. Wisc. Madison.

[16] E. Di Benedetto, Regularity results for the porous media equation. Ann. Mat. Pura Appl. (4) 121, 249-262 (1979).

[17] E. Di Benedetto, Continuity of weak solution to certain degenerate parabolic equations, Tech. Sum. Rep. 2124, M.R.C. Univ. of Wisc. Madison (1981).

[18] C.H. Lee, Etude de la classe des opérateurs m-accrétifs dans $L^1(\Omega)$ et accrétifs dans $L^\infty(\Omega)$. These 3e cycle. Paris VI. (1977).

[19] M.E. Rose, Numerical methods for a porous medium equation. Thesis, Univ. of Chicago (1978).

[20] L. Véron, Coercivité et propriétés régularisantes des semi-groupes non-linéaires dans les espaces de Banach, Publ. Math. Fac. Sc. Besançon 3 (1977).

[21] W.P. Ziemer, Interior and boundary continuity of weak solutions of degenerate parabolic equations. Preprint.

P. Bénilan
Departement de Mathématiques
Université de Franche Comté
25030 Besançon Cedex
France

F BERNIS

Compactness of the support in convex and non-convex fourth-order elasticity problems

0. INTRODUCTION

We consider the following Problem $\alpha\beta\Gamma$ or the Γ-Problem, taken from non-linear beam theory:

$$\text{minimize } J(u) = \frac{1}{2}\int_0^\infty (u''(x))^2\,dx + \int_0^\infty \Gamma(u(x))dx$$

in the set of locally integrable functions such that

$$u'' \in L^2(\mathbb{R}^+), \quad \int_0^\infty |\Gamma(u(x))|dx < \infty,$$

and verify the boundary conditions $u(0) = \alpha$, $u'(0) = \beta$.

Existence is standard if Γ is lower semicontinuous, $\Gamma \geqslant 0$ and the minimization set is non-empty. Uniqueness is well-known if Γ is convex (even if Γ is not strictly convex), but it does not hold in general for non-convex Γ.

When $\Gamma \in C^1(\mathbb{R})$, the associated Euler differential equation is

$$u^{iv}(x) + \Gamma'(u(x)) = 0.$$

If Γ is neither C^1 nor convex (for Γ convex see below, where we quote Bidaut-Véron), the validity and meaning of this equation is to be clarified. The 'fourth-order' of the title refers to the Euler equation.

When we set $\Gamma(s) = |s|^r$, $r > 0$, we shall speak of Problem $\alpha\beta r$ or the power problem. So Problem $\alpha\beta r$ is convex only if $r \geqslant 1$.

We present in the next sections our results [6], [7] about compactness of the support, validity of the Euler equation and regularity. For the power problem we also deal with uniqueness in the non-convex case ($0 < r < 1$).

Problem $\alpha\beta r$ with $r = 1$ (and a small modification) is to be found in Berkovitz and Pollard [4], [5]. See [5] for previous work. Redheffer [26] completes its study and applies the similarity method. Hestenes and Redheffer [15], [16] generalize it to include weight functions and terms with intermediate derivatives. Our work is, in many respects, on the line of the methods of these papers, specially [15] and [26], but our results are not implied by them except for $r = 1$.

Bidaut-Véron [8] - [10] studies the Γ-problem with Γ convex, $\Gamma(s) \geqslant C|s|$, $\Gamma(0) = 0$ and proves compactness of the support in dimension N with spherical symmetry. On the other hand, she establishes in [11] regularity and the relation of the minimization problem with the Euler equation in multivalued sense for any lower semicontinuous (finite or not) convex Γ. For $\Gamma(s) = |s|$, this has already been done by Ekeland and Teman [14].

Theorem 6 below about compactness of the support covers a wide class of convex (and non-convex) functions which do not verify the inequality $\Gamma(s) \geqslant C|s|$.

Compactness of the support in *second* order problems has been studied, in any dimension, by many authors, starting with Brézis [12], the essential tool being comparison principles. These methods do not work for fourth-order problems, mainly because comparison principles involve more boundary conditions and are valid in fewer situations. Nevertheless, we also use comparison (or positivity) principles in a more indirect way.

The second-order analogy of Problem $\alpha\beta r$ with $0 \leqslant r \leqslant 1$ ($r = 0$ means a limiting case) is studied, in dimension N, by Caffarelli [13], Alt and Caffarelli [1] and Phillips [23].

Some words about notation. Derivatives are in the sense of distributions, so that (in dimension one) "f' locally integrable" is equivalent to "f locally absolutely continuous". By u we mean a solution of Problem $\alpha\beta\Gamma$ or $\alpha\beta r$, unless otherwise stated. $a_1,\ldots,a_2,\ldots$ will be the consecutive zeros of u, and a_∞ the extreme of its support; a_∞ may be finite or not.

We suppose that $\alpha > 0$ (with any real β) or $\alpha = 0$ and $\beta > 0$, so that $u(x) > 0$ near $x = 0$. This does not cause loss of generality so long as the hypotheses about $\Gamma(s)$ are symmetric with respect to $s = 0$.

1. THE POWER PROBLEM AND THE SIMILARITY METHOD

The following theorems are proved in [7].

THEOREM 1. The solution of Problem $\alpha\beta r$ is unique if $r > (2/3)-\varepsilon$ (with any α,β) or if $\beta \geqslant 0$ (with any $r > 0$). In any case, there are at most two solutions.

THEOREM 2. The support of any solution is compact if $0 < r < 2$.

THEOREM 3. If $r > (2/3)-\varepsilon$,

(a) The solution u of Problem $\alpha\beta$ can be obtained from the solution U of Problem ($\alpha = 0$, $\beta = 1$) through one translation and one similarity:

$$u(x) = \tau U(\sigma x+s) \quad \text{for all } x \geqslant 0;$$

(b) U consists of an infinite sequence of contiguous similar arches of alternating sign:

$$-U(x+a_1) = \mu U(\lambda x) \text{ for all } x \geqslant 0.$$

THEOREM 4. Regularity of u.

(a) For all $r > 0$, $u''' \in L^1(\mathbb{R}^+)$ and therefore u'' is globally absolutely continuous in $\bar{\mathbb{R}}^+ = [0,\infty)$.
(b) $u''' \in L^\infty(R^+)$ if and only if $r \geqslant 2/3$,
(c) $u^{iv} \in L^1(R^+)$ if and only if $r > 2/3$. Then $u''' \in AbsC(\bar{\mathbb{R}}^+)$.
(d) $u^{iv} \in L^\infty(R^+)$ if and only if $r \geqslant 1$.
(e) $u^v \in L^1(R^+)$ if and only if $r > 1$. Then $u^{iv} \in AbsC(\bar{\mathbb{R}}^+)$.

(Hölder regularity results are to be found in [6]. They are just analogous to the second-order N-dimensional ones of [23].)

The core of the similarity method is, of course, Theorem 3. This method was applied in [26] to Problem $\alpha\beta r$ with $r = 1$. The idea is the following. Take point (b). If $U'(a_1) \neq 0$, then μ and λ can be chosen in such a way that $\mu U(\lambda x)$ verifies the same Euler equation and the same boundary conditions as $-U(x+a_1)$, so that they are equal if uniqueness is known. For $r > 1$, this argument becomes a proof with little supplementary work. For $0 < r < 1$ it is necessary to work with the functional $J(u)$ instead of the singular Euler equation; but the latter (see Theorem 5) is also needed in order to obtain uniqueness. On the other hand, the question whether or not $U'(a_1)$ is zero is related to regularity at a_∞ and the comments of Section 3 (Paragraph 9) apply.

Uniqueness and regularity when $\beta < 0$ and $r < 2/3$ admit of deeper study [6]. When r is small enough, Problem $\alpha\beta r$ "behaves like" the "$r = 0$" limiting case: U has only one arch and there are two different solutions for some (α,β). This limiting case is explicitly solved.

We quote [2] and [29] as general references about the similarity method in ordinary and partial differential equations.

2. RESULTS FOR THE Γ-PROBLEM

In all the theorems of this section we make the following hypotheses on Γ:

(a) Positive definitedness. $\Gamma(0) = 0$ and $\inf_{|t|\geqslant|s|} \Gamma(t) > 0$ if $s \neq 0$.

(b) Half-monotony. Γ is non-decreasing in $[0,\infty)$ and non-increasing in $(-\infty,0]$.

(c) Absolute continuity. $\Gamma \in AbsC_{loc}(R)$.

When taken with (b), (a) simplifies to $\Gamma(0) = 0$ and $\Gamma(s) > 0$ if $s \neq 0$. No hypothesis on convexity is made.

THEOREM 5. (Euler equation)

(a*) $\Gamma'(u(x))$ is defined a.e.w. in $(0,a_\infty)$ and belongs to $L^1_{loc}(0,a_\infty)$.

(b*) In the space of Radon measures $M(0,a_\infty)$:

$$u^{iv} + \Gamma'(u) + \Delta = 0$$

where Δ is an at most countable sum of Dirac masses placed at the zeros of u', with $u\Delta \geqslant 0$ and therefore $uu^{iv} \leqslant 0$. $\Delta = 0$ in the situation of Theorem 7(b).

For Problem αβr with $0 < r < 1$ we have

$$u^{iv} + \Gamma'(u) = 0 \text{ in } L^1_{loc}[0,a_\infty).$$

It is to be noted that, when $0 < r \leqslant 2/3$, neither $u^{iv} \in L^1_{loc}(R^+)$ nor $u^{iv} \in M(R^+)$, because of the singularity at a_∞.

THEOREM 6. The support of u is compact if

$$|\Gamma'(s)| \geqslant C|s|^{r-1} \text{ a.e.w.}$$

in some neighbourhood of $s = 0$ with $C > 0$ and $0 < r < 2$.

For the case in which this inequality holds globally, we give in [6] explicit bounds of the support depending exclusively on the data of the problem.

THEOREM 7. Regularity.

(1) With only the hypotheses (a), (b) and (c) on Γ, we have that u''' is continuous

where $u' \neq 0$ and

$$u^{iv} \in M(0,a_\infty), \quad u''' \in L^1(R^+) \text{ and } u'' \in AbsC(\bar{R}^+).$$

(2) If (in addition to (a), (b) and (c)), $\Gamma' \in L^q_{loc}(R-\{0\})$ with $q > 2$, then

$$u^{iv} \in L^1_{loc}[0,a_\infty) \text{ and } u'''(x)$$

is continuous for $x \neq a_\infty$.

(3) If (in addition to (a), (b) and (c)) $\Gamma' \in L^q_{loc}(R)$ with $q > 3$, then

$$u^{iv} \in L^1(R^+) \text{ and } u''' \in AbsC(\bar{R}^+).$$

(4) If $\Gamma \in Lip_{loc}(R)$, then $u^{iv} \in L^\infty(R^+)$.

When (3) holds u has an infinite number of arches.

Note that the origin is excluded in the hypothesis of (2). So $\Gamma(s) = |s|^r$ verifies (2) for all $r > 0$. On the other hand, $q > 3$ of (3) means $r > 2/3$ when $\Gamma(s) = |s|^r$.

3. GUIDE TO THE PROOFS FOR THE Γ-PROBLEM

The following steps are to be found in [6]. They supply the above-mentioned bounds of the support and constitute also a second proof of Theorems 2 and 4, now without similarity arguments.

We remark that the proofs about the Euler equation and regularity are much easier if Γ is a locally Lipschitz function (and still easier if $\Gamma \in C^1(R)$). On the contrary, the proof on compactness of the support becomes more difficult (as explained in Paragraph 7 below) when Γ is C^1 at the origin.

1. The positive definitedness hypothesis implies, by a theorem of Redheffer and Walter [28], that all the functions of the minimization set verify

$$\lim_{x\to\infty} u(x) = \lim_{x\to\infty} u'(x) = 0.$$

This theorem of Redheffer and Walter can be considered as a generalization of the Hadamard-Littlewood three-derivatives theorem (see results and references in [27], [28] and [3]) and also as a generalization of the interpolation inequalities of Nirenberg [22].

2. As in [4], if $u(k) = u'(k) = 0$, then $u(x) = 0$ for all $x \geqslant k$. This

implies that u has an at most countable sequence of contiguous arches of alternating sign.

3. From the half-monotony hypothesis it is proved that u^{iv} is a non-positive measure where $u > 0$, and a non-negative one where $u < 0$. So u'' is continuous where $u \neq 0$.

4. u has arbitrarily big zeros. From 3 and 1.

5. Euler equation. The main point is to establish, where $u' \neq 0$, a lemma of derivation of parametric integrals of the type

$$\frac{d}{d\lambda} \int_c^d \Gamma(u(x)+\lambda\phi(x))dx = \int_c^d \Gamma'(u(x)+\lambda\phi(x))\phi(x)dx.$$

For this lemma only the absolute continuity hypothesis on Γ is needed. The points where $u' = 0$ and $u \neq 0$ are dealt with in 3 and 2.

6. Following a method of [15], we obtain from $uu^{iv} \leqslant 0$ that

$$u'(a_n)\, u''(a_n) < 0 \tag{1}$$

Here a first study of the behaviour of u at a_∞ is needed.

7. Compactness of the support. Applying to each arch of u a comparison principle for the operator $y \to y^{iv}$ (with certain boundary conditions) and using (1), we prove that

$$|u'(a_{n+1})| < \frac{1}{2}\, |u'(a_n)| \tag{2}$$

$$a_{n+1}-a_n \leqslant \text{constant}\ |u'(a_n)|^{(2-r)/(2+r)} \tag{3}$$

and therefore $a_\infty = \Sigma(a_{n+1}-a_n) < \infty$ if $r < 2$.

When $0 < r < 1$, the comparison principle can be applied directly (to each arch) because we compare with the equation $y^{iv} = \mp 1$. The idea of obtaining (2) and (3) from (1) and the equation $y^{iv} = \mp 1$ is in [15]. The idea of combining this with a fourth-order comparison principle (in dimension N with spherical symmetry) is in [10]. The preceding steps (specially Euler equation) are needed to apply these ideas to non-convex Γ.

When $1 < r < 2$, we compare in fact with the equation $y^{iv} + y = 0$, and we circumvent in [6] the difficulty that the operator $y \to y^{iv} + y$ does not verify positivity properties in arbitrary intervals.

One-dimensional positivity (or comparison) principles for linear m-order differential operators are contained in, or easily implied by, the work of Pólya [25] (See more references in [3].) For the operator $y \to y^{iv}$ there are short proofs (see for example [6], [7]) in which y^{iv} is allowed to be a measure.

8. Continuity of u''' at the isolated zeros of u' (Point (2) of Theorem 7) is dealt with by another lemma of the type of Paragraph 5.

9. Continuity of u'' and u''' at the extreme of the support. We face the following difficulty (when Γ' is unbounded at the origin). If we consider a test function ϕ identically null beyond a_∞, we do not obtain enough information through differentiation of the functional. But if ϕ is not identically null beyond a_∞, then differentiation of the functional seems to be completely meaningless. What we do is to work directly with the functional, without any kind of differentiation.

Continuity of u''' at a_∞ involves the proof that u has an infinite number of arches and the bound:

$$|u'''(a_n)| \leqslant \text{constant } |u'(a_n)|^{(3r-2)/(r+2)} \tag{4}$$

where r is here such that $\Gamma(s) \leqslant \text{constant } |s|^r$. Note that $\Gamma' \in L^q$ implies $r = 1-(1/q)$.

Geometric bounds like (2) and (4) are the key to prove global absolute continuity and global integrability. They supply also a shorter proof of compactness of the support when $2/3 < r \leqslant 1$.

We finally comment that for the power problem we have developed a second, very different method to deal with the extreme of the support. It is based on continuity in the parameter r.

4. THE PHYSICAL MODEL

Our physical model is the beam version of certain models for plates to be found in Il'yushin [17], Kachanov [18] and Langenbach [20], [21]. The basic lines for a direct treatment of beams can be seen in Section 60 of Pisarenko et al. [24]. The general hypotheses of these models are 'geometric linearity' (or 'small deformations') and non-linear elastic laws. The latter appear not only by themselves as in [24], but also as one of the approaches to plasticity: see f.e. in [19] the deformation theory of plasticity and Hencky's relations. We remark that in von Karman's plates the hypotheses are just 'opposite':

non-linear geometry ("big deformations") and linear elastic Hooke's law.

We undertake a brief description of our physical model. The equilibrium equation of the beam is

$$\frac{d^2}{dx^2}\alpha(u''(x)) + \gamma(u(x)) = f(x)$$

which is associated with the energy functional

$$\int_0^c A(u''(x))dx + \int_0^c \Gamma(u(x))dx - \int_0^c f(x)\, u(x)dx$$

with

$$\alpha(s) = A'(s) \text{ and } \gamma(s) = \Gamma'(s).$$

The applied force per unit length is given by f. We prove in [6] that the solutions have compact support for some non-compact support f.

The term $\gamma(u(x))$ represents a non-linear elastic link, foundation or forces field. For example

$$\gamma(s) = |s|^{r-1}\text{sgn } s.$$

If we consider electrical forces, we obtain models with $r < 1$ and, in general, with non-convex Γ. The half-monotony hypothesis on Γ means that these forces are attractive.

Geometric linearity leads to identify u" with the curvature.

The function $\alpha(s)$ is directly related with the stress-strain law of the beam material; $\alpha(u''(x))$ represents the bending moment. Non-linear power-like laws (which lead to $A(s) = |s|^p$, $p > 1$) are considered in strength of materials - even in small deformations - for materials such as cast iron and stone. See again Section 60 of [24], who just applies these laws to beams. Plastic-rigid models [19] correspond to the $p = 1$ limiting case. When $p = 2$ the material of the beam follows linear Hooke's law. So in the Γ-problem of the present article non-linearity is only in the forces field.

For general p we have also obtained in [6] the compactness of the support property if $0 < r < p$.

On the other hand, we recall that the Berkovitz and Pollard work [4] comes from an optimal filter problem which physically has nothing to do with beams.

REFERENCES

[1] H. Alt and L.A. Caffarelli, A variational problem with a free boundary, J. für die Reine und Ang. Math. (to appear).

[2] G.I. Barenblatt, Similarity, Self-Similarity and Intermediate Asymptotics, Consultants Bureau, New York (1979).

[3] E. Beckenbach and R. Bellman, Inequalities, Springer, Berlin (1961).

[4] L.D. Berkovitz and H. Pollard, A non-classical variational problem arising from an optimal filter problem, Arch. Rat. Mech. and Anal. 26, 281-304 (1967).

[5] L.D. Berkovitz and H. Pollard, A non-classical variational problem arising from an optimal filter problem, II. Arch. Rat. Mech. and Anal. 38, 161-172 (1970).

[6] F. Bernis, Thesis, 1982.

[7] F. Bernis, Compactness of the support in convex and non-convex fourth order elasticity problems. J. of Non-Linear Anal.6(11), 1221-1243 (1982).

[8] M.F. Bidaut-Véron, Propriété de support compact de la solution d'une équation aux dérivées partielles non linéaire d'ordre 4, C.R. Acad. Sc. Paris 287, 1005-1008 (1978).

[9] M.F. Bidaut-Véron, Compacité du support de la solution d'une inéquation variationnelle d'ordre 4 dans R^N, Publ. Math. Univ. Pau, 1-33 (1977).

[10] M.F. Bidaut-Véron, Principe de maximum et support compact pour une classe d'équations elliptiques non linéaires d'ordre 4, Publ. Math. Univ. Pau 1-18 (1979-80).

[11] M.F. Bidaut-Véron, Equations elliptiques fortement non linéaires dans des domaines non bornés, Publ. Math. Univ. Pau 1-69 (1980).

[12] H. Brézis, Solutions with compact support of variational inequalities, Uspekhi Mat. Nauk. 129, 103-108 (1974).

[13] L.A. Caffarelli, Compactness methods in free boundary problems, Comm. in PDE, 5(4), 427-448 (1980).

[14] I. Ekeland and R. Teman, Analyse convexe et problèmes variationnels, Dunod, Paris (1974).

[15] M. Hestenes and R. Redheffer, On the minimization of certain quadratic functionals I, Arch. Rat. Mech. and Anal. 56, 1-14 (1974).

[16] M. Hestenes and R. Redheffer, On the minimization of certain quadratic functionals II, Arch. Rat. Mech. and Anal. 56, 15-33 (1974).

[17] A.A. Il'yushin, Plasticity, Gostekhizdat, Moscow (1948) French translation: Eyrolles, Paris (1956).

[18] L.M. Kachanov, Some Problems of Creep Theory. GITTL, Leningrad-Moscow (1949).

[19] L.M. Kachanov, Foundations of the Theory of Plasticity, North-Holland, Amsterdam (1971). First Russian edition in 1948.

[20] A. Langenbach, Elastisch-plastische Deformationen von Platten, Zeitschrift für angew. Math. und Mech. 41, 126-134 (1961).

[21] A. Langenbach, Monotone Potentialoperatoren in Theorie und Anwendung, Springer, Berlin (1977).

[22] L. Nirenberg, On elliptic partial differential equations. Annalli della Scuola Norm. Sup. Pisa, XIII, 115-162 (1959).

[23] D. Phillips, A minimization problem and the regularity of solutions in the presence of a free boundary, Preprint Purdue Univ., West Lafayette (1981).

[24] G.S. Pisarenko, A.P. Yákovlev and V.V. Matvéev, Handbook of Strength of Materials. Mir, Moscow (1975).

[25] G. Pólya, On the mean value theorem corresponding to a given linear homogeneous differential equation, Trans. Am. Math. Soc. 24, 312-324 (1922).

[26] R. Redheffer, On a non-linear functional of Berkovitz and Pollard, Arch. Rat. Mech. and Anal. 50, 1-9 (1973).

[27] R. Redheffer, A note on the Littlewood three-derivatives theorem, J. London Math. Soc. 9(2), 9-15 (1974).

[28] R. Redheffer and W. Walter, Inequalities involving derivatives, Pacific J. Math. 85(1) 165-178 (1979).

[29] L. Sédov, Similitude et dimensions en mécanique, Mir, Moscow (1972 and 1977).

F. Bernis
Departamento de Matematicas
Escuela de Ingenieros de Telecomunicacion,
APDO. 30.002,
Barcelona-34
Spain

M-F BIDAUT-VÉRON

On the solutions of some nonlinear elliptic equations of order $2m$

1. INTRODUCTION

Let Ω be a regular open subset of $\mathbb{R}^N$, possibly unbounded, $m \in \mathbb{N}$ and $p \in [1,+\infty[$. Here we study the variational inequality

$$\left.\begin{array}{l} u - \Phi \in W_0^{m,p}(\Omega), \\ \langle Au-f, v-u\rangle + J(v) - J(u) \geqslant 0, \quad \forall v \in \Phi + W_0^{m,p}(\Omega), \end{array}\right\} \tag{1.1}$$

where $f \in W^{-m,p'}(\Omega)$ $(p' = p/(p-1))$, $\Phi \in W^{m,p}(\Omega)$; A is an operator of the calculus of variations of order 2m, from $W^{m,p}(\Omega)$ to $W^{-m,p'}(\Omega)$:

$$u \mapsto Au = \sum_{|\alpha|\leqslant m} (-1)^{|\alpha|} D^{\alpha}(A_{\alpha}(\cdot,(u,D^1u,\ldots,D^{m-1}u),D^m u)), \tag{1.2}$$

with classical assumptions for the $A_\alpha((x,\eta,\zeta) \mapsto A_\alpha(x,\eta,\zeta))$: they are Carath-eodory functions, with

$$|A_\alpha(x,\eta,\zeta)| \leqslant c_1(|\eta|^{p-1} + |\zeta|^{p-1}) + s_1(x), \text{ where } c_1 > 0,\ s_1 \in L^{p'}(\Omega), \tag{1.3}$$

$$\sum_{|\alpha|=m} (A_\alpha(x,\eta,\zeta) - A_\alpha(x,\eta,\bar\zeta))(\zeta_\alpha - \bar\zeta_\alpha) > 0, \text{ if } \zeta \neq \bar\zeta; \tag{1.4}$$

and J is the function on $W^{m,p}(\Omega)$

$$v \mapsto J(v) = \int_\Omega j(\cdot,v)dx \quad (+\infty \text{ if not finite}), \tag{1.5}$$

where $(x,r) \mapsto j(x,r)$ is a nonnegative normal convex integrand in $\Omega \times \mathbb{R}$. A particular case is the unilateral constraint problem: $u \geqslant \Psi$ in Ω.

Some examples prove that for $f \in L^{p'}(\Omega)$, Problem (1.1) may not be equivalent to

$$\left.\begin{array}{l} u - \Phi \in W_0^{m,p}(\Omega), \quad Au \in L^1_{loc}(\Omega), \\ Au(x) + \beta(x,u(x)) \ni f(x), \text{ a.e. in } \Omega, \end{array}\right\} \tag{1.6}$$

where $\beta = \partial j$, subdifferential with respect to r. In fact solutions of (1.1) have not enough regularity.

First we study the existence of solutions of (1.1) under weak assumptions of coercivity. Then we notice that (1.1) is equivalent to

$$f - Au \in \partial J_{\Phi}(u-\Phi), \tag{1.7}$$

where ∂J_{Φ} is the subdifferential of the function J_{Φ} defined on $W_0^{m,p}(\Omega)$ by

$$U \mapsto J_{\Phi}(U) = J(U+\Phi). \tag{1.8}$$

We give a characterization of that subdifferential under suitable assumptions on j, and deduce the interpretation of inequality (1.1): the solutions u of (1.1) satisfy some generalized equations, where Au is a measure on Ω. By this way we obtain existence results for Problem (1.6) when $Au \in L^1_{loc}(\Omega)$.

2. EXISTENCE RESULTS FOR PROBLEM (1.1)

For any subset P of some set E, we denote by I_P the indicator function of P : $I_P(x) = 0$ if $x \in P$, $+\infty$ if not.

We suppose for A a semi-coercivity condition: setting $\xi = (\eta,\zeta)$,

$$\sum_{|\alpha|\leqslant m} A_\alpha(x,\eta,\zeta)\xi_\alpha \geqslant c_2 \sum_{|\alpha|=m} |\zeta_\alpha|^p - s_2(x), \quad \text{where} \qquad c_2 > 0,\ s_2 \in L^1(\Omega). \tag{2.1}$$

Then inequality (1.1) admits a solution when the following condition is satisfied: there exists $v_0 \in D(J)$ such that $v_0 - \Phi \in W_0^{m,p}(\Omega)$ and

$$\langle Au, u-v_0\rangle + J(v) - \langle f, u-\Phi\rangle \to +\infty, \quad \text{as}$$

$$\|u\|_{W^{m,p}(\Omega)} \to +\infty,\ u \in D(J),\ u-\Phi \in W_0^{m,p}(\Omega). \tag{2.2}$$

Indeed the operator A is pseudomonotone, from a recent result of Browder [7], then (2.2) is a variant of a well-known result applied to the function $J + I_{\Phi+W_0^{m,p}(\Omega)}$. This simplifies the proofs given in [1].

Now Condition (2.2) is satisfied if Ω is bounded, or A has a greater coercivity. In the general case we prove in [1] the following result:

THEOREM 2.1. Under the assumptions (1.2) to (1.5), we suppose that

$$j(x,0) = 0 \quad \text{a.e. in } \Omega, \tag{2.3}$$

$$\Phi \in D(J) \text{ and } \Phi \text{ has a bounded support,} \tag{2.4}$$

and there exist $k > 0$ and $\rho > 0$ such that, for almost all $x \in \Omega$,

$$j(x,r) \geq k\ |r|^p, \quad \forall r \in [-\rho, +\rho] \text{ with } j(x,r) < +\infty. \tag{2.5}$$

Then inequality (1.1) admits at least one solution.

Sketch of the proof: For every $u \in D(J)$, let $u_\rho = \text{sign}^0 u(|u|-\rho)^+$, $u'_\rho = u-u_\rho$; conditions (2.3) and (2.5) imply that $u_\rho \in L^1(\Omega)$, $u'_\rho \in L^p(\Omega)$ and

$$J(u) \geq k(\rho^{p-1}\|u_\rho\|_{L^1(\Omega)} + \|u'_\rho\|^p_{L^p(\Omega)}). \tag{2.6}$$

Now we notice that every $v \in L^1(\mathbb{R}^N)$ with its first derivatives in $L^p(\mathbb{R}^N)$ satisfies $v \in L^p(\mathbb{R}^N)$. We deduce that for any $\varepsilon > 0$, there exists $c_\varepsilon > 0$ such that, for every $u \in D(J)$,

$$\|u\|_{L^p(\Omega)} \leq \varepsilon\|u_\rho\|_{L^1(\Omega)} + 2\|u'_\rho\|_{L^p(\Omega)} + c_\varepsilon \sum_{|\alpha|=m} \|D^\alpha u\|_{L^p(\Omega)}. \tag{2.7}$$

Then with (2.1) we choose a suitable ε and deduce (2.2) when $\Phi \equiv 0$. In the general case we consider $U = u - \Phi$ and use a type of Poincaré inequality in $\Omega \cap \text{supp}\ \Phi$.

Remark 2.1. Condition (2.4) may be weakened in a local condition, in the neighbourhood of the boundary $\partial\Omega$ of Ω.

3. DESCRIPTION OF THE SUBDIFFERENTIAL ∂J_Φ

We study the subdifferential of J_Φ defined by (1.5), (1.8) in two important cases.

(I) First case ($j(x,r) = j_1(r) + j_2(x,r)$, j_2 finite)

Here we suppose that $j = j_1 + j_2$, where

$$j_1 : \mathbb{R} \to [0,+\infty] \text{ is a nonnegative l.s.c. function, with } j_1(0) = 0; \tag{3.1}$$

$j_2: \Omega \times \mathbb{R} \to [0,+\infty)$ is a nonnegative *real* normal convex integrand, with

$$x \mapsto j_2(x,r) \in L^1_{loc}(\Omega), \quad \forall r \in \mathbb{R}, \tag{3.2}$$

$$x \mapsto j_2(x,0) \in L^1(\Omega) \tag{3.3}$$

(notice that we do not suppose $j_2(x,0) = 0$, but $0 \in D(J)$). In some cases (3.2) will be strengthened in

$$x \mapsto j_2(x,r) \in L^\infty(\Omega), \quad \forall r \in \mathbb{R}, \tag{3.4}$$

or

$$x \to j_2(x,r) \in L^1(\Omega), \quad \forall r \in \mathbb{R}. \tag{3.5}$$

We set $\beta_1 = \partial j_1$, $\beta_2 = \partial j_2$ (hence $\beta = \partial j = \beta_1 + \beta_2$), $\overline{D(j_1)} = [a,b]$ $(-a,b \leqslant +\infty)$. Then $J = J^1 + J^2$, where J^i is associated to j_i as J is to j by (1.5) for $i = 1,2$.

Remark 3.1. Under the assumption (3.3) we can formulate (3.2), (3.4), (3.5) in an equivalent way in terms of β_2^o.

Our study of the subdifferential extends the results of Brézis [4] (case $j_2 \equiv 0$, $0 \in (a,b)$, $\Phi \equiv 0$, Ω bounded), those of Grun-Rehomme [9] (case $m = 1$), and our preceding results [2] (case $j_2(x,0) = 0$). The essential tool is the approximation of truncation within $W_0^{m,p}(\Omega)$ introduced by Hedberg [10]. In order to extend the proofs of [2] we use (3.3) and the following obvious inequality: for almost all $x \in \Omega$,

$$0 \leqslant j_2(x,r) \leqslant j_2(x,s) + j_2(x,t) \text{ if } s \leqslant r \leqslant t. \tag{3.6}$$

We deduce as in [2] the approximation property:

LEMMA 3.1. Under the assumptions (3.1) to (3.3), for any $U \in D(J_o)$, there exists a sequence $\phi_n \in \mathcal{D}(\Omega) \cap D(J_o)$ such that $\phi_n \to U$ in $W^{m,p}(\Omega)$ and $j(\cdot,\phi_n) \to j(\cdot,U)$ in $L^1(\Omega)$.

Then we study the domain of the conjugate function J_Φ^* of J_Φ in $W^{-m,p'}(\Omega)$:

THEOREM 3.1. Under the assumptions (3.1) to (3.3) with $\Phi \in D(J)$, let $T \in D(J_\Phi^*)$. Then

(a) T is a measure on Ω. So it admits the Lebesgue decomposition:

$$T = \gamma + S, \text{ where } \gamma \in L^1_{loc}(\Omega) \text{ and } S \text{ is a singular measure;} \tag{3.7}$$

we call σ the density of S with respect to its absolute value:

$$S = \sigma|S| \ , \quad \sigma \in L^\infty(\Omega,|S|). \tag{3.8}$$

(b) S is bounded when Φ verifies

$$a < \text{ess inf}_\Omega \ \Phi \leqslant \text{ess sup}_\Omega \ \Phi < b; \tag{3.9}$$

then $\gamma \in L^1(\Omega) + L^\infty(\Omega)$ (resp. $L^1_{loc}(\bar{\Omega})$) when j_2 satisfies (3.4) (resp. (3.5)).

(c) if Φ verifies

$$\left.\begin{array}{l} a < \text{ess inf}_\Omega \ \Phi \leqslant \text{ess sup}_\Omega \ \Phi < b, \text{ if } 0 \in (a,b), \\ \text{ess sup}_\Omega \ \Phi < b, \text{ when } a = 0 \ (a < \text{ess inf}_\Omega \ \Phi, \ \text{ when } b = 0), \end{array}\right\} \tag{3.10}$$

and

$$\text{there exists } \alpha > 1 \text{ such that } \alpha\Phi \in D(J), \tag{3.11}$$

then for any $V \in D(J_\Phi)$, we have

$$\gamma.V \in L^1(\Omega), \ \ \sigma.V \in L^1(\Omega,|S|), \tag{3.12}$$

$$\langle T,V\rangle = \int_\Omega \gamma.V \, dx + \int_\Omega \sigma.V d|S|. \tag{3.13}$$

<u>Remark 3.2</u>. These results hold in particular when $\Phi \equiv 0$ (then (3.10) and (3.11) are obviously satisfied). In the general case, (3.10) implies $\Phi \in L^\infty(\Omega)$, hence (3.10) implies (3.11) when j_2 verifies (3.5), or (3.4) with $\Phi \in L^1(\Omega)$ (use (3.6) in the first case, and

$$j_2(\cdot,\alpha\Phi) - j_2(\cdot,\Phi) = \int_\Phi^{\alpha\Phi} \beta_2^o(\cdot,r)dr$$

in the second one).

<u>Sketch of the proof of Theorem 3.1</u>. We argue as in [2], Propositions 4.2, 4.3:

Step 1 : case $\Phi \equiv 0$. We prove that for any compact $K \subset \Omega$ there exists $c(K) > 0$ such that, for every $\phi \in \mathcal{D}(\Omega)$ with support in K,

$$|\langle T,\phi\rangle| \leqslant c(K)\ \|\phi\|_{L^\infty(\Omega)}; \tag{3.14}$$

hence T is a measure on Ω. Then for any measure $T \in W^{-m,p'}(\Omega)$ we prove that

$$J_0^*(T) = \int_\Omega j^*(\cdot,\gamma)dx + \int_\Omega I^*_{[a,b]}(\sigma)\ d|S|,\ \text{in } [-J(0),+\infty], \tag{3.15}$$

where γ, S, σ are defined by (3.7), (3.8), and j^*, $I^*_{[a,b]}$ are the conjugate functions of j, $I_{[a,b]}$ with respect to r. From Lemma 3.1 and (3.15) we prove (3.12) and the following inequalities: for any $v \in D(J_0)$,

$$\left.\begin{aligned} 0 &\leqslant \int_\Omega (j(\cdot,v)+j^*(\cdot,\gamma)-\gamma v)dx \leqslant J_0(v) + J_0^*(T) - \langle T,v\rangle, \\ 0 &\leqslant \int_\Omega (I_{[a,b]}(v)+I^*_{[a,b]}(\sigma)-\sigma v)d|S| \leqslant J_0(v) + J_0^*(T) - \langle T,v\rangle. \end{aligned}\right\} \tag{3.16}$$

Identity (3.13) follows, from the results of [6].

Step 2 : general case. We apply the results of Step 1 to the restriction of J to $W_0^{m,p}(\Omega')$, where Ω' is any open set strictly contained in Ω, and deduce (a) by localization. Assumption (3.9) allows us to improve (3.14) and obtain (b). Under the assumptions of (c), we prove that $T \in D(J_0^*)$: the essential point is to verify that for the functions $\xi = \xi(\Omega,\Omega') \in C^\infty(\bar{\Omega})$ with $\xi(\Omega) \subset [0,1]$, $\xi/\Omega' \equiv 1$, $\xi \equiv 0$ in a neighbourhood of $\partial\Omega$, $\langle T,\xi\Phi\rangle$ is bounded above; now this comes from

$$(\alpha-1)\ \langle T,\xi\Phi\rangle \leqslant J_\Phi^*(T) + J((\alpha-1)\xi\Phi+\Phi) \leqslant J_\Phi^*(T) + J(\Phi) + J(\alpha\Phi). \tag{3.17}$$

Then we can go to the limit in the inequalities (3.16) applied to $\xi(V + \Phi)$ for $V \in D(J_\Phi)$ and deduce (3.12), (3.13).

Now we give a characterization of the subdifferential:

THEOREM 3.2. Under the assumptions (3.1) to (3.3) with $\Phi \in D(J)$, let $U = u - \Phi \in D(J_\Phi)$ and $T \in W^{-m,p'}(\Omega)$.

Then $T \in \partial J_\Phi(U)$ if and only if: T is a measure on Ω, and, defining γ, S, σ by (3.7), (3.8), we have

$$\gamma = \gamma_1 + \gamma_2, \qquad \gamma_i \in L^1_{loc}(\Omega) \text{ for } i = 1,2, \tag{3.18}$$

$$\gamma_1(x) \in \beta_1(u(x)), \text{ a.e. in } \Omega, \tag{3.19}$$

$$\gamma_2(x) \in \beta_2(x,u(x)), \text{a.e. in } \Omega, \tag{3.20}$$

$$\sigma(x) \in \partial I_{[a,b]}(u(x)), \quad |S| \text{ - a.e. in } \Omega. \tag{3.21}$$

Sketch of the proof: (i) Suppose $T \in \partial J_\Phi(U)$; then we have only to prove (3.18) to (3.21) in Ω'; we use inequalities of type (3.16) in Ω', applied to functions $\xi(\Omega',\Omega")u$, where $\Omega"$ is any open set strictly contained in Ω', and let $\Omega"$ grow to Ω'; hence we get (3.19)-(3.21) and $j^*(\cdot,\gamma) \in L^1(\Omega')$; then $j_1^*(\gamma_1)$, $j_2^*(\cdot,\gamma_2) \in L^1(\Omega')$, since

$$j^*(\cdot,\gamma) = j_1^*(\gamma_1) + j_2^*(\cdot,\gamma_2) \geqslant j_2^*(\cdot,\gamma_2) \geqslant -j_2(\cdot,0). \tag{3.22}$$

This implies $\gamma_i v \in L^1(\Omega')$ for any $v \in D(J)$ for $i = 1,2$; hence (3.18) holds (consider $\gamma_i u$).
(ii) Suppose (3.18) to (3.21). Notice that for almost all $x \in \Omega$,

$$\gamma(x)r \geqslant \gamma_2(x)r \geqslant j_2(x,r) - j_2(x,0) \geqslant -j_2(x,0), \quad \forall r \in \mathbb{R}; \tag{3.23}$$

so with (3.3) we can apply the results of [6] to $\xi(\Omega,\Omega')u$ and get $\gamma u \in L^1(\Omega')$, $\sigma u \in L^1(\Omega',|S|)$. To prove that $T \in \partial J_\Phi(U)$ we use functions $\xi(\Omega',\Omega")u$ as in [2] and let $\Omega"$ grow to Ω.

Remark 3.3. In the same way as in (i) we can prove that, for $T \in \partial J_\Phi(U)$, assumptions (3.10), (3.11) imply

$$\gamma_i V \in L^1(\Omega), \quad \forall V \in D(J_\Phi), \text{ for } i = 1,2; \tag{3.24}$$

indeed they imply $T \in D(J_0^*)$, hence $j^*(\cdot,\gamma) \in L^1(\Omega)$; from (3.22) $\gamma_i(V + \Phi) \in L^1(\Omega)$; we deduce $\gamma_i\Phi \in L^1(\Omega)$ and (3.24).

Remark 3.4. Conditions $\Phi \in D(J)$ and (3.10), (3.11) can be weakened in local conditions in the neighbourhood of $\partial\Omega$; see [2] and use (3.3).

(II) Second case ($j(x,r) = j_2(x,r)$ if $r \geq \Psi(x)$, $+\infty$ if not)

Here we suppose that $j = j_\Psi + j_2$, where j_2 is defined as in the first case, Ψ is a given function in $W^{m,p}(\Omega)$ and

$$j_\Psi(x,r) = 0 \text{ if } r \geq \Psi(x), \quad +\infty \text{ if not;} \tag{3.25}$$

so that $J = I_C + J^2$, where C is the convex set of constraints

$$C = \{u \in W^{m,p}(\Omega) \mid u(x) \geq \Psi(x), \text{ a.e. in } \Omega\} . \tag{3.26}$$

Notice that j satisfies the first case's conditions when Ψ is a nonpositive constant. Now we can also reduce the study of ∂J_Φ to the first case when Ψ is a bounded function in the domain of J^2, with a suitable change of variables:

PROPOSITION 3.1. Under the assumptions (3.2), (3.3), (3.25) with $\Psi \in L^\infty(\Omega) \cap D(J_2)$, let $\tilde{j} = \tilde{j}_1 + \tilde{j}_2$, with $\tilde{j}_1 = I_{[0,+\infty[}$ and, for almost all $x \in \Omega$,

$$\tilde{j}_2(x,r) = j_2(x,r + \Psi(x)), \quad \forall r \in \mathbb{R}; \tag{3.27}$$

let $\tilde{J}$, $\tilde{J}^1$, $\tilde{J}^2$ the functions of $W^{m,p}(\Omega)$ associated with $\tilde{j}$, $\tilde{j}_1$, $\tilde{j}_2$ by (1.5).

Then $\tilde{j}_1$ satisfies (3.1), $\tilde{j}_2$ is a nonnegative normal convex integrand and satisfies (3.2), (3.3); and we have

$$\partial J_\Phi = \partial \tilde{J}_{\Phi-\Psi}. \tag{3.28}$$

(the important point is $x \mapsto \tilde{j}_2(x,0) = j_2(x,\Psi(x)) \in L^1(\Omega)$).

Now we can apply the results of this section to the interpretation of inequality (1.1).

4. GENERALIZED EQUATIONS OF PROBLEM (1.1)

(I) First case ($j = j_1 + j_2$)

From Theorem 3.2 we deduce the following result:

COROLLARY 4.1. Assume (1.2)-(1.5), (3.1)-(3.3) and $\Phi \in D(J)$. Then Problem (1.1) is equivalent to the following problem:

$$
\left.\begin{aligned}
&u - \Phi \in W^{m,p}(\Omega),\\
&\text{and there exist } \gamma_1, \gamma_2 \in L^1_{loc}(\Omega) \text{ and two nonnegative singular}\\
&\text{measures } S_b, S_a, \text{ concentrated on the coincidence sets}\\
&\{x \in \Omega | u(x) = b\}, \{x \in \Omega | u(x) = a\}, \text{ such that}\\
&\qquad f - Au = \gamma_1 + \gamma_2 + S_b - S_a,\\
&\qquad \gamma_1(x) \in \beta_1(u(x)), \gamma_2(x) \in \beta_2(x,u(x)), \text{ a.e. in } \Omega.
\end{aligned}\right\} \quad (4.1)
$$

Remark 4.1. Let us give some properties of any solution u:

(i) under (3.9), S_a, S_b are bounded;

(ii) under (3.10), (3.11), we have, for any $v \in D(J) \cap (\Phi + W_0^{m,p}(\Omega))$,

$$\gamma_i(v-\Phi) \in L^1(\Omega) \text{ for } i = 1,2; \ (v-\Phi) \in L^1(\Omega, S_b + S_a); \quad (4.2)$$

$$\langle f-Au, v-\Phi\rangle = \int_\Omega \gamma_1(v-\Phi)dx + \int_\Omega \gamma_2(v-\Phi)dx + \int_\Omega (v-\Phi)dS_b - \int_\Omega (v-\Phi)dS_a; \quad (4.3)$$

in particular, if u, $\hat{u}$ are two solutions for data f, $\hat{f}$, we have, with obvious notation,

$$
\begin{aligned}
\langle f-\hat{f}-(Au-A\hat{u}), u-\hat{u}\rangle &= \int_\Omega (\gamma_1-\hat{\gamma}_1)(u-\hat{u})\,dx + \int_\Omega (\gamma_2-\hat{\gamma}_2)(u-\hat{u})\,dx + \ldots \\
&+ \int_\Omega (u-\hat{u})\,dS_b + \int_\Omega (\hat{u}-u)\,dS_a + \int_\Omega (\hat{u}-u)\,d\hat{S}_b + \int_\Omega (u-\hat{u})\,d\hat{S}_a,
\end{aligned} \quad (4.4)
$$

where each integral of the right-hand side is nonnegative.

(II) Second case ($j = j_\Psi + j_2$)

Here we consider a unilateral constraint problem, where j_2 satisfies (3.2) (3.3) and j_Ψ is given by (3.25). To simplify we suppose $\Phi \equiv 0$.

Then Problem (1.1) has the following form:

$$
\left.\begin{aligned}
&u \in C_o,\\
&\langle Au-f, v-u\rangle + J^2(v) - J^2(u) \geqslant 0, \quad \forall v \in C_o,
\end{aligned}\right\} \quad (4.5)
$$

where C_o is the convex set

$$C_o = \{u \in W_0^{m,p}(\Omega) | u(x) \geq \Psi(x), \text{ a.e. in } \Omega\}. \tag{4.6}$$

We suppose that

$$\text{ess sup}_V \Psi \leq 0, \text{ where } V \text{ is a neighbourhood of } \partial\Omega, \tag{4.7}$$

so that C_o is nonempty (see [2]).

COROLLARY 4.2. Assume (1.2)-(1.5), (3.2), (3.3) and $\Psi \in L^\infty(\Omega) \cap D(J^2)$, with (4.7). Then

(i) Problem (4.5) is equivalent to the following problem:

$$\left.\begin{array}{l} u \in C_o, \\ \text{there exist } \gamma_1, \gamma_2 \in L^1_{loc}(\Omega) \text{ and a singular measure } S \\ \text{such that} \\ f - Au = \gamma_1 + \gamma_2 + S, \\ \gamma_2(x) \in \beta_2(x,u(x)), \text{ a.e. in } \Omega, \\ \text{and } \gamma_1, S \text{ are nonpositive, concentrated on the coincidence} \\ \text{set } \{x \in \Omega | u(x) = \Psi(x)\}. \end{array}\right\} \tag{4.8}$$

(ii) Assume that

$$\text{there exists } \theta < 0 \text{ such that } \theta\Psi \in D(J^2), \tag{4.9}$$

then Problem (4.5) is also equivalent to the problem

$$\left.\begin{array}{l} u \in C_o, \\ \text{and there exists } \gamma_2 \in L^1_{loc}(\Omega) \text{ such that } \gamma_2 u \in L^1(\Omega) \text{ and} \\ \gamma_2(x) \in \beta_2(x,u(x)), \text{ a.e. in } \Omega, \\ \langle Au-f,\phi-u\rangle + \int_\Omega \gamma_2(\phi-u)\, dx \geq 0, \quad \forall\phi \in C_o \cap \mathcal{D}(\Omega). \end{array}\right\} \tag{4.10}$$

Remark 4.2. Furthermore any solution u of (4.5) satisfies:

(i) S is bounded if $\text{ess sup}_V \Psi < 0$.

This property extends a result of Caffarelli and Friedman [8] (where $m = 2$, A is the bi-Laplacian and $j_2 \equiv 0$).

(ii) Under the assumption (4.9), we have, for any $v \in C_o \cap D(J^2)$,

$$\gamma_i v \in L^1(\Omega) \ (i = 1,2), \ v \in L^1(\Omega,|S|), \tag{4.11}$$

$$\langle f-Au,v\rangle = \int_\Omega \gamma_1 v \, dx + \int_\Omega \gamma_2 v \, dx - \int_\Omega v \, d|S|; \tag{4.12}$$

if u, $\hat{u}$ are two solutions for data f, $\hat{f}$, then

$$\langle f-\hat{f}-(Au-A\hat{u}),u-\hat{u}\rangle = \int_\Omega (\gamma_1-\hat{\gamma}_1)(u-\hat{u}) \, dx + \int_\Omega (\gamma_2-\hat{\gamma}_2)(u-\hat{u}) \, dx + \ldots$$

$$+ \int_\Omega (\hat{u}-u) d|S| + \int_\Omega (u-\hat{u}) \, d|\hat{S}|, \tag{4.13}$$

where each integral of the right-hand side is nonnegative. These results improve our previous ones [2] (where $j_2 \equiv 0$) and some results of Boccardo and Giachetti [3]. Notice that assumptions (3.5) and $\Psi \in L^\infty(\Omega)$ imply $\Psi \in D(J^2)$, (4.9) and $\gamma_2 \in L^1(\Omega)$ (use $\gamma_2 u \in L^1(\Omega)$).

Sketch of the proof of Corollary 4.2: (i) This comes directly from Section 3.

(ii) Any solution of (4.5) is a solution of (4.10), from (4.11), (4.12) and sign properties of γ_1,S given in (4.8). In the converse let u be a solution of (4.10) and let $v \in C_o \cap D(J^2)$; notice that for every $\phi \in C_o \cap L^\infty(\Omega)$ with compact support in Ω, we have $\phi \in D(J^2)$ and

$$\langle Au-f,\phi-u\rangle + J^2(\phi) - J^2(u) \geqslant 0. \tag{4.14}$$

Then consider ξv, where $\xi = \xi(\Omega,\Omega')$ and $\Omega' \supset (\Omega-\bar{V})$ is an open set strictly contained in Ω: $\xi v \in C_o \cap D(J^2)$; applying Lemma 3.1 to $\xi(v-\Psi)$ and the function $(x,r) \mapsto \tilde{j}_1(r) + j_2(x,r + \xi(x)\Psi(x))$, we deduce from (4.14):

$$\langle Au-f,\xi v-u\rangle + J^2(\xi v) - J^2(u) \geqslant 0. \tag{4.15}$$

We let grow Ω' to Ω and obtain (4.5).

5. EXISTENCE RESULTS FOR PROBLEM (1.6)

From Section 4 we see that, when $f \in L^{p'}(\Omega)$, solutions of Problem (1.1) are solutions of Problem (1.4) if and only if the singular measure S vanishes identically. First we give an obvious consequence:

COROLLARY 5.1. Under the assumptions (1.2)-(1.5), (3.2), (3.3), $j_1 \equiv 0$, let $\Phi \in D(J^2)$ and $f \in L^{p'}(\Omega)$. Then Problems (1.1) and (1.6) are equivalent.

Hence we obtain for Problem (1.6) the existence results of Section 2, and the properties of Section 4 (Remark 4.1); this improves some results of Brézis and Browder [5] in the convex case.

Now we prove in [1] the following regularity result for the case of *second-order* operators when j does not depend on x:

THEOREM 5.1. We assume (1.2)-(1.5) with m = 1, $j \equiv j_1$ given by (3.1), $\Phi \equiv 0$, $f \in L^{p'}(\Omega)$. Then Problems (1.1) and (1.6) are equivalent. Furthermore any solution u satisfy $Au \in L^{p'}(\Omega)$.

Idea of the proof: We consider any solution u of (1.1) and notice that u is the unique solution of an implicit problem of the same type. Then we use a method of Yosida adapted to $L^p(\Omega)$ to approximate this last problem; the approximate solution u_λ satisfies $Au_\lambda \in L^{p'}(\Omega)$, and we get a priori estimates using positivity properties of second-order operators, which allows us to go to the limit. The converse comes from Corollary 4.1.

This theorem extends the usual results where p = 2 and A is the Laplacian.

REFERENCES

[1] M.F. Bidaut-Véron, Variational inequalities of order 2m in unbounded domains, Nonlinear Anal., T.,M. and A. 6(3),253-269 (1982).

[2] M.F. Bidaut-Véron, Nonlinear elliptic equations of order 2m and sub-differentials, J.O.T.A. 40(3), 405-432 (1983).

[3] L. Boccardo and D. Giachetti, Strongly nonlinear unilateral problems, to appear.

[4] H. Brézis, Intégrales convexes dans les espaces de Sobolev, Isr. J. of Math. 13 (1), 9-23 (1972).

[5] H. Brézis and F.E. Browder, Strongly nonlinear elliptic boundary value problems, Ann. Sc. Norm. Sup. Pisa 5, 587-603 (1978).

[6] H. Brézis and F.E. Browder, Some properties of higher order Sobolev spaces, to appear in J. de Math. Pures et Appl.

[7] F.E. Browder, Pseudomonotone operators and nonlinear elliptic boundary value problems on unbounded domains, Proc. Nat. Acad. Sci. 74, 2659-2661 (1977).

[8] L. Caffarelli and A. Friedman, The obstacle problem for the biharmonic operator, Ann. Sc. Norm. Sup. Pisa 6, 151-184 (1979).

[9] M. Grun-Rehomme, Caractérisation du sous-différentiel d'intégrales convexes dans les espaces de Sobolev, J. de Math. Pures et Appl. 56, 149-156 (1977).

[10] L.I. Hedberg, Two approximations problems in function spaces, Ark. för Mat. 16, 51-81 (1978).

M.-F. Bidaut-Véron
Département de Mathématiques,
Faculté des Sciences,
Parc de Grandmont,
37200 Tours,
France

H BRÉZIS

Nonlinear elliptic equations involving measures

1. INTRODUCTION

Assume $\Omega \subset \mathbb{R}^N$ is a bounded smooth domain such that $0 \in \Omega$. Consider the following problem:

$$\left.\begin{aligned} -\Delta u + |u|^{p-1}u &= f(x) \quad \text{in } \Omega, \\ u &= 0 \quad \text{on } \partial\Omega, \end{aligned}\right\} \tag{1}$$

where $1 < p < \infty$ and f is a given function in L^1, or more generally f is a measure. It is known (see Brézis and Strauss [4]) that given $f \in L^1(\Omega)$, Problem (1) has a unique solution in some appropriate class. More precisely $u \in L^p(\Omega)$ and we have

$$\left.\begin{aligned} -\int u\ \Delta\phi + \int |u|^{p-1}u\phi = \int f\phi \\ \forall \phi \in C^2(\bar{\Omega}) \text{ with } \phi = 0 \text{ on } \partial\Omega. \end{aligned}\right\} \tag{2}$$

Moreover $u \in W_0^{1,q}(\Omega)$ for all $q < N/(N-1)$. We turn now to the case where f is a measure. Our main results are the following:

THEOREM 1. (Bénilan-Brézis): Assume $p \geqslant N/(N-2)$ $(N \geqslant 3)$ and $f = \delta$. Then there is no solution of (1). More precisely there is no function $u \in L^p_{loc}(\Omega)$ such that

$$-\Delta u + |u|^{p-1}u = \delta \text{ in } \mathcal{D}'(\Omega). \tag{3}$$

THEOREM 2. (Bénilan-Brézis): Assume $p < N/(N-2)$ (any $1 < p < \infty$ when $N = 1$, $N = 2$). Then, for any bounded measure $f \in M(\Omega)$ there is a unique solution of (1) in the sense of (2).

Theorem 1 is a straightforward consequence of:

THEOREM 3. Assume $p \geqslant N/(N-2)$. Let $f \in L^1(\Omega)$ and suppose $u \in L^p_{loc}(\Omega \setminus \{0\})$

satisfies:

$$-\Delta u + |u|^{p-1}u = f \text{ in } \mathcal{D}'(\Omega \setminus \{0\}). \tag{4}$$

Then $u \in L^p_{loc}(\Omega)$ and we have

$$-\Delta u + |u|^{p-1}u = f \text{ in } \mathcal{D}'(\Omega). \tag{5}$$

Theorem 3 extends an earlier result of Brézis and Véron [5] (dealing with the case where f is bounded). Clearly Theorem 3 implies Theorem 1 (choose $f = 0$). Solving (1) for a measure f a natural method consists of solving (1) for a sequence of smooth functions (f_n) such that $f_n \to f$. Our next result describes what happens to (u_n) as $n \to \infty$.

THEOREM 4. Assume $p \geqslant N/(N-2)$. Let $f \in L^1(\Omega)$ and let (f_n) be a sequence on $L^1(\Omega)$ such that

$$\int_{\Omega \setminus B_r(0)} |f_n - f| \xrightarrow[n\to\infty]{} 0 \text{ for each } r > 0. \tag{6}$$

Let u_n denote the solution of:

$$\left.\begin{aligned} -\Delta u_n + |u_n|^{p-1}u_n &= f_n \text{ in } \Omega, \\ u_n &= 0 \text{ in } \partial\Omega. \end{aligned}\right\} \tag{7}$$

Then $u_n \to u$ in $L^p(\Omega \setminus B_r(0))$ for each $r > 0$, where u is the (unique) solution of (1).

In particular if (f_n) is any sequence in $L^1(\Omega)$ such that $\operatorname{supp} f_n \subset B(0,\frac{1}{n})$, then $u_n \to 0$. This result is rather surprising; for example when $f_n \to \delta$ we would expect that $u_n \to u$, where u satisfies

$$-\Delta u + |u|^{p-1}u = \delta. \tag{8}$$

But we know that such a u does not exist. By contrast note that if $p < N/(N-2)$ it is indeed the case that $u_n \to u$ where u satisfies (8) (see Section 3).

Further interesting results concerning Problem (1) have recently been obtained by Baras and Pierre [1].

Finally we mention that there are analogous results for nonlinear parabolic equations (see Brézis and Friedman [3]).

2. SKETCH OF THE PROOF OF THEOREM 3

For simplicity - but this is not essential - we assume that

$$\Omega = \{x \in \mathbb{R}^3; \quad |x| < 1\} \text{ and } p = 3.$$

Let v be the solution of

$$\left.\begin{aligned} -\Delta v &= |f| \text{ on } \Omega \\ v &= 0 \quad \text{on } \partial\Omega , \end{aligned}\right\}$$

so that $v \in M^3(\Omega)$ (see Bênilan et al. [2]).

Step 1. We have

$$|u(x)| \leqslant \frac{3}{|x|} + v(x) \text{ for a.e. } x \text{ with } |x| < \frac{1}{2}. \tag{9}$$

Indeed fix $x_0 \in \mathbb{R}^3$ with $0 < |x_0| < \frac{1}{2}$ and let $0 < R < |x_0|$. In the ball $B_R(x_0)$ we define

$$U(x) = \frac{3R}{R^2 - |x-x_0|^2} + v(x),$$

(such a construction is inspired by the paper of Loewner and Nirenberg [7]). An easy computation shows that

$$-\Delta U + U^3 \geqslant |f| \text{ on } B_R(x_0). \tag{10}$$

Comparing (10) with the equation

$$-\Delta u + u^3 = f \text{ in } \mathcal{D}'(\Omega \setminus \{0\}), \tag{11}$$

we obtain

$$-\Delta(u-U) + u^3 - U^3 \leqslant 0 \text{ in } \mathcal{D}'(B_R(x_0)). \tag{12}$$

We deduce from Kato's inequality (see [6]) that

$$-\Delta(u-U)^+ + (u^3-U^3)^+ \leqslant 0 \text{ in } \mathcal{D}'(B_R(x_0)) \tag{13}$$

and in particular we find

$$-\Delta(u-U)^+ \leqslant 0 \text{ in } \mathcal{D}'(B_R(x_0)). \tag{14}$$

Since, formally, $(u-U)^+ = 0$ on $\partial B_R(x_0)$ we conclude from the maximum principle that

$$u(x) \leqslant U(x) \text{ for a.e. } x \in B_R(x_0) \tag{15}$$

Finally we let $x \to x_0$ (provided x_0 is a Lebesgue point for v) and we obtain (9).

In order to justify (15) we may proceed as follows. Define w(x) on $\mathbb{R}^3$ by

$$w = \begin{cases} (u-U)^+ & \text{on } B_R(x_0) \\ 0 & \text{otherwise.} \end{cases}$$

We claim that

$$-\Delta w \leqslant 0 \text{ in } \mathcal{D}'(\mathbb{R}^3), \tag{16}$$

so that $w \leqslant 0$ a.e. on $\mathbb{R}^3$.

Indeed let $\phi \in \mathcal{D}_+(\mathbb{R}^3)$ and let (ζ_n) be a sequence of smooth functions such that $0 \leqslant \zeta_n \leqslant 1$,

$$\zeta_n(x) = 1 \text{ for } |x-x_0| < R - \frac{2}{n}$$
$$\zeta_n(x) = 0 \text{ for } |x-x_0| > R - \frac{1}{n},$$
$$|\nabla\zeta_n(x)| \leqslant Cn, |\Delta\zeta_n(x)| \leqslant Cn^2 \text{ for } R - \frac{2}{n} < |x-x_0| < R - \frac{1}{n}.$$

We have

$$-\int w\Delta(\phi\zeta_n) \leqslant 0. \tag{17}$$

In order to deduce (16) from (17) we must check that

$$\int w \, \nabla\phi \, \nabla\zeta_n \to 0 \text{ as } n \to \infty \tag{18}$$

$$\int w\phi \quad \Delta\zeta_n \to 0 \text{ as } n \to \infty . \tag{19}$$

We shall, for instance, verify (19). We have

$$\left|\int w\phi \, \Delta\zeta_n\right| \leqslant Cn^2 \int |w| \text{ (where the right-hand integration is carried out over } R - \tfrac{2}{n} < |x-x_0| < R - \tfrac{1}{n})$$

$$\leqslant Cn^2 \int_{\substack{B_R(x_0) \\ [u>n]}} \leqslant C \int_{\substack{B_R(x_0) \\ [u>n]}} |u|^3 \to 0 \text{ as } n \to \infty$$

since $u \in L^3(B_R(x_0))$ (note that $U(x) \sim n$ on the set $[R - \frac{2}{n} < |x-x_0| < R - \frac{1}{n}]$).

Step 2. We have $u \in L^3_{loc}(\Omega)$. From (11) and Kato's inequality we deduce that

$$-\Delta |u| + |u|^3 \leq |f| \text{ in } \mathcal{D}'(\Omega \setminus \{0\}). \tag{20}$$

Choose a sequence (ζ_n) in $\mathcal{D}_+(\Omega \setminus \{0\})$ such that

$$\begin{aligned} \zeta_n(x) &= 0 \quad \text{for } |x| < \tfrac{1}{n} \\ \zeta_n(x) &= 1 \quad \text{for } \tfrac{2}{n} < |x| < \tfrac{1}{2} \\ \zeta_n(x) &= \zeta(x) \text{ for } |x| > \tfrac{1}{2} \\ |\Delta\zeta_n(x)| &\leq Cn^2 \quad \text{for } \tfrac{1}{n} < |x| < \tfrac{2}{n}, \end{aligned}$$

where $\zeta(x)$ is a fixed function in $\mathcal{D}_+(\Omega)$ (independent of n). It follows from (20) that

$$\int |u|^3 \zeta_n \leq \int |u| \, |\Delta\zeta_n| + \int |f| \, \zeta_n \leq Cn^2 \int_{[\frac{1}{n}<|x|<\frac{2}{n}]} |u| + C.$$

From Step 1 we know that $u \in M^3(|x| < \frac{1}{2})$ and thus $\int_{[\frac{1}{n}<|x|<\frac{2}{n}]} |u| \leq C/n^2$. Hence $\int |u|^3 \zeta_n$ remains bounded as $n \to \infty$ and we conclude that $u \in L^3_{loc}(\Omega)$.

Step 3. We have

$$-\Delta u + u^3 = f \text{ in } \mathcal{D}'(\Omega).$$

Choose a sequence (ζ_n) of smooth functions such that

$$\begin{aligned} &\zeta_n(x) = 0 \quad \text{for } |x| < \tfrac{1}{n} \\ &\zeta_n(x) = 1 \quad \text{for } |x| > \tfrac{2}{n} \\ &|\nabla\zeta_n(x)| \leq Cn, \quad |\Delta\zeta_n(x)| \leq Cn^2 \text{ for } \tfrac{1}{n} < |x| < \tfrac{2}{n}. \end{aligned}$$

Given $\phi \in \mathcal{D}(\Omega)$ we must check that

$$-\int u\Delta\phi + \int u^3\phi = \int f\phi.$$

We already know, by (4), that

$$-\int u\Delta(\phi\zeta_n) + \int u^3\phi\zeta_n = \int f\phi\zeta_n.$$

Using Step 2 and dominated convergence it remains to verify that as $n \to \infty$

$$\int u\ \nabla\phi\ \nabla\zeta_n \to 0 \text{ and } \int u\phi\ \Delta\zeta_n \to 0.$$

Let us consider, for example, the second property. We have

$$\left|\int u\phi\ \Delta\zeta_n\right| \leqslant Cn^2 \int_{[\frac{1}{n}<|x|<\frac{2}{n}]} |u| \leqslant C\Big[\int_{[\frac{1}{n}<|x|<\frac{2}{n}]} |u|^3\Big]^{1/3} \to 0$$

since $u \in L^3_{loc}(\Omega)$.

3. SKETCH OF THE PROOF OF THEOREM 2

Let (f_n) be a sequence of smooth functions such that $f_n \longrightarrow f$ for the weak * topology of measures and so $\|f_n\|_{L^1} \leqslant C$. Let u_n be the solution of

$$-\Delta u_n + |u_n|^{p-1} u_n = f_n \text{ on } \Omega, \quad u_n = 0 \text{ on } \partial\Omega. \tag{21}$$

We have the following estimates:

$$\left.\begin{aligned} \|\Delta u_n\|_{L^1} &\leqslant C \\ \|u_n\|_{L^p} &\leqslant C \end{aligned}\right\} \tag{22}$$

(In order to establish (22), we multiply (21) by $\theta_j(u_n)$ where $\theta_j(t) \to \text{sign}(t)$ as $j \to \infty$.) Finally we pass to the limit on (21) with the help of the following compactness lemma:

LEMMA 1. The set

$$\{v \in W_0^{1,1}(\Omega);\ \Delta v \in L^1(\Omega) \text{ and } \|\Delta v\|_{L^1} \leqslant 1\}$$

is relatively compact in $L^q(\Omega)$ for each $q < N/(N-2)$.

4. SKETCH OF THE PROOF OF THEOREM 4

Since $f_n \to f$ in $L^1(\Omega \setminus B_r(0))$ for each $r > 0$, we may assume, modulo a subsequence, that

$$|f_n| \leq g \text{ on } \Omega,$$

for some fixed function g such that $g \in L^1(\Omega \setminus B_r(0))$ for each $r > 0$. The same argument as in Section 2 (Step 1) shows that

$$\int_\omega |u_n| \leq C_\omega \quad \text{for each } \omega \subset\subset \Omega \setminus \{0\}. \tag{23}$$

Multiplying (7) by $\theta_j(u_n)\zeta$ where θ_j is a smooth approximation of sign, we obtain

$$-\int |u_n| \, \Delta\zeta + \int |u_n|^p \zeta \leq \int |f_n| \zeta$$
$$\forall \zeta \in C^2(\bar{\Omega}), \; \zeta \geq 0 \text{ on } \Omega. \tag{24}$$

Choosing in (24) a smooth function ζ such that $\zeta(x) = 0$ for $|x| < r/2$ and $\zeta(x) = 1$ for $|x| > r$ we see that

$$\int_{\Omega \setminus B_r(0)} |u_n|^p \leq C_r \quad \forall r > 0. \tag{25}$$

Using again (7) with (25) we obtain

$$\int_{\Omega \setminus B_r(0)} |\Delta u_n| \leq C_r. \tag{26}$$

It follows from (23) and (26) (as in Lemma 1) that (u_n) is relatively compact in $L^1(\omega)$ for each $\omega \subset\subset \Omega \setminus \{0\}$. In particular we may assume that

$$u_n \to u \text{ in } L^1(\omega) \quad \forall \omega \subset\subset \Omega \setminus \{0\}. \tag{27}$$

On the other hand we have (as in (24))

$$-\int |u_n - u_m| \Delta\zeta + \int \left| |u_n|^{p-1} u_n - |u_m|^{p-1} u_m \right| \zeta \leq \int |f_n - f_m| \zeta$$
$$\forall \zeta \in C^2(\bar{\Omega}), \quad \zeta \geq 0 \text{ on } \Omega. \tag{28}$$

Choosing for ζ the same function as above we deduce from (27) and (28) that

$$u_n \to u \text{ in } L^p(\Omega \setminus B_r^{(0)}) \quad \forall r > 0. \tag{29}$$

Finally, we multiply (7) by $\phi \in C^2(\bar{\Omega})$ with $\phi = 0$ on $\partial\Omega$:

$$-\int u_n \, \Delta\phi + \int |u_n|^{p-1} u_n \, \phi = \int f_n \phi. \tag{30}$$

If we assume further that $\phi \equiv 0$ near 0 we may pass to the limit in (30) using (29) and we obtain

$$-\int u \, \Delta\phi + \int |u|^{p-1} u \, \phi = \int f\phi, \tag{31}$$
$$\forall \phi \in C^2(\bar{\Omega}),\ \phi = 0 \text{ on } \partial\Omega,\ \phi \equiv 0 \text{ near } 0.$$

In particular u satisfies, $u \in L^p_{loc}(\Omega \setminus \{0\})$ and $-\Delta u + |u|^{p-1}u = f$ in $\mathcal{D}'(\Omega \setminus \{0\})$. Applying Theorem 3 we conclude that $u \in L^p(\Omega)$. By the same method as in Section 2, Step 3, we deduce from (31) that

$$-\int u \, \Delta\phi + \int |u|^{p-1} \, u\phi = \int f\phi$$
$$\forall \phi \in C^2(\bar{\Omega}),\ \phi = 0 \text{ on } \partial\Omega.$$

In other words, u is the unique solution of (1).

REFERENCES

[1] P. Baras and M. Pierre, Singularités éliminables d'équations élliptiques semi-linéaires (to appear).

[2] Ph. Bênilan, H. Brézis and M. Crandall, A semilinear elliptic equation in $L^1(\mathbb{R}^N)$, Ann. Sc. Norm. Pisa 2, 523-555 (1975).

[3] H. Brézis and A. Friedman, Nonlinear parabolic equations involving measures as initial conditions, J. Math. Pures et Appl. (to appear).

[4] H. Brézis and W. Strauss, Semilinear, second order elliptic equations in L^1, J. Math. Soc. Japan, 25, 565-590 (1973).

[5] H. Brézis and L. Véron, Removable singularities for some nonlinear elliptic equations, Arch. Rat. Mech. Anal. 75, 1-6 (1980).

[6] T. Kato, Schrödinger operators with singular potentials, Israel J. Math. 13, 135-148 (1972).

[7] C. Loewner and L. Nirenberg, Partial differential equations invariant under conformal or projective transformations, in Contributions to Analysis, Academic Press, 245-272 (1974).

H. Brézis
Université Pierre et Marie Curie
Paris VI, France

A GUTIÉRREZ CARDONA

On the invertibility of operators related to the boundary value problems of elastostatics on C^1-domains

Let D be a bounded C^1-domain of $\mathbb{R}^3$. We consider the equations of elasticity

$$A(\partial_X)u \equiv \mu\, \Delta u + (\lambda+\mu)\nabla \operatorname{div} u = 0, \text{ on } D \tag{1}$$

where u is the displacement vector, $u = (u_1, u_2, u_3)$ and λ and μ, the Lamé constants, are positive.

We define a double layer potential, related to these equations, for any $f = (f_1, f_2, f_3)$, with $f_j \in L^p(\partial D)$, $1 < p < \infty$, $j = 1,2,3$, in the following way:

$$L_1(f)(X) = \int_{\partial D} [T(\partial_Q, N_Q)\, \Gamma(X-Q)]' f(Q)\, d\sigma(Q),\ X \notin \partial D.$$

Here $\Gamma(X)$ is a fundamental solution of $A(\partial_X)u = 0$ given by

$$\Gamma(X) = \frac{A}{4\pi r} I + \frac{C}{4\pi r^3} X \otimes X$$

with

$$A = \frac{1}{2}\left|\frac{1}{\mu} + \frac{1}{\lambda+2\mu}\right|, \quad C = \frac{1}{2}\left|\frac{1}{\mu} - \frac{1}{\lambda+2\mu}\right| .$$

$r = |X|$, I is the identity 3×3 matrix, the tensor product $X \otimes X$ is given by the matrix with components $(X \otimes X)_{ij} = x_i x_j$, and the prime represents the transpose of the preceding matrix.

The operator $T(\partial_Q, N_Q)$ acting on Γ is defined by the matrix

$$T(\partial_X, N(X)) = (T_{ij}(\partial_X, N(X))),$$

where

$$T_{ij}(\partial_X, N(X)) = \lambda N_i(X) \frac{\partial}{\partial x_j} + \mu N_j(X) \frac{\partial}{\partial x_i} + \mu \delta_{ij} \frac{\partial}{\partial N(X)}$$

with

$$\frac{\partial}{\partial N(X)} = \langle N(X), \nabla_X \rangle,$$

$N(X)$ is an arbitrary unit vector and $\langle \cdot,\cdot \rangle$ is the usual inner product of vectors in $\mathbb{R}^3$. By N_Q we denote the unit exterior normal to ∂D at the point Q.

It is clear that the potential $L_1(f)$ is a solution of equation (1). Besides there exists the limit of $L_1(f)(X)$ when X approaches the point $P \in \partial D$ nontangentially, i.e. through the cones $\Gamma_\alpha^-(P) = \{x \in D, \langle P-X, N_P\rangle > \alpha|X-P|\}$ or

$$\Gamma_\alpha^+(P) = \{X \in \mathbb{R}^3 \setminus \bar{D}, \langle X-P, N_P\rangle > \alpha|X-P|\},$$

where α is a given real number, $0 < \alpha < 1$.

In fact

$$L_1(f)(X) \to \pm \frac{1}{2} f(P) + \int_{\partial D}^{*} [T(\partial_Q, N_Q)\Gamma(P-Q)]\,' f(Q)\, d\sigma(Q)$$

pointwise for almost every $P \in \partial D$, and in $L^p(\partial D)$, as $X \to P$, $X \in \Gamma_\alpha^\pm(P)$ respectively. The integral is defined as the limit, as $\varepsilon \to 0$, of integrals over $\partial D \setminus \rho_\varepsilon$, where ρ_ε represents the portion of ∂D whose projection on its tangent plane at P is a circle of radius ε. See Selvaggi and Sisto [4].

From now on and for the sake of simplicity we will use the following notation:

$$K(f)(P) = \int_{\partial D}^{*} [T(\partial_Q, N_Q)\Gamma(P-Q)]\,' f(Q)\, d\sigma(Q).$$

We will denote the adjoint of K by K^*.

Therefore it is clear that the Dirichlet problem on D, for data in $L^p(\partial D)$, can be solved if we are able to invert the operator $-\frac{1}{2} I + K$ on $L^p(\partial D)$, because the solution would be

$$u(X) = \int_{\partial D} [T(\partial_Q, N_Q)\,\Gamma(X-Q)]\,'(-\tfrac{1}{2} I + K)^{-1}(g)(Q)\, d\sigma(Q)$$

for any $g \in L^p(\partial D)$, given on the boundary.

If we define now the single-layer potential

$$L_2(f)(X) = \int_{\partial D} \Gamma(X-Q) f(Q)\, d\sigma(Q) \quad X \notin \partial D$$

for any $f \in L^p(\partial D)$, we obtain that

$$T(\partial_X, N_P)L_2(f)(X) \to (\mp\tfrac{1}{2} I + K^*)f(P)$$

pointwise for almost every $P \in \partial D$, and in $L^p(\partial D)$, when $X \to P$, $X \in \Gamma_\alpha^{\pm}(P)$, respectively. See Selvaggi and Sisto [4].

Since $T(\partial_P, N_P)$ is the boundary operator associated with the second boundary-value problem of elastostatics, the invertibility of the operator $(+\frac{1}{2} I + K^*)$ will solve this second problem on D, provided that the data verify the necessary condition for the existence of a solution, i.e. they belong to $R^p(\partial D)$, the subspace of $L^p(\partial D)$ consisting of those functions $g \in L^p(\partial D)$, which verify

$$\int_{\partial D} g \, d\sigma = 0, \quad \int_{\partial D} g(Q) \wedge Q \, d\sigma(Q) = 0.$$

In other words, in order to obtain a state of equilibrium the forces g must satisfy the requirement that their resultant and their resultant moment vanish.

From now on, $T(\partial_P, N_P)u(P)$ will be denoted by $B(u)(P)$. In order to prove the invertibility of the operator $-\frac{1}{2}I + K$ and $\frac{1}{2}I + K^*$, we need some previous theorems and a few definitions. But first note that if D is of class C^1, then for every $\varepsilon > 0$, we can find a finite number of spheres $(B(P_j, \delta_j))_{j=1}^{\ell} = \{B_j\}_{j=1}^{\ell}$, with $P_j \in \partial D$ such that $\partial D \subset \cup_j B_j$ and for every j there is a coordinate system, with the origin at P_j, such that

$$D \cap B_j = \{(x,t) \in \mathbb{R}^3 \mid t > \phi(x)\} \cap B_j$$

with $\phi_j \in C_0^1(\mathbb{R}^2)$, $\phi_j(0) = \frac{\partial \phi_j}{\partial x_i}(0) = 0$, $i = 1,2$, and $\max_{x \in \mathbb{R}^2} |\nabla \phi_j(x)| < \varepsilon$.

DEFINITION 1. For $1 < p < \infty$, $L_1^p(\mathbb{R}^2)$ denotes the space of functions $f \in L^p(\mathbb{R}^2)$ whose Jacobian matrix ∇f verifies

$$\|\nabla f(x)\| \in L^p(\mathbb{R}^2).$$

We set

$$\|f\|_{L_1^p(\mathbb{R}^2)} = \|f\|_{L^p(\mathbb{R}^2)} + \| \, \|\nabla f\| \, \|_{L^p(\mathbb{R}^2)}$$

DEFINITION 2. For $1 < p < \infty$, $L_1^p(\partial D)$ denotes the space of functions $f = (f_1, f_2, f_3)$ such that $f_k \in L^p(\partial D)$, $k = 1,2,3$ and for any covering $\{B_j\}_{j=1}^{\ell}$ of ∂D with the properties described in the definition of a C^1 domain, and for any $\psi \in C_0^1(B_j)$, the functions

$$\psi(x,\phi_j(x)f_k(x,\phi_j(x)) \equiv \psi\tilde{f}_k(x)$$

belong to $L_1^p(\mathbb{R}^2)$, $k = 1,2,3$.

For a fixed covering $\{B_j\}_{j=1}^{\ell}$ of ∂D and a partition of unity $\{\psi_j \in C_0^1(\mathbb{R}^2)\}$ subordinate to this cover we can define the norm

$$\|f\|_{L_1^p(\partial D)} = \|f\|_{L^p(\partial D)} + \sum_j \|\nabla\psi\tilde{f}_j\|_{L^p(\mathbb{R}^2)}.$$

Using a different covering we will obtain an equivalent norm (see Seeley [3]).

THEOREM 1. K is a compact operator in $L^p(\partial D)$.

Proof: The proof of this theorem follows the line of argument given in Theorem 1.3 of Fabes et al. [1], and we omit the details.

THEOREM 2. K is a compact operator in $L_1^p(\partial D)$.

Proof: See Theorem 1.6 of the same authors, [1].

THEOREM 3. If $f \in L_1^p(\partial D)$, $1 < p < \infty$, let us define

$$u(x) = \int_{\partial D} [T(\partial_Q,N_Q)\ \Gamma(X-Q)]'f(Q)\ d\sigma(Q), \quad X \notin \partial D.$$

Then given α, $0 < \alpha < 1$, there exists a $\delta = \delta_{\alpha,D} > 0$ such that the non-tangential maximal functions of ∇u, i.e.

$$(\nabla u)^*_{\alpha^-}(P) = \sup \{ \|\nabla u(X)\| \ \big|\ |X-P| < \delta,\ X \in \Gamma_\alpha^-(P)\}$$

and

$$(\nabla u)^*_{\alpha^+}(P) = \sup \{ \|\nabla u(X)\| \ \big|\ |X-P| < \delta,\ X \in \Gamma_\alpha^+(P)\}$$

belong to $L^p(\partial D)$ and verify

$$\|(\nabla u)^*_{\alpha^+}\|_{L^p(\partial D)} + \|(\nabla u)^*_{\alpha^-}\|_{L^p(\partial D)} \leq c\,\|f\|_{L^p_1(\partial D)}$$

Proof: We refer the reader to Theorem 1.7 of [1].

THEOREM 4. If $f \in L^p_1(\partial D)$, $1 < p < \infty$, and

$$u(X) = \int_{\partial D} [T(\partial_Q, N_Q)\,\Gamma(X-Q)]\,'f(Q)\,d\sigma(Q), \quad X \notin \partial D$$

then for almost every $P \in \partial D$, the following equality holds:

$$\lim_{X\to P,\ X\in\Gamma^+(P)} T(\partial_X, N_P)u(X) = \lim_{X\to P,\ X\in\Gamma^-(P)} T(\partial_X, N_P)u(X)$$

for any $0 < \alpha < 1$.

Proof: Because of Theorem 3 it will be sufficient to show the equality when $f \in C^1(\partial D)$. But in this case the theorem is a consequence of integration by parts and Theorem 3.1 of Selvaggi and Sisto [4].

THEOREM 5. If $\mathbb{R}^3 \setminus \bar{D}$ is connected, then $-\frac{1}{2}I + K$ is invertible on $L^p(\partial D)$, $1 < p < \infty$.

Proof: We show first that $-\frac{1}{2}I + K^*$ is injective. Let f be such that $f \in L^p(\partial D)$ and

$$(-\tfrac{1}{2}I + K^*)f = 0. \tag{2}$$

Then, as a consequence of (2), $f \in L^q(\partial D)$, for every q, $1 < q < \infty$. See Theorem 2.1 of [1]. For such f we define

$$L_2(f)(X) = \int_{\partial D} \Gamma(X-Q)\,f(Q)\,d\sigma(Q).$$

We use now Betti's formula

$$\int_\Omega E(u,v)dx = -\int_\Omega \langle u, A(\partial_X)v\rangle\,dx + \int_\Omega \langle u, B(v)\rangle\,d\sigma$$

with $\Omega \equiv \mathbb{R}^3 \setminus \bar{D}$ and $u = v = L_2(f)$.

Since

$$E(u,v) = \lambda \operatorname{div} u \cdot \operatorname{div} v + 2\mu \sum_{i,j}^{3} e_{ij}(u)\, e_{ij}(v)$$

with

$$e_{ij}(\omega) = \frac{\partial \omega_i}{\partial x_j} + \frac{\partial \omega_j}{\partial x_i},$$

we obtain that $E(L_2(f), L_2(f)) = 0$ a.e. on $\mathbb{R}^3 \setminus \bar{D}$. Therefore $L_2(f)(X) = a + BX$, where a is a constant vector and B a constant skew-symmetric matrix. Because of the condition at infinity, $L_2(f)(X) = 0$ on $\mathbb{R}^3 \setminus \bar{D}$. Since $L_2(f)(X)$ is a solution of $A(\partial_X)u = 0$, which verifies $L_2(f)\big|_{\partial D} = 0$, using the theorem of uniqueness of Friedrichs [2], we conclude that $L_2(f)(X) = 0$ in $\mathbb{R}^3$. Therefore

$$0 = \lim_{X \to P,\ X \in \Gamma_\alpha^-(P)} T(\partial_X, N_P) L_2(f)(X) = (\tfrac{1}{2} I + K^*) f(P)$$

and consequently $f = 0$ on ∂D, using (2).

<u>COROLLARY 6</u>. For $1 < p < \infty$, $-\frac{1}{2} I + K$ is invertible on $L_1^p(\partial D)$.

<u>Proof</u>: From the previous theorem we conclude that $-\frac{1}{2} I + K$ is injective on $L_1^p(\partial D)$. Since K is compact in $L_1^p(\partial D)$, the result follows using the Fredholm alternative theorem.

<u>THEOREM 7</u>. If $\mathbb{R}^3 \setminus \bar{D}$ is connected, then $\frac{1}{2} I + K^*$ is invertible on $R^p(\partial D)$, $1 < p < \infty$.

<u>Proof</u>: Because of Theorem 1, it is enough to show that $\frac{1}{2} I + K^*$ is injective in $R^p(\partial D)$. For this purpose it will be sufficient to prove that $\frac{1}{2} I + K$ is one-to-one on the dual space of $R^p(\partial D)$, i.e. on the quotient space $L^{p'}(\partial D)/(a+BP)$, where p' is the conjugate of p, a is any constant vector, B is any 3×3 constant and skew-symmetric matrix and $P \in \partial D$.

We assume that $f \in L^{p'}(\partial D)$, $1 < p' < \infty$, and

$$(\tfrac{1}{2} I + K) f = 0. \tag{3}$$

Then, using techniques of localization, it can be shown that such an f not only belongs to $L^{p'}(\partial D)$ but also to $L_1^q(\partial D)$ for every q, $1 < q < \infty$.

We define the double layer potential for this f,

$$L_1(f)(X) = \int_{\partial D} [T(\partial_Q, N_Q)\Gamma(X-Q)]'f(Q)\, d\sigma(Q), \quad X \notin \partial D$$

and use Betti's formula with $\Omega \equiv \mathbb{R}^3 \setminus \bar{D}$ and $u = v = L_1(f)$. We obtain that $L_1(f)(X) = 0$ on $\mathbb{R}^3 \setminus \bar{D}$.

From this fact and Theorem 4 we conclude that

$$0 = \lim_{X \to P,\ X \in \Gamma_\alpha^-(P)} T(\partial_X, N_P)\, L_1(f)(X) = B(L_1(f))(P).$$

Using Betti's formula again, but with $\Omega \equiv D$ and $u = v = L_1(f)$, we have

$$L_1(f)(X) = a + BX$$

with a and B as defined for the dual space of $R^p(\partial D)$. Since, because of (3), a.e. in D,

$$\lim_{X \to P,\ X \in \Gamma_\alpha^-(P)} L_1(f)(X) = -f(P),$$

we have

$$-f(P) = a + BP, \quad P \in \partial D$$

and this proves the theorem.

REFERENCES

[1] E.B. Fabes, M. Jodeit and N.M. Riviere, Potential techniques for boundary value problems on C^1-domains, Acta Math. 141 165-186 (1978).

[2] K.O. Friedrichs, On the boundary-value problems of the theory of Elasticity and Körn's inequality, Annals of Math. 48, 441-471 (1947).

[3] R.T. Seely, Singular integrals on compact manifolds, Am. J. Math. 81 (3), 658-690 (1959).

[4] R. Selvaggi and I. Sisto, Regolarità di certe trasformazioni integrali relative ad aperti di classe C^1, Napoli, Rendiconti, Accad. Sc. Fis. Matematiche, s. IV, 45, 393-410 (1978).

A. Gutiérrez Cardona
Universidad Autónoma de Madrid,
Matematicas
Madrid-34,
Spain

J CARRILLO

An evolution free boundary problem: filtrations of a compressible fluid in a porous medium

Our purpose in this paper is to prove the existence of a solution to the problem of filtrations of a compressible fluid through a porous medium. In this part we shall mainly use the techniques of Gilardi [5] and some techniques for bounding solutions of elliptic problems. In this paper we shall also prove a regularity result for solutions.

1. STATEMENT OF THE PROBLEM

We consider a connected open bounded set Ω of $\mathbb{R}^2$ with Lipschitz boundary $\partial\Omega$. Moreover two relatively open subsets Γ_1 and Γ_2 of $\partial\Omega$ are given such that $\Gamma_1 \cap \Gamma_2 = \emptyset$ and $\bar{\Gamma}_1 \cup \bar{\Gamma}_2 = \partial\Omega$.

Physically Ω represents the section of a porous medium which separates two or more reservoirs containing a compressible (or incompressible) fluid and Γ_2 is the pervious part and Γ_1 the impervious part of $\partial\Omega$.

Our purpose is to study the problem of finding the pressure in Ω at any time $t \in [0,T]$ when the levels of fluid of reservoirs vary with respect to t.

The equation in the unknown function u is given by

$$\Delta u + (\partial y - \partial t)\chi - \nu u_t = 0 \quad \text{in} \quad \Omega \times]0,T[, \tag{1.1}$$

where

$$\Delta = \frac{\partial^2}{\partial n^2} + \frac{\partial^2}{\partial y^2},$$

$\chi(t)$ is the characteristic function in space variables of the wet region at time t and ν is a nonnegative constant (storativity coefficient); the case $\nu = 0$ is related to an incompressible fluid and has been studied by Gilardi [5]. (The case where Ω is rectangular has been studied by Torelli and Friedmann [4], [6], [7].)

With (1.1), the unknown function u (in (1.1) χ also is unknown) satisfies the following conditions:

$$u = \phi \quad \text{on} \quad \Gamma_2 \times]0,T[. \tag{1.2}$$

Here ϕ is a given function which represents the pressure on $\Gamma_2 \times]0,T[$:

$$\frac{\partial u}{\partial n} + \chi \cos ny = 0 \text{ on } \Gamma_1 \times]0,T[\tag{1.3}$$

$$\frac{\partial u}{\partial n} + \chi \cos ny \leqslant 0 \text{ on } \Gamma_2 \times]0,T[\setminus \operatorname{supp} \phi \tag{1.4}$$

1.3 and 1.4 mean that the normal velocity of the fluid vanishes on $\Gamma_1 \times]0,T[$ (because Γ_1 is impervious) and is nonnegative on $(\Gamma_2 \times]0,T[) \setminus \operatorname{supp} \phi$ (because the fluid can leave the porous medium via those parts of Γ_2 where the pressure is zero).

Finally we have also an initial condition:

$$\nu u(0) = \nu\phi(0) \text{ and } \chi(0) = \chi([\phi(0) > 0]). \tag{1.5}$$

Before stating a weak formulation of this problem, we define

$$Q = \Omega \times]0,T[$$

$$\Sigma_1 = \Gamma_1 \times]0,T[$$

$$\Sigma_2 = \Gamma_2 \times]0,T[$$

$$\Omega_t = \Omega \times \{t\} \quad \text{for any } t \in [0,T],$$

and we assume that ϕ is a Lipschitz continuous and nonnegative function on $\mathbf{R}^3$; then $\chi[\phi(0) > 0]$ is the characteristic function of an open subset of Ω.

Now we shall state a weak formulation of this problem:

Problem 1. Find a pair (u,g) satisfying:

$$u \in L^2(0,T; H^1(\Omega)); \quad g \in L^\infty(Q) \tag{1.6}$$

$$u \geqslant 0, \ 0 \leqslant g \leqslant 1, \ u(1-g) = 0 \text{ a.e. in } Q \tag{1.7}$$

$$u = \phi \quad \text{on } \Sigma_2 \tag{1.8}$$

$$\left.\begin{array}{l} \int_Q \{\nabla u \, \nabla \xi + g\xi_y - (g+\nu u)\xi_t\} \leqslant 0 \\ \text{for every smooth function satisfying: } \xi(0) = \xi(T) = 0 \\ \xi = 0 \text{ on } \Sigma_2 \cap \operatorname{supp} \phi, \ \xi \geqslant 0 \text{ on } \Sigma_2 \setminus \operatorname{supp} \phi \end{array}\right\} \tag{1.9}$$

$$g(0) + \nu u(0) = \chi[\phi(0) > 0] + \nu\phi(0). \tag{1.10}$$

In order to solve Problem 1, we shall approximate this by elliptic problems in the same way as Gilardi does.

2. APPROXIMATING PROBLEMS: EXISTENCE AND UNIQUENESS

For each $\varepsilon > 0$ we define:

(a) a function ϕ_ε on $\bar{Q}$ satisfying:

$$\phi_\varepsilon \in H^1(Q), \quad \phi_\varepsilon(0) \in C^\infty(\Omega); \tag{2.1}$$

$$\phi_\varepsilon = \phi \text{ on } \partial Q - \Omega_0, \; 0 \leqslant \phi_\varepsilon \leqslant 2\|\phi\|_{L^\infty(Q)} \text{ on } \Omega_0 \tag{2.2}$$

$$\|\phi_\varepsilon(0) - \phi(0)\|_{H^1(\Omega)} \leqslant \varepsilon; \; \|\phi_\varepsilon - \phi\|_{H^1(Q)} \leqslant c\varepsilon, \tag{2.3}$$

where c is a constant.

In order to find a function ϕ_ε satisfying (2.1), (2.2) and (2.3), we choose a function $\psi \in C^\infty(\Omega)$ such that:

$$\psi = \phi(0) \text{ on } \partial\Omega ;$$

and we choose a function $w_\varepsilon \in C_0^\infty(\Omega)$ such that

$$\|\phi(0) - \psi - w_\varepsilon\|_{H^1(\Omega)} \leqslant \varepsilon;$$

then we define $\phi_\varepsilon(0) = w_\varepsilon + \psi$; $\phi_\varepsilon = \phi$ on $\partial Q - \Omega_0$ and we define ϕ_ε in Q as a function of $H^1(Q)$ such that:

$$\|\phi_\varepsilon - \phi\|_{H^1(Q)} \leqslant C\, \|\phi_\varepsilon - \phi\|_{H^{\frac{1}{2}}(\partial Q)} \leqslant c\, \|\phi_\varepsilon - \phi\|_{H^1(\partial Q)} \leqslant c\varepsilon.$$

(b) a function H_ε on $\mathbb{R}$, satisfying:

$$H_\varepsilon \in C^\infty(\mathbb{R}) \tag{2.4}$$

$$H_\varepsilon(r) = 1 \text{ if } r \geqslant \varepsilon^{1/3}, \quad H_\varepsilon(0) = 0 \tag{2.5}$$

$$0 \leqslant H'_\varepsilon \leqslant \frac{2}{\varepsilon^{1/3}} . \tag{2.6}$$

(c) Problem (2.3). Find a function u_ε satisfying:

$$u_\varepsilon \in H^1(Q) \tag{2.7}$$

$$u_\varepsilon = \phi_\varepsilon \text{ on } \Sigma_2 \cup \Omega_0 \tag{2.8}$$

$$\left.\begin{array}{l}\int_Q \{\nabla u_\varepsilon \nabla \xi + \varepsilon u_{\varepsilon_t} \xi_t + H_\varepsilon(u_\varepsilon)\xi_y - G_\varepsilon(u_\varepsilon)\xi_t\} + \int_\Omega G_\varepsilon(u_\varepsilon(T))\xi(T) = 0 \\ \text{for every } \xi \in H(Q),\ \xi = 0 \text{ on } \Sigma_2 \cup \Omega_0,\end{array}\right\} \tag{2.9}$$

where $G_\varepsilon(r) = H_\varepsilon(r) + \nu r \quad \forall r \in \mathbb{R}$.

First, we shall prove a comparison result:

<u>LEMMA 2.1</u>. Let v_1 and v_2 be two functions satisfying (2.7) and (2.9) and such that $v_1 \geqslant v_2$ on $\Sigma_2 \cup \Omega_0$. Then we have:

$$v_1 \geqslant v_2 \quad \text{a.e. in } Q. \tag{2.10}$$

<u>Proof</u>. We suppose that (2.10) is not true, then for each $\delta > 0$ we define a function

$$\xi_\gamma = \frac{(v_1 - v_2 - \delta)^+}{v_1 - v_2} \quad \text{a.e. in } Q.$$

Then, obviously $\xi_\delta \in H^1(Q)$ and $\xi_\delta = 0$ on $\Sigma_2 \cup \Omega_0$ such that:

$$0 = \int_Q \{\nabla(v_2 - v_1)\nabla\xi_\delta + \varepsilon(v_2 - v_1)_t(\xi_\delta)_t + [H_\varepsilon(v_2) - H_\varepsilon(v_1)](\xi_\delta)_y$$

$$- [G_\varepsilon(v_2) - G_\varepsilon(v_1)](\xi_\delta)_t\} + \int_\Omega [G_\varepsilon(v_2(T)) - G_\varepsilon(v_1(T)]\xi_\delta(T)$$

$$0 = \delta\int_{[v_2 - v_1 > \delta]} \left\{\left|\frac{\nabla(v_2 - v_1)}{v_2 - v_1}\right|^2 + \varepsilon\left|\frac{(v_2 - v_1)_t}{v_2 - v_1}\right|^2\right.$$

$$\left. + [H_\varepsilon(v_2) - H_\varepsilon(v_1)]\frac{(v_2 - v_1)_y}{(v_2 - v_1)^2} - [G_\varepsilon(v_2) - G_\varepsilon(v_1)]\frac{(v_2 - v_1)_t}{(v_2 - v_1)^2}\right\}$$

$$+ \int_\Omega [G_\varepsilon(v_2(T) - G_\varepsilon(v_1(T))] \frac{(v_2(T) - v_1(T) - \delta)^+}{v_2(T) - v_1(T)}.$$

Obviously, the last integral is nonnegative, and we have:

$$\int_{[v_2-v_1 > \delta]} \left\{\left|\frac{\nabla(v_2-v_1)}{v_2-v_1}\right|^2 + \varepsilon\left|\frac{(v_2-v_1)t}{v_2-v_1}\right|^2\right\}$$

$$\leqslant -\int_{[v_2-v_1 > \delta]} \left\{[H_\varepsilon(v_2)-H_\varepsilon(v_1)]\frac{(v_2-v_1)_y}{(v_2-v_1)^2} - [G_\varepsilon(v_2)-G_\varepsilon(v_1)]\frac{(v_2-v_1)_t}{(v_2-v_1)^2}\right\}$$

$$\leqslant \int_{[v_2-v_1 > \delta]} \left\{\frac{2}{\varepsilon^{1/3}}\left|\frac{(v_2-v_1)_y}{v_2-v_1}\right| + \left(\nu + \frac{2}{\varepsilon^{1/3}}\right)\left|\frac{(v_2-v_1)_t}{v_2-v_1}\right|\right\},$$

from which we easily deduce

$$\int_{[v_2-v_1 > \delta]} \left|\frac{\nabla(v_2-v_1)}{v_2-v_1}\right|^2 + \left|\frac{(v_2-v_1)_t}{v_2-v_1}\right|^2 \leqslant C(\varepsilon), \tag{2.11}$$

where $C(\varepsilon)$ does not depend on δ.

Now, from (2.11) we have:

$$\int_Q \left|\nabla\log\left(1 + \frac{(v_2-v_1-\delta)^+}{\delta}\right)\right|^2 + \left|\frac{\partial}{\partial t}\log\left(1 + \frac{(v_2-v_1-\delta)^+}{\delta}\right)\right|^2 \leqslant C(\varepsilon)$$

and from Poincaré's inequality we get:

$$\int_Q \left|\log\left(1 + \frac{(v_2-v_1-\delta)^+}{\delta}\right)\right|^2 \leqslant K(\varepsilon) \tag{2.12}$$

where $K(\varepsilon)$ does not depend on δ. Then when $\delta \to 0$ we get a contradiction, so the proof is complete. (See [2]).

<u>COROLLARY 2.1</u>. Problem 2.3 has no more than one solution.

<u>Proof</u>. This is an obvious consequence of Lemma 2.1.

<u>COROLLARY 2.2</u>. For some M_ε we have:

$$0 \leqslant u_\varepsilon \leqslant M_\varepsilon.$$

<u>Proof</u>. The first inequality is obvious since 0 satisfies (2.7) and (2.9) and $u_\varepsilon = \phi_\varepsilon \geqslant 0$ on $\Sigma_2 \cup \Omega_0$. In order to prove the second inequality we define a function $w \in C^\infty(\bar{Q})$:

$$w(x,y,t) = v(y) \quad \forall(x,y,t) \in \bar{Q},$$

where v is the unique solution of

$$\begin{cases} v'(y) = -H_\varepsilon(v(y)) \text{ in } \mathbb{R} \\ v(y_M) = \|\phi_\varepsilon\|_{L^\infty(\delta Q)} \end{cases}$$

where $y_M = \sup\{y \in \mathbb{R} \mid (x,y) \in \Omega \text{ for some } x\}$.

Obviously, w satisfies (2.7) and (2.9) $w \geq \|\phi_\varepsilon\|_{L^\infty(\delta Q)} \geq \phi_\varepsilon$ on $\Sigma_2 \cup \Omega_0$, so that $w \geq u_\varepsilon$ on a.e. in Q, and then:

$$0 \leq u_\varepsilon \leq M_\varepsilon = \|w\|_{L^\infty(Q)} = \|v\|_{L^\infty(P_y\Omega)}$$

where $P_y\Omega$ is the projection of Ω on Oy.

Remark 2.1. Let us define $\bar{G}_\varepsilon$ as:

$$\bar{G}_\varepsilon(r) = \begin{cases} G_\varepsilon(M_\varepsilon) & \text{if } r \geq M_\varepsilon \\ G_\varepsilon(r) & \text{if } 0 \leq r \leq M_\varepsilon \\ 0 & \text{if } r \leq 0 \end{cases}$$

and $H_\varepsilon^+ = \text{Max}(H_\varepsilon, 0)$.

Then, from Corollary 2.2, if there exists a function u_ε satisfying (2.7), (2.8) and (2.9), u_ε also satisfies:

$$\left.\begin{aligned} &\int_Q \{\nabla u_\varepsilon \nabla\xi + \varepsilon u_{\varepsilon_t}\xi_t + H_\varepsilon^+(u_\varepsilon)\xi_y - \bar{G}_\varepsilon(u_\varepsilon)\xi_t\} + \int_\Omega \bar{G}_\varepsilon(u_\varepsilon(T)\xi(T) = 0 \\ &\text{for every } \xi \in H^1(\Omega),\ \xi = 0 \text{ on } \Sigma_2 \cup \Omega_0 \end{aligned}\right\} \quad (2.13)$$

Reciprocally, if there exists a function v_ε satisfying (2.7), (2.8) and (2.13), then in the same way that we prove Lemma 2.1 and Corollaries 2.1 and 2.2, we can prove equivalent results (since the proof of Lemma 2.1 is only grounded on the Lipschitz continuity of H_ε and G_ε and here, H_ε^+ and $\bar{G}_\varepsilon$ are also Lipschitz continuous); in particular $0 \leq v_\varepsilon \leq M_\varepsilon$, where M_ε represents the same number as in Corollary 2.2. Then, obviously v_ε also satisfies (2.9) and from uniqueness we have $v_\varepsilon = u_\varepsilon$.

Now we can prove

THEOREM 2.1. Problem 2.3 has one and only one solution.

Proof. Uniqueness has been proved in Corollary 2.1.

In order to prove the existence of a solution we shall take into account Remark 2.1, and then we shall prove the existence of a function satisfying (2.7), (2.8) and (2.13). We consider the operator $T_\varepsilon : H^1(Q) \to H^1(Q)$ which associates with $v \in H^1(Q)$ the unique solution w of the problem:

$$w \in H^1(Q) \tag{2.14}$$

$$w = \phi_\varepsilon \text{ on } \Sigma_2 \cup \Omega_0 \tag{2.15}$$

$$\left.\begin{aligned} &\int_Q \{\nabla w \nabla \xi + \varepsilon w_t \xi_t + H_\varepsilon^+(v)\xi_y - \bar{G}_\varepsilon(v)\xi_t\} + \int_\Omega \bar{G}_\varepsilon(v_\varepsilon(T))\xi(T) = 0 \\ &\text{for every } \xi \in H^1(Q),\ \xi = 0 \text{ on } \Sigma_2 \cup \Omega_0. \end{aligned}\right\} \tag{2.16}$$

Since H_ε^+ and $\bar{G}_\varepsilon$ are Lipschitz continuous, we prove easily that T_ε is a continuous mapping from $H^1(Q)$ with the weak topology into $H^1(Q)$ with the strong topology and since H_ε^+ and $\bar{G}_\varepsilon$ are bounded we get easily:

$$||w||_{H^1(Q)} \leqslant C(\varepsilon),$$

where $C(\varepsilon)$ does not depend on v (see Gilardi [5]). Then T_ε maps $H^1(Q)$ into the ball

$$B = \{w \in H^1(Q) / \ ||w||_{H^1(Q)} \leqslant C(\varepsilon)\}$$

which is compact in $H^1(Q)$ with respect to the weak topology. Thus we can apply the Tychonov theorem (see Baiocchi and Capelo [1]) to deduce the existence of a fixed point to T_ε. Then the fixed point obviously satisfies (2.7), (2.8) and (2.13), and from Remark 2.1, this also satisfies (2.9), so the proof is complete.

We have also:

PROPOSITION 2.1. The solution u_ε of Problem 2.3 satisfies

$u_\varepsilon \in C^\infty(\Omega \times [0,T])$.

Proof. See Gilardi [5].

3. EXISTENCE OF A SOLUTION TO THE PROBLEM 2.3

In this section u_ε represents the solution of Problem 2.3.

PROPOSITION 3.1. For some M and for any $\varepsilon \in]0,1]$ we have

$$0 \leqslant u_\varepsilon \leqslant M \text{ in } Q. \tag{3.1}$$

Proof. We define a function $w \in C^\infty(\bar{Q})$:

$$w(x,y,t) = v(y) \text{ in } \bar{Q}$$

where v is the unique solution of

$$\begin{cases} v'(y) = -1 \text{ in } \mathbb{R} \\ v(y_M) = \max\left(2\|\phi\|_{L^\infty(Q)},1\right); \end{cases}$$

then $H_\varepsilon(w) = 1$ in Q and w satisfies (2.7) and (2.9), and $w \geqslant \phi_\varepsilon$ on $\Sigma_2 \cup \Omega_0$. Therefore, from Lemma 2.1, we have

$$0 \leqslant u_\varepsilon \leqslant M = \|w\|_{L^\infty(Q)} = \|v\|_{L^\infty(P_y\Omega)}.$$

PROPOSITION 3.2. For some K and for any $\varepsilon \in]0,1]$ we have:

$$\int_Q \{|\nabla u_\varepsilon|^2 + \varepsilon|u_{\varepsilon_t}|^2\} \leqslant K \tag{3.2}$$

Proof. See Gilardi [5].

Applying techniques of Gilardi we have also:

PROPOSITION 3.3. Let Θ be a nonnegative function of $D(\Omega)$, then for some $C(\Theta)$ and for any $\varepsilon \in]0,1]$ we have:

$$\int_\Omega \Theta|\varepsilon u_{\varepsilon_t}(0)|^2 \leqslant C(\Theta)\varepsilon^{1/6} \tag{3.3}$$

Proof. See Gilardi [5].

Finally, we have

PROPOSITION 3.4. Let Ω' be an open subset of Ω satisfying $\bar{\Omega}' \subset \Omega$, then for some $C(\Omega')$ and for any $\varepsilon \in]0,1]$ we have

$$\nu\int_{\Omega'\times]0,T[} |u_{\varepsilon_t}|^2 \leqslant C(\Omega') \tag{3.4}$$

Proof. First we shall prove that for every nonnegative $\Theta \in D(\Omega)$ there exists a constant $K(\Theta)$ such that:

$$\int_Q H_\varepsilon'(u_\varepsilon)|u_{\varepsilon_y}|^2\Theta \leqslant K(\Theta). \tag{3.5}$$

In order to prove this we define a function $v_\varepsilon = \min\,(u_\varepsilon, \varepsilon^{1/3})$, then we have

$$0 \leqslant \int_Q H_\varepsilon'(u_\varepsilon)|u_{\varepsilon_y}|^2\Theta \leqslant \frac{2}{\varepsilon^{1/3}} \int_Q u_{\varepsilon_y} v_{\varepsilon_y}\Theta$$

$$= \frac{2}{\varepsilon^{1/3}} \int_Q [-\nabla u_\varepsilon \nabla\Theta v_\varepsilon - u_{\varepsilon_x} v_{\varepsilon_x}\Theta - u_{\varepsilon_t} v_{\varepsilon_t}\Theta - H_\varepsilon(u_\varepsilon)v_{\varepsilon_y}\Theta + v_\varepsilon\Theta_y)$$

$$+ G_\varepsilon(u_\varepsilon)v_{\varepsilon_t}\Theta] - \frac{2}{\varepsilon^{1/3}}\int_\Omega \varepsilon u_{\varepsilon_t}(0)v_\varepsilon(0)\Theta$$

$$- \frac{2}{\varepsilon^{1/3}}\int_\Omega G_\varepsilon(u_\varepsilon(T))v_\varepsilon(T)\Theta \quad + \frac{2}{\varepsilon^{1/3}}\int_\Omega G_\varepsilon(u_\varepsilon(0))v_\varepsilon(0)\Theta.$$

From (3.3) the second integral is obviously bounded. From boundedness of the data on Ω_0 the last integral is bounded. Moreover we have

$$\int_Q G_\varepsilon(u_\varepsilon)v_{\varepsilon_t}\Theta - \int_\Omega G(u_\varepsilon(T))v_\varepsilon(T)\Theta \leqslant \int_Q G_\varepsilon(v_\varepsilon)v_{\varepsilon_t}\Theta - \int_\Omega G_\varepsilon(v_\varepsilon(T))v_\varepsilon(T)\Theta$$

$$= (*) = \int_Q \frac{\partial}{\partial t}\hat{G}_\varepsilon(v_\varepsilon)\Theta - \int_\Omega G_\varepsilon(v_\varepsilon(T))v_\varepsilon(T)$$

where $\hat{G}_\varepsilon(r) = \int_0^r G_\varepsilon(s)\,ds$ and G_ε satisfies

$$\hat{G}_\varepsilon(r) \leqslant rG_\varepsilon(r) \quad \forall r \geqslant 0$$

then we have

$$(*) = \int_\Omega \{G_\varepsilon(v_\varepsilon(T))\Theta - G_\varepsilon(v_\varepsilon(T))v_\varepsilon(T)\Theta\} - \int_\Omega G_\varepsilon(v_\varepsilon(0)) \ \Theta \leqslant 0.$$

Finally we have

$$\int_Q H_\varepsilon(u_\varepsilon)(v_{\varepsilon_y}) = \int_Q H_\varepsilon(v_\varepsilon)(v_{\varepsilon_y}) = \int_Q \frac{\partial}{\partial y}(\hat{H}_\varepsilon(v_\varepsilon)\Theta) - H_\varepsilon(v_\varepsilon)\Theta_y$$

$$= - \int_Q \hat{H}_\varepsilon(v_\varepsilon)\Theta_y$$

where

$$\hat{H}_\varepsilon(r) = \int_0^r H_\varepsilon(s)ds$$

and satisfies $H_\varepsilon(r) \leqslant r \quad \forall r \geqslant 0$. Then

$$0 \leqslant \int_Q \{H'_\varepsilon(u_\varepsilon)|u_{\varepsilon_y}|^2\Theta \leqslant \frac{2}{\varepsilon^{1/3}} \int_Q - \nabla u_\varepsilon \nabla\Theta v_\varepsilon + \hat{H}_\varepsilon(v_\varepsilon)\Theta_y - H_\varepsilon(u_\varepsilon)v_\varepsilon\Theta_y\}$$

$$\leqslant K(\Theta),$$

which proves (3.5).

In order to prove (3.4), consider a function $\Theta \in D(\Omega)$ and satisfying $\Theta = 1$ in Ω'; then

$$\nu\int_Q u^2_{\varepsilon_t}\Theta^2 = \int_Q \{\varepsilon u_{\varepsilon_{tt}} u_{\varepsilon_t}\Theta^2 + \Delta u_\varepsilon u_{\varepsilon_t}\Theta^2 + H_\varepsilon(u_\varepsilon)_y u_{\varepsilon_t}\Theta^2 - H_\varepsilon(u_\varepsilon)_t u_{\varepsilon_t}\Theta^2\}$$

$$= - \frac{1}{2} \int_\Omega \varepsilon|u_{\varepsilon_t}(0)|^2\Theta^2 - \int_Q \nabla u_\varepsilon \ \nabla u_{\varepsilon_t}\Theta^2 - 2 \int_Q \nabla u_\varepsilon \nabla\Theta \ \Theta u_{\varepsilon_t}$$

$$+ \int_Q H'_\varepsilon(u_\varepsilon)\Theta^2[u_{\varepsilon_y} u_{\varepsilon_t} - u^2_{\varepsilon_t}]$$

$$\leqslant \frac{1}{2} \int_\Omega |\nabla u_\varepsilon(0)|^2\Theta^2 - 2\int_Q \nabla u\nabla\Theta \ \Theta u_{\varepsilon_t} + \int_Q H'_\varepsilon(u_\varepsilon)\Theta^2[u_{\varepsilon_y} u_{\varepsilon_t} - u^2_{\varepsilon_t}]^+,$$

but

$$\int_Q H'_\varepsilon(u_\varepsilon)\Theta^2 [u_{\varepsilon_y} u_{\varepsilon_t} - u^2_{\varepsilon_t}]^+ \leq \int_Q H'_\varepsilon (u_\varepsilon)\Theta^2 |u_{\varepsilon_y}|^2 \leq K(\Theta^2);$$

hence obviously

$$\nu\int_{\Omega'\times]0,T[} |u_{\varepsilon_t}|^2 \leq \nu\int_Q |u_{\varepsilon_t}|^2\Theta^2 \leq C(\Theta) = C(\Omega'),$$

which completes the proof.

Now we can prove the main result of this paper:

THEOREM 3.1. There exists at least one solution (u,g) to Problem 1. Moreover, u satisfies:

(i) $u \in L^\infty(Q)$

(ii) $u \in H^1(\Omega' \times]0,T[)$ for every open subset Ω' of Ω satisfying $\bar{\Omega}' \subset \Omega$

(iii) $\nu u(0) = \nu\phi(0)$.

Proof. The case $\nu = 0$ has been proved by Gilardi.
The case $\nu > 0$:

From Propositions 3.1 and 3.2 we can extract a subsequence of ε, still denoted by ε, such that

$$u_\varepsilon \rightharpoonup u \text{ in } L^2(0,T;\ H^1(\Omega)) \tag{3.6}$$

and

$$H_\varepsilon(u_\varepsilon) \to g \text{ in } L^\infty(Q) \text{ weak } *. \tag{3.7}$$

Moreover, from Proposition 3.4 and from (3.6),

$$u_\varepsilon \rightharpoonup u \text{ in } H^1(\Omega' \times (0,T)) \tag{3.8}$$

for every open subset Ω' of Ω satisfying $\bar{\Omega}' \subset \Omega$.

Then we deduce $\nu u(0) = \nu\phi(0)$ a.e. in Ω, and also

$$0 \leq \int_{(\Omega' \times (0,T))} u(1-g) = \lim_{\varepsilon\to 0} \int_{(\Omega' \times (0,T))} u_\varepsilon(1-H_\varepsilon(u_\varepsilon)) \leq \lim_{\varepsilon\to 0} \varepsilon^{1/3}\ |Q| = 0.$$

Then, obviously, (u,g) satisfies (1.6), (1.7) and (1.8).

In order to prove (1.9) from (3.2) and (3.6) we deduce that

$$\sqrt{\varepsilon}\, u_{\varepsilon_t} \longrightarrow 0 \text{ in } L^2(Q) \tag{3.9}$$

and then, choosing a function ξ as in (1.9), we have

$$\int_Q (\nabla u\, \nabla\xi + g\xi_y - (g+\nu u)\xi_t) = \lim_{\varepsilon\to 0} \int_Q (\nabla u_\varepsilon \nabla\xi + \varepsilon u_{\varepsilon_t}\xi_t + H_\varepsilon(u_\varepsilon)\xi_y - G_\varepsilon(u_\varepsilon)\xi_t)$$

$$= \lim_{\varepsilon\to 0} [\int_Q - (\Delta u_\varepsilon + \varepsilon u_{\varepsilon_{tt}} + H_\varepsilon(u_\varepsilon)_y - G_\varepsilon(u_\varepsilon)_t)\xi$$

$$+ \int_{\Sigma_2 \setminus \mathrm{supp}} [\frac{\partial u_\varepsilon}{\partial n} + H_\varepsilon(u_\varepsilon) \cos (n,y)]\xi]$$

$$= \lim_{\varepsilon\to 0} \int_{\Sigma_2 \setminus \mathrm{supp}} \frac{\partial u_\varepsilon}{\partial n}\xi \leqslant 0,$$

since $u_\varepsilon = 0$ on $\Sigma_\varepsilon \setminus \mathrm{supp}\, \phi = \Sigma_2 \setminus \mathrm{supp}\, \phi_\varepsilon$ and $u_\varepsilon \geqslant 0$ in Q. Finally, for every $\Theta \in D(\Omega)$ and for every $\psi \in D(]0,T[)$ we have

$$0 = \int_Q [\nabla u_\varepsilon\, \nabla\Theta + H_\varepsilon(u_\varepsilon)\Theta_y]\psi + \int_Q (\varepsilon u_{\varepsilon_t} - G_\varepsilon(u_\varepsilon))\Theta \cdot \psi_t$$

$$= \int_0^T \psi \int_\Omega [\nabla u_\varepsilon\, \nabla\Theta + H_\varepsilon(u_\varepsilon)\Theta_y] + \int_0^T \psi_t \int_\Omega (\varepsilon u_{\varepsilon_t} - G_\varepsilon(u_\varepsilon))\Theta$$

$$= \int_0^T \psi \int_\Omega [\nabla u_\varepsilon\, \nabla\Theta + H_\varepsilon(u_\varepsilon)\Theta_y] - \langle \frac{d}{dt} \int_\Omega (\varepsilon u_{\varepsilon_t} - G_\varepsilon(u_\varepsilon))\Theta, \psi \rangle_{D',D}$$

and then, obviously

$$\frac{d}{dt} \int_\Omega (\varepsilon u_{\varepsilon_t} - G_\varepsilon(u_\varepsilon))\Theta \in L^2(0,T)$$

and

$$F_\varepsilon(t) = \int_\Omega (\varepsilon u_{\varepsilon_t} - G_\varepsilon(u_\varepsilon))\Theta \in H^1(0,T).$$

In the same way we prove that $F(t) = \int_\Omega -(g + \nu u)\Theta \in H^1(0,T)$, and then from (3.6), (3.7) and (3.9) we easily get

$$F_\varepsilon \rightharpoonup F \text{ in } H^1(0,T),$$

which implies

$$F_\varepsilon(0) \to F(0).$$

But

$$F_\varepsilon(0) = \int_\Omega [\varepsilon u_{\varepsilon_t}(0) - G_\varepsilon(u_\varepsilon(0))]\Theta$$

and also we have

$$\int_\Omega \varepsilon u_{\varepsilon_t}(0)\Theta \to 0 \text{ when } \varepsilon \to 0,$$

$$\int_\Omega \nu u_\varepsilon(0)\Theta \to \int_\Omega \nu\phi(0)\Theta = \int_\Omega \nu u(0)\Theta,$$

and

$$\int_\Omega H_\varepsilon(u_\varepsilon(0))\Theta = \int_\Omega [H_\varepsilon(u_\varepsilon(0)) - H_\varepsilon(\phi(0))]\Theta + \int_\Omega H_\varepsilon(\phi(0))\Theta.$$

Moreover

$$\left|\int_\Omega [H_\varepsilon(u_\varepsilon(0)) - H_\varepsilon(\phi(0))]\Theta\right| \leqslant \frac{2}{\varepsilon^{1/3}} \int_\Omega [u_\varepsilon(0) - \phi(0)]\Theta \leqslant \frac{2}{\varepsilon^{1/3}} \|\Theta\|_{L^2(\Omega)}$$

and

$$\int_\Omega H_\varepsilon(\phi(0))\Theta \to \int_\Omega \chi[\phi(0) > 0]\Theta.$$

Then we deduce

$$F(0) = -\int_\Omega [g(0) + \nu u(0)]\Theta = -\int_\Omega [\chi[\phi(0) > 0] + \nu\phi(0)]\Theta,$$

leading to (1.10):

$$g(0) + \nu u(0) = \chi[\phi(0) > 0] + \nu\phi(0) \quad \text{a.e. in } \Omega,$$

which completes the proof.

REFERENCES

[1] C. Baiocchi and A. Capelo, Disequazioni variazionali quasivariazionali. Applicazioni a problemi di frontiera libera, Vols. 1, 2, Pitagora Editrice, Bologna (1978).

[2] H. Brézis, D. Kinderlehrer and G. Stampacchia, Sur une nouvelle formulation du problème de l'écoulement à travers une digue, C.R.A. Sci. Paris, Série A 287, 711-714 (1978).

[3] J. Carrillo and M. Chipot, On the Dam problem, to appear in the Journal of Differential Equations

[4] A. Friedmann and A. Torelli, A free boundary problem connected with non-steady filtrations in porous media, Non-linear Analysis 1, 503-545 (1977).

[5] G. Gilardi, A new approach to evolution free boundary problems, comm. in Partial Diff. Equat. 4(10), 1099-1122 (1979).

[6] A Torelli, On a free boundary problem connected with a non steady filtration phenomenon, Ann. Scuola Norm. Sup. Pisa, 4(4), 33-59 (1977).

[7] A. Torelli, On a free boundary problem connected with a non-steady filtration phenomenon of compressible fluid, to appear.

J. Carrillo
Departamento de ecuaciones funcionales,
Universidad Complutense,
Madrid-3,
Spain

A CASAL & A SOMOLINOS

Entrainment in a nonlinear functional differential equation—applications

1. INTRODUCTION

In this work we point out some mathematical techniques which have been useful in the study of a particular mathematical model which tries to clarify some experimental facts of entrainment of the hypocotyls of Helianthus annuus (Sunflower), when an external periodic perturbation is applied.

As discussed elsewhere (see Casal and Somolinos [5]), we can consider as a suitable mathematical model for such a motion, the equation

$$\ddot{x}(t) + (a/r)\,\dot{x}(t) + (b/r)[x(t-r) + (1/6)x^3(t-r)] = \nu f \cos \omega t, \quad (1.1)_f$$

which is accurate enough within the range of experimental values, given, for example, in Andersen and Johnsson (1). In this equation, $x(t)$ is the angle of the plant with the vertical, $r > 0$ is the geotropic reaction time, a, b, f are positive constants, and ν is a small positive parameter. It has been proved previously (see Grafton [9], Casal and Somolinos [4] and Somolinos [21]) that the equation $(1.1)_0$ has periodic solutions for a range of values of the parameters a, b, r, with a frequency σ_0 (depending on them) which will be called, from now on, the free running frequency. When a periodic external perturbation is applied, it is interesting to study the response of the plant for different frequencies of the perturbation. The experiments mentioned above show the existence of an interval around the free running frequency such that if the forcing frequency is inside, the motion is entrained to a periodic one with the same frequency as the forcing term, and if the forcing frequency is outside that interval, some other phenomena, such as beats, appear.

Although equation $(1.1)_f$ is infinite dimensional, some known results, [3], [7], [10], allow us to carry the whole study in the framework of a nonlinear ordinary differential equation. Then we study this equation along the path, for example, of [7]. What happens is, essentially, the following: when we perturb with a $2\pi/\omega$-periodic forcing function, the already existing $2\pi/\sigma_0$-periodic orbit and the equilibrium point (zero) of the unperturbed equation

$(1.1)_0$ become a torus and a $2\pi/\omega$-periodic solution, respectively, and the stability properties of these new solutions depend upon the forcing frequency When this varies and approaches the free running frequency from either side, the torus, which corresponds to the experimental beats, becomes unstable and jumps back to the $2\pi/\omega$-periodic solution, which represents the entrained motion, for a certain range of forcing frequencies. This change of stability leads us to a satisfactory explanation of the experimental facts.

2. PRELIMINARIES

Equation $(1.1)_f$ can be written as a system of first-order retarded functional differential equations,

$$\dot{x}(t) = Ax(t) + Bx(t-r) + f(t,x_t), \tag*{$(2.1)_f$}$$

where

$$x = \begin{bmatrix} x_1 \\ x_2 \end{bmatrix}, \; A = \begin{bmatrix} 0 & 1 \\ 0 & -a/r \end{bmatrix}, \; B = \begin{bmatrix} 0 & 0 \\ -b/r & 0 \end{bmatrix}, \; f(t,x_t) = \begin{bmatrix} 0 \\ (b/6r)x^3(t-r)+f(t) \end{bmatrix}.$$

and where x_t is, as usually in functional differential equations (see, for example, Hale [10]), an element of the Banach space C of continuous functions mapping the interval $[-r,0]$ into R^2 with the topology of uniform convergence, and defined as $x_t(\theta) = x(t+\theta)$, $-r \leq \theta \leq 0$, or as an equation

$$L(x_t) \equiv \ddot{x}(t) + (a/r)\dot{x}(t) + (b/r)x(t-r) = f(t,x_t) \tag*{$(2.2)_f$}$$

with

$$f(t,x_t) = (b/6r)x^3(t-r) + f(t).$$

For our purposes, the following results, on the variation of the spectrum of L as the parameter a varies, will be useful:

<u>LEMMA</u> (Somolinos) Let ζ be the root of $(1/br)\sigma^2 = \cos\sigma$ in $(0,\pi/2)$. For fixed r and b, there exists an a_0, $2br/\pi < a_0 < br$ such that $a_0 = (br/\zeta)\sin\zeta$ and

(i) if $a > a_0$, all solutions of the characteristic equation of L,

$$\Delta(\lambda) \equiv \lambda^2 + (a/r)\lambda + (b/r)\exp(-\lambda r) = 0, \tag{2.3}$$

have negative real parts;

(ii) if $a = a_0$, there is a conjugate pair of pure imaginary roots $\lambda_0 \pm i\sigma_0$, and all characteristic roots $\lambda_j \neq \lambda_0$ satisfy $\lambda_j \neq m\lambda_0$ for any integer m;

(iii) there exists an $\varepsilon > 0$ and a complex function $\lambda(a)$ with continuous derivative $\lambda'(a)$ defined in $[a_0-\varepsilon, a_0 + \varepsilon]$. The function $\lambda(a)$ satisfies $\Delta(\lambda(a)) = 0$, $\lambda(a_0) = \pm i\sigma_0$, and $\mathrm{Re}\,\lambda'(a_0) \neq 0$;

(iv) if $a < a_0$, there are precisely two roots of (2.3) with $\mathrm{Re}\,\lambda > 0$ and $-\pi/r < \mathrm{Im}\,\lambda < \pi/r$.

The results of the above lemma correspond to the questions we have raised about our own problem, and allow us to study them by analogy for an ordinary differential equation. Since there is only a pair of eigenvalues of L with zero or positive real part, we can consider in C the generalized eigenspace P associated with this pair. When $a = a_0$, these eigenvalues are pure imaginary, and a basis for P is given by

$$\Phi(\theta) = \begin{bmatrix} \cos\sigma_0\theta & \sin\sigma_0\theta \\ -\sin\sigma_0\theta & \cos\sigma_0\theta \end{bmatrix}.$$

We consider the bilinear form

$$(\psi\cdot\phi) = \psi(0)\phi(0) + \int_{-r}^{0}\int_{0}^{\theta} \psi(\xi-\theta)[d\eta(\theta)]\phi(\xi)\, d\xi$$

where η is the absolutely continuous kernel of the Riesz representation of the linear functional L in C, form which is defined for $\phi \in C$ and $\psi \in C([0,r], R^{2*})$, R^{2*} being the space of 2-dimensional row vectors.

We can define the adjoint equation and the generalized eigenspace P* associate with the same pair of eigenvalues. It is possible to get a basis Ψ, for P* such that $(\Psi,\Phi) = I$. Later we will give those values of that matrix that are relevant for our purposes.

With this basis, we can decompose the space C as follows:

$$C = P \oplus Q,$$

$$P = \{\phi \text{ in } C: \phi = \Phi b \text{ for some vector } b\},$$

$$Q = \{\phi \text{ in } C: (\Psi,\phi) = 0\},$$

and, therefore, for any ϕ in C,

$$\phi = \phi^P + \phi^Q,$$

$$\phi^P = \Phi(\Psi,\phi).$$

The dimension of P is two, and Q is infinite dimensional.

It is known (Hale [10]) that if, in the original equation, we make

$$x_t = x_t^P + x_t^Q,$$

in the subspace P we have

$$x_t^P = \Phi y(t)$$

and the original retarded functional differential equation with the initial condition $x_0 = \phi$ is equivalent to

$$\dot{y}(t) = \Lambda_0 y(t) + \Psi(0)f(t,\Phi y(t) + x_t^Q) \qquad (2.4)$$

$$x_t^Q = T(t-\sigma)\phi^Q + \int_\sigma^t T(t-s)X_0^Q f(s,x_s^P + x_s^Q)\, ds \qquad (2.5)$$

where

$$\Lambda_0 = \begin{bmatrix} 0 & \sigma_0 \\ -\sigma_0 & 0 \end{bmatrix},$$

and both T and X_0^Q are defined in Hale [10]. We do not need the explicit expression for them now, because the relevant approximation and the study of the stability properties of the periodic solution can be obtained by letting $x_t^Q = 0$. We will justify this approximation later on: it is based on the type of nonlinearity, and on the fact that all the eigenvalues corresponding to Q have negative real parts, and it will be clear after the scaling we will introduce and use in the following section.

We will set $x_t^0 = 0$. The equations (2.4) and (2.5) become uncoupled, and we need study only the ordinary differential equation

$$\dot{y} = \Lambda_0 y + \Psi(0)f(t,\Phi y), y(t) = \text{col }(y_1(t), y_2(t)).$$

We obtain its explicit expression by some computation, and we get

$$\dot{y}(t) = \Lambda_0 y + \begin{pmatrix} \alpha_1^0 \\ \gamma_1^0 \end{pmatrix} (\ell_0(y(t)))^3 + \begin{pmatrix} \alpha_0^0 \\ \gamma_0^0 \end{pmatrix} f(t) \tag{2.6}$$

for a_0, where

$$\begin{aligned}
\alpha_0^0 &= \frac{1}{D_1} [-b/2r)\{(1/\sigma_0) \sin \sigma_0 r - r \cos \sigma_0 r\}] \\
\gamma_0^0 &= \frac{1}{D_1} [1-(b/2) \sin \sigma_0 r] \qquad (2.7)\\
\alpha_1^0 &= -(b^2/12r^2 D_1)\{(1/\sigma_0) \sin \sigma_0 r - r \cos \sigma_0 r\} \\
\gamma_1^0 &= (b/6rD_1)\{1-(b/2) \sin \sigma_0 r\}
\end{aligned}$$

$$\left.\begin{aligned} \alpha_2^0 &= \cos \sigma_0 r \\ \gamma_2^0 &= -\sin \sigma_0 r \end{aligned}\right\} \ell_0(y(t)) = \alpha_2^0\, y_1(t) + \gamma_2^0\, y_2(t)$$

$$\begin{aligned}
D_1 &= 1 + (b^2/4) - a_0\sigma_0(1 + (a_0\sigma_0/4(\sigma_0 r)^2)) \\
&= -(1/4(\sigma_0 r)^2)\{(a_0\sigma_0 - \rho_+)(a_0\sigma_0 - \rho_-)\}
\end{aligned}$$

$$\rho_\pm = 2(\sigma_0 r)\,[-(\sigma_0 r) \pm \sqrt{(\sigma_0 r)^2 + 1 + (b/2)^2}].$$

If a varies, the system (2.6) varies also because then the matrix of the linear part has two complex eigenvalues with real part different from zero, and because all the coefficients depending on a are no longer taken at the particular value $a = a_0$.

If we want to take into account this variation of a we will consider the same form as for the system (2.6) but just drop the upper index zero as follows:

$$\dot{y}(t) = \Lambda y(t) + \begin{pmatrix}\alpha_1\\ \gamma_1\end{pmatrix} (\ell(y(t)))^3 + \begin{pmatrix}\alpha_0\\ \gamma_0\end{pmatrix} f(t) \tag{2.8}$$

where

$$\Lambda = \begin{bmatrix} \mu & \sigma \\ -\sigma & \mu \end{bmatrix}$$

and the coefficients $\alpha_0, \gamma_0, \alpha_1, \gamma_1, \alpha_2, \gamma_2$ (in ℓ) correspond to the variation of the parameter a.

3. ENTRAINMENT

According to the experimental evidence we are trying to clarify we will assume in the following the nonresonance condition $\omega \neq \sigma$. The experiments in Andersen and Johnsson [1] were always conducted with forcing frequencies different from the free running frequency, although they could be close to it. This assumption mathematically implies that, for every fixed μ (even $\mu = 0$), the system (2.8) is noncritical with respect to the class of $(2r/\omega)$-periodic functions. Then, for each μ, the result in, for example, Hale ([11], Ch. IV, Thm. 2.1), can be applied to (2.8) and it gives the existence of a $(2\pi/\omega)$-periodic solution, unique in a sufficiently small neighbourhood of zero, and for a sufficiently small range of values of ν. It is not difficult to see in the above-mentioned result that this periodic solution is analytic with respect to the parameter ν. This fact allows us to write this solution in the form

$$\psi(t) = \nu\psi_1(t) + \nu^2\psi_2(t) + \ldots$$

in which we can try to obtain an approximation of it, by assuming

$$\psi_1(t) = \begin{bmatrix} \alpha & \beta \\ \gamma & \delta \end{bmatrix} \begin{bmatrix} \sin \omega t \\ \cos \omega t \end{bmatrix} .$$

This first approximation is also an exact solution of the linear equation

$$\dot{y}(t) = \Lambda y(t) + f(t).$$

After some computation, we obtain

$$\alpha = -\frac{f}{D_2}\{\alpha_0\mu(\omega^2 + \sigma^2 + \mu^2) + \gamma_0\sigma[\omega^2 - (\sigma^2 + \mu^2)]\}$$

$$\beta = \frac{f}{D_2}\{\alpha_0\omega(\omega^2 - \sigma^2 + \mu^2) - 2\gamma_0\omega\sigma\mu\} \tag{3.1}$$

$$\gamma = -\frac{f}{D_2}\{\alpha_0\sigma(\sigma^2 - \omega^2 + \mu^2) + \gamma_0\mu(\omega^2 + \sigma^2 + \mu^2)\}$$

$$\delta = \frac{f}{D_2}\{2\alpha_0\sigma\omega\mu + \gamma_0\omega(\omega^2 - \sigma^2 + \mu^2)\},$$

where

$$D_2 = [\omega^2 - (\sigma^2 + \mu^2)]^2 + 4\mu^2\omega^2. \tag{3.2}$$

Thus, in order to have this solution also defined for $\mu = 0$, we use again the nonresonance condition $\omega \neq \sigma$.

To study the stability properties and the existing $(2\pi/\omega)$-periodic solution, we will consider more closely the neighbourhood of this function $\psi(t)$. We make the transformation of variables $y(t) = z(t) + \psi(t)$. Since the equation (2.8) is of the form

$$\dot{y} = L(\mu)y + N(y) + f,$$

where L and N are the linear and nonlinear part, respectively, after this transformation we obtain

$$\dot{z} = L(\mu)z + N(z+\psi) - N(\psi). \tag{3.3}$$

The function $z = 0$ is a solution of (3.3) (corresponding to $y = \psi(t)$) and its stability in the above equation will give that of $\psi(t)$. The equation (3.3) is

$$\dot{z}(t) = \begin{pmatrix} \mu & \sigma \\ -\sigma & \mu \end{pmatrix} z + \begin{pmatrix} \alpha_1 \\ \gamma_1 \end{pmatrix} \{[\ell(z+\psi)]^3 - [\ell(\psi)]^3\} \tag{3.4}$$

which can be written as

$$\dot{z}(t) = L(0)z + \mu Iz + \begin{pmatrix} \alpha_1 \\ \gamma_1 \end{pmatrix} \{3\ell(z)\ell^2(\psi) + 3\ell^2(z)\ell(\psi) + \ell^3(z)\} \tag{3.5}$$

We will change to polar coordinates to study the behaviour near ψ, but first we rescale the equation in such a way that nothing important is lost but everything becomes more clear. Thus, the scaling has to be done in a critical way, and in this respect it is useful (see Carr [3]) to consider the first of the equations (3.5) in polar coordinates (without scaling). Taking $z_1 = r\cos\theta$, $z_2 = r\sin\theta$,

$$\dot{r} = \mu r + \{O(r^3) + O(\nu r^2) + O(\nu^2 r)\}.$$

Then, if there is a bounded solution (periodic, for example), when r attains its maximum we have $\dot{r} = 0$, and if we do not want to lose any information about this solution, all the terms in the right-hand side have to be of the same order. A possible choice to meet these requirements is to have r of the order of ν and μ of the order of ν^2, so we will make the transformation

$$z = \nu\bar{z}$$
$$\mu = \nu^2\bar{\mu}.$$

After this scaling, the above equation (3.5) becomes

$$\dot{\bar{z}} = L(0)\bar{z} + \nu^2\bar{\mu}I\bar{z} + \begin{pmatrix}\alpha_1\\ \gamma_1\end{pmatrix} \nu^2\{\ell^3(\bar{z}) + 3\ell^2(\bar{z})\ell(\psi_1) + 3\ell(\bar{z})\ell^2(\psi_1)\} + O(\nu^3) \tag{3.5'}$$

We will change to polar coordinates and average with respect to t. To do so, we need the supplementary assumption

$$|m\sigma + n\omega| \neq 0, \quad |m| + |n| \leqslant 4. \tag{3.6}$$

and we get

$$\dot{\rho} = \nu\{[\bar{\mu} + \frac{3}{4}(\alpha_1\alpha_2 + \gamma_1\gamma_2)(\alpha_2^2(\alpha+\beta)^2 + \gamma_2^2(\gamma+\delta)^2)]\rho$$

$$+ [\frac{3}{8}(\alpha_2^2 + \gamma_2^2)(\alpha_1\alpha_2 + \gamma_1\gamma_2)]\rho^3\} + O(\nu^3)$$

$$= \nu^2\{(\bar{\mu} + k_1)\rho + k_2\rho^3\} + O(\nu^3). \tag{3.7}$$

The sign of the term $(\alpha_1\alpha_2 + \gamma_1\gamma_2)$ appears, then, as a crucial one for the discussion of the above equation. Although the computation of that term is difficult in the general case, $\bar{\mu} \neq 0$, since we are only interested in the discussion of its sign, this can be done by continuity from the case $\bar{\mu} = 0$.

Taking into account the values of α_1^0, α_2^0, γ_1^0, γ_2^0 given by (2.7), we obtain

$$D_3 = (\alpha_1\alpha_2 + \gamma_1\gamma_2)_{\mu=0} = \frac{1}{12rD_1}\{(a_0\sigma_0)^2 - (a_0\sigma_0)(2+\sigma_0) + \sigma_0^2(\sigma_0 r)^2\}$$

$$= \frac{1}{12rD_1}[(a_0\sigma_0) - \tau_+][(a_0\sigma_0) - \tau_-] \qquad (3.8)$$

where D_1 is given by (2.7). The roots τ_+ and τ_- of $12rD_1D_3$ are positive and negative, respectively, and with (3.8) it is easy to discuss the sign of D_3. Considering $a_0\sigma_0 > 0$, there is always an interval [0,c), where D_3 is negative, with $c = \min(\rho_+,\tau_+)$. In the case in which $\tau_+ < \rho_+$, $(\alpha_1\alpha_2 + \gamma_1\gamma_2)(\tau_+) = 0$, for example. Typical experimental values (see Klein [16] or Andersen and Johnsson [1]) are, for example, $a_0 = 4.7$, $\sigma_0 r = 1.5$, $r \simeq 40$, and then $\sigma_0 = 0.04$, so we have $(a_0\sigma_0) << \tau_+ < \rho_+$, and the value of D_3 is negative.

On the other hand, the sign of $(\bar{\mu} + k_1)$ depends on the values of $\bar{\mu}$ (positive or zero), the sign of the term just discussed, which is negative, and the sign of the term $\alpha_2^2(\alpha+\beta)^2 + \gamma_2^2(\gamma+\delta)^2$. This term is small if the forcing frequency, ω, is far from the free running frequency, but becomes bigger as ω approaches σ. So, there must be an interval around the free running frequency such that if the forcing frequency is out of it, $\bar{\mu} + k_1 > 0$, and if it is inside it, $\bar{\mu} + k_1 \lessdot 0$.

For the experimental conditions, we have $k_2 < 0$ and, if $\bar{\mu} + k_1 > 0$, we know (from the implicit function theorem, as can be seen, for example, in Chow and Mallet-Paret [7]) that there is a unique periodic solution for the averaged equation (3.7), of frequency close to σ_0, and also one close to the free running frequency, and a torus for the full equation. Moreover, because of the sign of the coefficients, both are stable. This torus corresponds to the experimental beats, which occur when the forcing frequency is far from the free running one, outside the above-mentioned interval. When the forcing frequency approaches the other one, the coefficient $\bar{\mu} + k_1$ vanishes for, say, $\omega = \omega_1$, and the torus loses its stability, which means that the beats disappear, and the only stable solution for the averaged solution is the zero one, which corresponds to the $(2\pi/\omega)$-periodic solution for the full equation (3.5'). Then, the torus jumps back to the $(2\pi/\omega)$-periodic solution ψ. The coefficients of ψ, given by (3.1) become important in this interval because, even though ν is a small parameter, the term $(1/D_2)$ is larger. This $(2\pi/\omega)$-periodic solution, which now is stable, corresponds to the entrainment of the motion of the plant to the frequency of the forcing term. The same thing happens where the forcing frequency approaches the free running one from the other side, and there is a ω_2 in which $\bar{\mu} + k_1$ becomes zero. These two numbers ω_1 and ω_2 determine an interval in which the phenomenon of entrainment takes place.

All this study corresponds to equation (2.4) with $x_t^Q = 0$. It remains to justify this, which can now be done by considering the equation (2.5). It can be written (see Hale [10]) as

$$x_t^Q = \int_{-\infty}^{t} T(t-s)X_0^Q \, f(s, x_s^P + x_s^Q) \, ds.$$

Since $f(s, x_s)$ is the function in (2)(f), rescaling as before $x_t = \nu \bar{x}_t$ we obtain

$$\bar{x}_t^Q = \nu \int_{-\infty}^{t} T(t-s)X_0^Q \, f(s, \nu^2(\bar{x}_t^P + \bar{x}_t^Q)) \, ds$$

because the terms with x_s in f are cubics. This expression, the known exponential estimates for $T(t-s)X_0^Q$ and the fact that all the eigenvalues corresponding to Q have negative real parts, gives $|\bar{x}_t^Q| \leqslant \Omega|\nu|$, where Ω is a constant. Thus, this justifies taking $x_t^Q = 0$ in the neighbourhood of P and then in (2.4). So, there exists an annulus surrounding the periodic solution, which is also positively invariant because the term k_2 is negative.

ACKNOWLEDGEMENT

The authors want to express their gratitude to Professor Jack K. Hale, whose suggestions and help led to improvements in the original conception and in the results of this work. Also, they want to thank Professor Jean Mawhin for useful discussions.

REFERENCES

[1] H. Andersen and A. Johnsson, Entrainment of geotropic oscillations in hypocotyls of Helianthus annuus. An experimental and theoretical investigation, Physiol. Plant. 26: Part I, 44-51; Part II, 52-61 (1972).

[2] N. Bogolioubov and I. Mitropolski, Les méthodes asymptotiques en théorie des oscillations non linéaires, Gauthier-Villars, Paris (1962).

[3] J. Carr, Applications of Centre Manifold Theory. Lecture Notes, Lefschetz Center for Dynamical Systems, Division of Applied Mathematics, Brown University (June 1979).

[4] A. Casal and A. Somolinos, Estudio analítico y numérico de la ecuación del girasol. Rev. de la Real Academia de Ciencias Exactas, Físicas y Naturales, 73, 503-519 (1979).

[5] A. Casal and A. Somolinos, Forced oscillations for the sunflower equation, entrainment. Nonlinear Analysis, Theory, Methods and Applications 6(4), 397-414 (1982).

[6] S. Chow and J.K. Hale, Methods of Bifurcation Theory (to be published).

[7] S. Chow and J. Mallet-Paret, Integral averaging and bifurcation, J. Differential Equations 26, 112-159 (1977).

[8] R. Grafton, A periodicity theorem for autonomous functional equations, J. Differential Equations 6, 87-109 (1969).

[9] R. Grafton, Periodic solutions of certain Liénard equations with delay, J. Differential Equations 11, 519-527 (1972).

[10] J.K. Hale, Theory of Functional Differential Equations, Appl. Math. Sci. Vol. 3, 2nd edn. Springer-Verlag, Berlin and New York (1977).

[11] J.K. Hale, Ordinary Differential Equations, Wiley-Interscience, New York (1969); 2nd edn., Krieger (1980).

[12] J.K. Hale, Oscillations in Nonlinear Systems, McGraw-Hill, New York (1963).

[13] J.K. Hale, Nonlinear oscillations in equations with delays, 157-185 in Nonlinear Oscillations in Biology (ed. F. Hoppenstadt), Lectures in Applied Mathematics, Vol. 17 (1979).

[14] J.K. Hale, Topics in Dynamic Bifurcation Theory, NSF-CBMS Conference, Arlington, Texas (June 1980).

[15] D. Israelson and A. Johnsson, A theory for circumnutations in Helianthus Annuus, Physiol. Plant, 20, 957-976 (1967).

[16] E. Klein, An application of non-linear retarded differential equations to the circumnutation of plants. Master thesis, Division of Applied Mathematics, Brown University, June 1972.

[17] J. Mawhin, Degré topologique et solutions périodiques des systèmes différentiels non linéaires, Bull. Soc. R. Sci. Liège, 38, 308-398 (1969).

[18] J. Mawhin, Periodic solutions of nonlinear functional differential equations, J. Differential Equations, 10, 240-261 (1971).

[19] C. Perelló, Periodic solutions of differential equations with time lag containing a small parameter, J. Differential Equations, 4, 160-175 (1968).

[20] N. Rouche and J. Mawhin, Equations Différentielles Ordinaires, Tome 2: Stabilité et solutions périodique, Masson et Cie., Paris (1973).

[21] A. Somolinos, Periodic solutions of the sunflower equation: $\ddot{x} + (a/r)\dot{x} + (b/r) \sin x(t-r) = 0$. Quarterly Appl. Math. 35, 465-478 (1978).

[22] H. Stech, The Hopf bifurcation: a stability result and application. J. Math. Anal. and Appl. 71, 525-546 (1979).

A. Casal
Facultad de Matemáticas,
Universidad Complutense
Madrid-3,
Spain

A. Somolinos
Division of Applied Mathematics
Brown University
Providence, R.I. 02912,
USA

and Universidad de Alcalá
Alcalá de Henares,
Madrid
Spain

T CAZENAVE

Stability and instability of stationary states in nonlinear Schrödinger equations

1. INTRODUCTION

Nonlinear Schrödinger equations arise in various domains of mathematical physics. These equations are generally of the form

$$i \frac{\partial \phi}{\partial t} + \Delta\phi + F(\phi) = 0 \text{ in } R \times R^N \tag{1}$$

where F is a local or nonlocal nonlinearity.

Various examples can be found in Bialynicki-Birula and Mycielski [4], Hartree [10], Kelley [11], Lam et al. [12], Lions [13], Suydam [17]. F is generally the gradient of a functional $G \in C^1(H^1(R^N),R)$ and one has, at least formally, the two following conservation laws:

$$\int_{R^N} |\phi(t,x)|^2 \, dx = \text{constant} \tag{2}$$

$$E(\phi(t,\cdot)) = \text{constant} \tag{3}$$

where $E(u) = \frac{1}{2} \int_{R^N} |\nabla u(x)|^2 \, dx - G(u)$.

Considering the above conservation laws, it is natural to solve (1) in $H^1(R^N)$, that is to consider solutions $\phi \in C(a,b,H^1(R^N))$. (1) may possess stationary states, that is solutions of the form $\phi(t,x) = e^{i\lambda t}u(x)$. Those solutions are of particular interest from both the physical and mathematical point of view. They are given by solutions of the following problem:

$$-\Delta u + \lambda u = F(u) \text{ in } R^N, \; u \in H^1(R^N). \tag{4}$$

In this paper, we study the problem of the stability or instability for the stationary states of (1). For the sake of simplicity we shall consider a local, parameterized nonlinearity of the form $F(u) = |u|^{p-1}u$. Hence we consider the following nonlinear Schrödinger equation:

$$i \frac{\partial \phi}{\partial t} + \Delta\phi + |\phi|^{p-1}\phi = 0 \text{ in } R \times R^N \tag{S}$$

and the associated nonlinear scalar field equation

$$- \Delta u + \lambda u = |u|^{p-1}u \text{ in } R^N, \ u \in H^1(R^N). \qquad (E_\lambda)$$

It is known (see Ginibre and Velo [8]) that if $1 < p < (N+2)/(N-2)$ ($1 < p < \infty$ if $N = 1,2$) then for any $\phi_0 \in H^1(R^N)$ there is a unique local solution ϕ of (S), $\phi \in C(-\varepsilon, +\varepsilon, H^1(R^N))$ such that $\phi(0) = \phi_0$. In addition ϕ satisfies the following conservation laws:

$$\int_{R^N} |\phi(t,x)|^2 \, dx = \text{constant} \qquad (5)$$

$$E(\phi(t,\cdot)) = \text{constant} \qquad (6)$$

where

$$E(u) = \frac{1}{2} \int_{R^N} |\nabla u(x)|^2 \, dx - \frac{1}{p+1} \int_{R^N} |u(x)|^{p+1} \, dx. \qquad (7)$$

Furthermore, if $1 < p < 1+4/N$, then all the solutions of (S) exist globally (see [8]) and if $1 + 4/N \leqslant p < (N+2)/(N-2)$, then certain solutions of (S) blow up in a finite time (see Glassey [9], Strauss [16], and Cazenave [5]).

The stationary problem (E_λ) can be solved for $\lambda > 0$ and $1 < p < (N+2)/(N-2)$. More precisely (see Berestycki and Lions [3]) (E_λ) possesses a ground state, that is a nontrivial solution u_λ of (E_λ) that minimizes S_λ among all nontrivial solutions of (E_λ), where

$$S_\lambda(u) = E(u) + \frac{\lambda}{2} \int_{R^N} |u(x)|^2 \, dx.$$

A ground state is, up to a translation and a phase change, a spherically symmetric, nonnegative, nonincreasing solution. In part 2 we prove the stability of stationary states of (1) associated to ground states of (E_λ) when $1 < p < 1 + 4/N$, and in part 3 we prove the instability of the same stationary states when $1 + 4/N < p < (N+2)/(N-2)$.

More general and detailed results on this subject can be found in Berestycki and Cazenave [1], [2], Cazenave [6], Cazenave and Lions [7].

2. A STABILITY RESULT

To avoid technicalities we give a stability result in the case $N \leq 3$.

THEOREM 1. Assume $N \leq 3$ and $1 < p < 1 + \frac{4}{N}$. Let $\lambda > 0$ and u_λ be a ground state of (E_λ). Then the associated stationary state of (S), $\psi(t,x) = e^{i\lambda t} u_\lambda(x)$, is stable in the following sense:

For any $\varepsilon > 0$ there is $\delta > 0$ such that if $\phi_0 \in H^1(R^N)$ satisfies $\|\phi_0 - u_\lambda\|_{H^1(R^N)} \leq \delta$, then there is $\theta(t) \in R$ and $y(t) \in R^N$ such that the solution ϕ of (S) with $\phi(0) = \phi_0$ satisfies

$$\|\phi(t,\cdot) - e^{i\theta(t)} u_\lambda(\cdot - y(t))\|_{H^1(R^N)} \leq \varepsilon \text{ for } t \in R.$$

Remark 2. Simple considerations show (see [7]) that neither $\theta(t)$ nor $y(t)$ can be dropped in Theorem 1.

Proof of Theorem 1. We apply a general method of [7] and compactness results of Lions [14] and [15].

Let u_λ be a ground state of (E_λ). We set

$$\mu = \int_{R^N} |u_\lambda|^2 \, dx \quad \text{and} \quad I_\mu = \inf_{\substack{u \in H^1(R^N) \\ \int_{R^N} |u|^2 dx = \mu}} E(u) .$$

It is proved in [16] that $I_\mu > -\infty$ and that $\Sigma_\mu \neq \emptyset$, where

$$\Sigma_\mu = \{u \in H^1(R^N) \,\Big|\, \int_{R^N} |u|^2 \, dx = \mu,\ E(u) = I_\mu\}.$$

Furthermore we have $I_\mu = E(u_\lambda)$ and $\Sigma_\mu = \{e^{i\theta} u_\lambda(\cdot - y)\}_{\substack{\theta \in R \\ y \in R^N}}$

Theorem 1 is then an immediate consequence of the following:

PROPOSITION 3. S_μ is stable in the following sense:

For any $\varepsilon > 0$ there is $\delta > 0$ such that if $\phi_0 \in H^1(R^N)$ satisfies $d(\phi_0, \Sigma_\mu) \leq \delta$, then the solution ϕ of (S) with $\phi(0) = \phi_0$ satisfies $d(\phi(t), \Sigma_\mu) \leq \varepsilon$ for

$t \in R$, where d denotes the distance in $H^1(R^N)$.

Proof. We argue by contradiction. If Proposition 2 is wrong, there is $(\phi_n^0)_{n\in N}$, $(t_n)_{n\in N}$ and $\varepsilon > 0$ such that:

$$d(\phi_n^0, \Sigma_\mu) \underset{n\to\infty}{\to} 0 \tag{8}$$

$$d(\phi_n(t_n), \Sigma_\mu) \geq \varepsilon \tag{9}$$

where ϕ_n is the solution of (S) such that $\phi_n(0) = \phi_n^0$.

It follows from (8) that $\int_{R^N} |\phi_n^0|^2 \, dx \underset{n}{\to} \mu$ and $E(\phi_n^0) \underset{n}{\to} I_\mu$. From (5), (6) we get

$$\int_{R^N} |\phi_n(t_n)|^2 \, dx \underset{n}{\to} \mu \quad \text{and } E(\phi_n(t_n)) \underset{n}{\to} I_\mu.$$

Hence, it follows from Lions [14] and [15] that $d(\phi_n(t_n), \Sigma_\mu) \underset{n}{\to} 0$, which contradicts (8).

Remark 4. We do not know about the stability of stationary states of (S) associated with bound states of (E_λ). It is expected that they are not stable in the sense of Theorem 1.

3. AN INSTABILITY RESULT

We state the following result of [2] with a sketch of the proof.

THEOREM 5. Assume $1 + 4/N < p < (N+2)/(N-2)$. Let $\lambda > 0$ and u_λ be a ground state of (E_λ). Then the associated stationary state $\psi(t,x) = e^{i\lambda t} u_\lambda(x)$ is unstable in the following sense:

For any $\varepsilon > 0$ there is $\phi_\varepsilon^0 \in H^1(R^N)$ with $\|\phi_\varepsilon^0 - u_\lambda\|_{H^1(R^N)} \leq \varepsilon$ such that the solution ϕ_ε of (S) satisfying $\phi_\varepsilon(0) = \phi_\varepsilon^0$ blows up in a finite time in $H^1(R^N)$.

Remark 6. We have no result concerning the instability of stationary states of (S) associated with bound states of (E_λ). It is probable that they are unstable in the sense of Theorem 5.

Remark 7. For technical reasons in the limiting case $p = 1 + 4/N$, it is possible to prove that all the stationary states of (S) are unstable in the sense of Theorem 5.

The proof of Theorem 5 is based upon

(i) a variational characterization of the ground states of (E_λ);

(ii) an identity of Glassey [9].

We first define $Q \in C^1(H^1(R^N),R)$ by

$$Q(u) = \int_{R^N} |\nabla u|^2 \, dx - \frac{N}{2} \left(\frac{p-1}{p+1}\right) \int_{R^N} |u|^{p+1} \, dx.$$

Then we set:

$$M = \{u \in H^1(R^N) \mid u \neq 0, \; Q(u) = 0\}$$

$$d_\lambda = \inf_{u \in M} S_\lambda(u)$$

$$K^\lambda = \{u \in H^1(R^N) \mid S_\lambda(u) < d_\lambda, \;\; Q(u) < 0\}.$$

We then state the following characterization of ground states of E_λ which is proved in [2].

PROPOSITION 8. Let $u \in H^1(R^N)$, $\lambda > 0$. Then u is a ground state of (E_λ) iff:

(i) $u \in M$

(ii) $S_\lambda(u) = d_\lambda$.

We deduce easily from (5), (6) the following.

PROPOSITION 9. K_λ is stable by the flow associated with (S).

On the other hand, we have (see [9], [8], [2])

PROPOSITION 10. Let $\phi_0 \in H^1(R^N)$ be such that $|\cdot|\phi_0(\cdot) \in L^2(R^N)$. Let $[0, Tmax[$ be the maximal interval of existence of the solution ϕ to (S) satisfying $\phi(0) = \phi_0$. Then:

(i) $|\cdot|\phi(t,\cdot) \in C([0,Tmax[, L^2(R^N))$

(ii) $\| \, |\cdot|\phi(t,\cdot)\|_{L^2(R^N)} \in C^2([0,Tmax[)$

(iii) $\frac{d^2}{dt^2} \| \ |\cdot| \phi(t,\cdot) \|^2_{L^2(R^N)} = 4Q(\phi(t))$ for $0 \leqslant t < Tmax$.

As a consequence, we have the following:

PROPOSITION 11. Let $\phi_0 \in K_\lambda$ be such that $|\cdot|\phi_0(\cdot) \in L^2(R^N)$. Then the solution ϕ of (S) satisfying $\phi(0) = \phi_0$ blows up in a finite time in $H^1(R^N)$.

Proof. From Propositions 9 and 10 we deduce

$$\frac{d^2}{dt^2} \| \ |\cdot| \phi(t,\cdot) \|^2_{L^2(R^N)} < 0 \text{ for } 0 \leqslant t < Tmax.$$

A more precise study (see [2]) shows that there is $\varepsilon > 0$ that

$$\frac{d^2}{dt^2} \| \ |\cdot| \phi(t,\cdot) \|^2_{L^2(R^N)} \leqslant -\varepsilon \text{ for } 0 \leqslant t < Tmax.$$

Since $\| |\cdot| \phi(t,\cdot) \|_{L^2(R^N)} \geqslant 0$ we get $Tmax < \infty$.

Proof of Theorem 5. We define ϕ^0_ε by $\phi^0_\varepsilon(x) = (1+\varepsilon)^{N/2} u_\lambda((1+\varepsilon)x)$. It is easily checked that $\phi^0_\varepsilon \to u_\lambda$ as $\varepsilon \to 0$ and that $\phi^0_\varepsilon \in K_\lambda$. Hence Theorem 5 is proved.

REFERENCES

[1] H. Berestycki and T. Cazenave, Instabilité des états stationnaires dans les équations de Schrödinger et de Klein-Gordon non linéaires, C.R. Acad. Sci. Paris 320, 489-492 (1981).

[2] H. Berestycki and T. Cazenave, Instability of stationary states in non-linear Schrödinger and Klein-Gordon equations (to appear).

[3] H. Berestycki and P.L. Lions, Nonlinear scalar fields equations. I & II (to appear, Arch. Rat. Mech. Anal.).

[4] I. Bialynicki-Birula and J. Mycielski, Nonlinear wave mechanics. Ann. of Phys. 100, 62-93 (1976).

[5] T. Cazenave, Equations de Schrödinger non lineaires en dimension deux. Proc. Roy. Soc. Edinburgh, 88, 327-346 (1979).

[6] T. Cazenave, Stable solutions of the logarithmic Schrödinger equation (to appear, Nonlinear Anal. T.M.A.).

[7] T. Cazenave and P.L. Lions, Orbital stability of standing waves for some nonlinear Schrödinger equations, Comm. Math. Phys. 85,549-561 (1982).

[8] J. Ginibre and G. Velo, On a class of nonlinear Schrödinger equations, I & II. J. Funct. Anal. 32, 1-71 (1979).

[9] R.T. Glassey, On the blowing-up of solutions to the Cauchy problem for nonlinear Schrödinger equations. J. Math. Phys. 18, 1794-1797 (1977).

[10] D. Hartree, The wave mechanics of an atom with a non Coulomb central field. I. Proc. Camb. Phil. Soc. 24, 89-132 (1968).

[11] P.L. Kelley, Self focusing of optical beams. Phys. Review Letters 15, 1005-1008 (1965).

[12] J.F. Lam, B. Lippmann and F. Tappert, Self-trapped laser beams in plasma, Phys. Fluids 20, 1176-1179 (1977).

[13] P.L. Lions, The Choquard equation and related equations. Nonlinear Anal. T.M.A. 4, 1063-1073 (1980).

[14] P.L. Lions, Principe de concentration en calcul des variations, C.R. Acad. Sci. Paris 294, 261-264 (1982) and this volume.

[15] P.L. Lions (to appear).

[16] W. Strauss, The nonlinear Schrödinger equation. In Contemporary Developments in Continuum Mechanics and Partial Differential Equations (ed. De la Penha-Medeiros), 452-465 Amsterdam, North Holland (1978).

[17] B.R. Suydam, Self-focusing of very powerful laser beams, Spec. Pub. Natn. Bur. Stand. 387, 42-48 (1973).

T. Cazenave
Laboratoire d'Analyse Numérique
Université Pierre et Marie Curie
4 Place Jussieu-75230
Paris Cedex 05
France

R CIPOLATTI & A DAMLAMIAN

Nonconvex duality for a free boundary problem: the plasma equilibrium

The following problem was introduced by Temam [8] as a simple model for the equilibrium of a confined plasma gas in a Tokamak machine: in a bounded domain Ω of $\mathbb{R}^2$ (more generally $\mathbb{R}^N$ although this has no physical meaning in general), find a solution of

$$\left.\begin{aligned} -\Delta u &= \lambda u^+ \quad \text{in } \Omega \\ u &= u_\Gamma \quad \text{an unknown constant on } \Gamma = \partial\Omega \\ -\int_\Gamma \frac{\partial u}{\partial n} &= I \quad \text{a given constant.} \end{aligned}\right\}$$

This problem has been studied extensively by several authors (see [8] and the bibliography therein, [1] and [6]) and in [3] one of the authors applied Toland's nonconvex duality method ([10], [11]) to prove the equivalence between the two apparently distinct variational formulations introduced in [1] and [9].

The purpose of this paper is to give a detailed proof of such a result in a more abstract and general setting.

Recently Heron and Sermange have studied the same problem with a slightly different approach [4].

1. THE PROBLEM

Let Ω be a bounded regular domain in $\mathbb{R}^N$ and Γ its boundary which is assumed to be Lipschitz continuous. Let ψ be an element of $H^{1/2}(\Gamma)$ whose harmonic extension to Ω is still denoted ψ.

Let ϕ be a convex continuous functional on $L^2(\Omega)$. The problem is to find u in $H^2(\Omega)$ such that

$$\left.\begin{aligned} & u \in H_0^1(\Omega) \oplus \mathbb{R}\psi \\ & -\Delta u \in \partial\phi(u) \\ & -\int_\Omega \Delta u.\psi = I, \text{ a given real number.} \end{aligned}\right\} \qquad \text{(P)}$$

Here $\partial\Omega$ is the subdifferential of ϕ on $L^2(\Omega)$. This operator can be multivalued but in this case it is everywhere defined since ϕ is continuous on $L^2(\Omega)$.

Remark 1. The situation can be generalized to the case where $\mathbb{R}\psi$ is replaced by a finite-dimensional subspace X of $H^1(\Omega)$ which is orthogonal to $H_0^1(\Omega)$, the problem being to find u such that:

$$\left.\begin{array}{l} u \in H_0^1(\Omega) \oplus x \\ -\Delta u \in \partial\phi(u) \\ \chi \mapsto \int_\Omega -\Delta u\ \chi \quad \text{is a given linear form on X.} \end{array}\right\}$$

Remark 2. More generally the operator $-\Delta$ can be replaced by any second-order symmetric operator A which is elliptic on $H_0^1(\Omega)$ so that the bilinear form a associated with A is an equivalent scalar product on $H_0^1(\Omega)$.

Even more generally A could be an operator of order 2m which is elliptic on $H_0^m(\Omega)$ (which would replace $H_0^1(\Omega)$) and with $L^2(\Omega)$ domain $H^{2m}(\Omega)$.

2. FIRST FORMULATION OF (P) AS A CRITICAL POINT PROBLEM

Let H be $H_0^1(\Omega) \oplus \mathbb{R}\psi$ on which the following scalar product is equivalent to that induced by $H^1(\Omega)$:

$$(u,v)_H = \int_\Omega \nabla u.\nabla v + u_\Gamma\ v_\Gamma,$$

where $u = \tilde{u} + u_\Gamma\psi, v = \tilde{v} + v_\Gamma\psi$, $\tilde{u}$, $\tilde{v}$ in $H_0^1(\Omega)$.

On H, define

$$E_1(v) = \frac{1}{2}\ |\nabla v|^2_{L^2(\Omega)} - \phi(v) - \frac{1}{2}|\nabla\psi|^2_{L^2}\ v_\Gamma^2 + Iv_\Gamma.$$

We have

PROPOSITION 1. Suppose ϕ is Gâteaux-differentiable on $L^2(\Omega)$. Then u is a solution of (P) if and only if u is a critical point of E_1 on H.

Proof. It is easy to check by Green's formula that E_1 is Gâteaux-differentiable on H and its derivative is given by

$$\langle E_1'(v),w\rangle = \int_\Omega (-\Delta v - \phi'(v))w + w_\Gamma(I + \int_\Gamma \frac{\partial v}{\partial n}\psi - v_\Gamma|\nabla\psi|^2_{L^2})$$

$$= \int_\Omega (-\Delta v - \phi'(v))w + w_\Gamma(I + \int_\Omega \Delta v.\psi).$$

In the general case for ϕ, we can write

$$E_1(v) = \frac{1}{2}|v|^2_H - F_1(v) = G_1(v) - F_1(v),$$

where

$$F_1(v) = \phi(v) + \frac{1}{2}(1 + |\nabla\psi|^2_{L^2})v^2_\Gamma - Iv_\Gamma$$

is continuous convex on H.

In Proposition 2 it will be shown that the elements of $\partial F_1(u)$ are the following linear forms on H:

$$w \mapsto \int_\Omega \theta.w + (1 + |\nabla\psi|^2_{L^2})u_\Gamma w_\Gamma - Iw_\Gamma$$

for every θ in $L^2(\Omega)$ such that $\theta \in \partial\phi(u)$.

Since $\partial G_1(u)$ is the linear form which corresponds to the scalar product with u in H, it is easy to check (with some application of Green's formula) that (P) is exactly equivalent to finding u in H such that $\partial G_1(u) \in \partial F_1(u)$, which we write as

$$\partial G_1(u) \cap \partial F_1(u) \neq \emptyset$$

Toland takes this last property as the definition of a critical point for $E_1 = G_1 - F_1$ (see [10], [11]).

Remark 3. Although the above definition seems to depend upon the decomposition of E_1 as a difference of two l.s.c. convex functions, in this particular case it is not so because there is an equivalent intrinsic expression (due to the fact that G_1 is itself everywhere Gâteaux-differentiable):

$$\sup_{t \to 0^+} \lim \frac{E_1(u + tw) - E_1(u)}{t} \leq 0$$

for all w in H.

3. SECOND FORMULATION OF (P) AS A CRITICAL POINT PROBLEM

As Toland noticed, the condition $v \in \partial G_1(u) \cap \partial F_1(u)$ is equivalent to $u \in \partial G_1^*(v) \cap \partial F_1^*(v)$, where G_1^* and F_1^* are the convex conjugates of G_1 and F_1.

From now on we shall insist in taking conjugation on H itself, that is by identifying H with its dual space. Therefore $\partial G_1 = Id_H$ so that $u = v$ above and the dual problem is written as $u \in \partial F_1^*(u)$. But writing it in such a way does not improve anything and it is more productive to view this as a critical-point problem for a new functional on H:

$$E_2(v) = F_1^*(v) - G_1^*(v) = F_1^*(v) - G_1(v),$$

for which Remark 3 holds also.

PROPOSITION 2.

$$F_1^*(u) = \phi^*(-\Delta u) + \frac{1}{2(1+|\nabla\psi|^2_{L^2})}\{u_\Gamma(1+|\nabla\psi|^2_{L^2}) + I + \int_\Omega \Delta u.\psi\}^2$$

with the understanding that $\phi^*(-\Delta u) = +\infty$ whenever $-\Delta u$ is not in $L^2(\Omega)$.

$$v \in \partial F_1(u) \Leftrightarrow \begin{cases} -\Delta v \in \partial\phi(u) \text{ in } L^2(\Omega) \\ v_\Gamma = u_\Gamma - \dfrac{1}{1+|\nabla\psi|^2_{L^2}}(I + \int_\Omega \Delta v.\psi) \end{cases}$$

Proof. By definition of the convex conjugate

$$F_1^*(u) = \sup_{v\in H}\{\int_\Omega \nabla u.\nabla v + u_\Gamma v_\Gamma + I v_\Gamma - \frac{1}{2}(1 + |\nabla\psi|^2_{L^2})v_\Gamma^2 - \phi(v)\}.$$

We claim that if $-\Delta u$ is not in $L^2(\Omega)$, then $F_1^*(u) = +\infty$. Indeed, if $-\Delta u$ is not in $L^2(\Omega)$, there exists a bounded sequence $\{v_n\}$ in $L^2(\Omega)$ consisting of smooth functions (v_n in $\mathcal{D}(\Omega)$) such that

$$\int_\Omega \nabla u.\nabla v_n \xrightarrow[n\to+\infty]{} +\infty.$$

Since ϕ is continuous on $L^2(\Omega)$, it is bounded on some neighbourhood of zero to which we can assume that the v_n belong. Thus going to the limit yields $F_1^*(u) = +\infty$.

For u in H with Δu in $L^2(\Omega)$ one has

$$\int_\Omega \nabla u . \nabla v = \int_\Omega -\Delta u . v + \int_\Gamma \frac{\partial u}{\partial n} v = \int_\Omega -\Delta u . v + v_\Gamma \int_\Gamma \frac{\partial u}{\partial n} \psi$$

so

$$F_1^*(u) = \sup_{v \in H} \{\int_\Omega -\Delta u . v - \phi(v) + v_\Gamma(u_\Gamma + I + \int_\Gamma \frac{\partial u}{\partial n} \psi)$$

$$- \frac{1}{2} (1 + |\nabla\psi|^2_{L^2}) v_\Gamma^2\}.$$

Now we claim that

$$F_1^*(u) = \sup_{w \in L^2(\Omega)} \{\int_\Omega -\Delta u . w - \phi(w)\}$$

$$+ \sup_{b \in \mathbb{R}} \{(u_\Gamma + I + \int_\Gamma \frac{\partial u}{\partial n} \psi) b - \frac{1}{2} (1 + |\nabla\psi|^2) b^2\}.$$

Indeed let w be in $L^2(\Omega)$ and b in $\mathbb{R}$. There is a sequence $\{w_n\}$ in $H_0^1(\Omega)$ such that w_n converges to $w - b\psi$ in $L^2(\Omega)$ so that $v_n = w_n + b\psi$ is in H and converges to w in $L^2(\Omega)$.

Passing to the limit in the definition of $F_1^*(u)$, since ϕ is continuous on $L^2(\Omega)$ we get:

$$F_1^*(u) \geqslant \int_\Omega -\Delta u . w - \phi(w) + (u_\Gamma + I + \int_\Gamma \frac{\partial u}{\partial n} \psi) b - \frac{1}{2} (1 + |\nabla\psi|^2_{L^2}) b^2.$$

This proves one inequality in the claim, but the converse inequality is obvious.

Finally, using standard duality arguments on $L^2(\Omega)$, we get

$$F_1^*(u) = \phi^*(-\Delta u) + \frac{1}{2(1+|\nabla\psi|^2_{L^2})} (u_\Gamma + I + \int_\Gamma \frac{\partial u}{\partial n} \psi)^2,$$

and we use

$$\int_\Gamma \frac{\partial u}{\partial n} \psi = u_\Gamma \int_\Gamma \psi \frac{\partial \psi}{\partial n} + \int_\Omega \Delta u . \psi = \int_\Omega \Delta u . \psi + |\nabla\psi|^2_{L^2} u_\Gamma .$$

In order to determine $\partial F_1(u)$, one now uses the equivalence

$$v \in \partial F_1(u) \Leftrightarrow F_1(u) + F_1^*(v) = (u,v)_H$$

which becomes, successively,

$$\phi(u) + \frac{1}{2}(1 + |\nabla\psi|^2_{L^2})u_\Gamma^2 - Iu_\Gamma + \phi^*(-\Delta v)$$

$$+ \frac{1}{2(1+|\nabla\psi|^2_{L^2})}\{v_\Gamma(1 + |\nabla\psi|^2_{L^2}) + I + \int_\Omega \Delta v.\psi\}^2$$

$$= \int_\Omega \nabla u.\nabla v + u_\Gamma v_\Gamma = \int_\Omega -\Delta v.u + \int_\Gamma \frac{\partial v}{\partial n}\psi u_\Gamma + v_\Gamma u_\Gamma$$

$$= \int_\Omega -\Delta v.u + u_\Gamma(v_\Gamma + \int_\Gamma \frac{\partial v}{\partial n}\psi)$$

$$= \int_\Omega -\Delta v.u + u_\Gamma(v_\Gamma(1 + |\nabla\psi|^2_{L^2}) + \int_\Omega \Delta v\,\psi).$$

Put

$$\phi(u) + \phi^*(-\Delta v) + \int_\Omega \Delta v.u = A$$

and

$$\frac{1}{2}(1 + |\nabla\psi|^2_{L^2})u_\Gamma^2 + \frac{1}{2(1+|\nabla\psi|^2)}\{v_\Gamma(1 + |\nabla\psi|^2_{L^2}) + I + \int_\Omega \Delta v.\psi\}^2$$

$$- Iu_\Gamma - v_\Gamma(u_\Gamma(1 + |\nabla\psi|^2_{L^2})) - u_\Gamma \int_\Omega \Delta v.\psi = B;$$

thus we have $A + B = 0$.

But A is always nonnegative by definition of the convex conjugate and

$$B = \frac{1}{2(1+|\nabla\psi|^2_{L^2})}\{(1 + |\nabla\psi|^2_{L^2})u_\Gamma - v_\Gamma(1+|\nabla\psi|^2_{L^2}) - I - \int_\Omega \Delta v.\psi\}^2$$

therefore $A = B = 0$, which implies that

$$-\Delta v \in \partial\phi(u)$$

and

$$v_\Gamma + \frac{1}{1+|\nabla\psi|^2_{L^2}}(I + \int_\Omega \Delta v.\psi) = u_\Gamma.$$

To determine $E_2(v)$ we write $v = \tilde{v} + v_\Gamma\psi$, so that

$$|\nabla v|^2_{L^2} = |\nabla\tilde{v}|^2_{L^2} + v_\Gamma^2|\nabla\psi|^2_{L^2} \text{ and } \Delta v = \Delta\tilde{v}.$$

Now

$$E_2(v) = \phi^*(-\Delta\tilde{v}) - \frac{1}{2}|\nabla\tilde{v}|^2_{L^2}$$

$$+ v_\Gamma(I + \int_\Omega \Delta\tilde{v}.\psi) + \frac{1}{2}\frac{(I + \int_\Omega \Delta\tilde{v}\ .\psi)^2}{1 + |\nabla\psi|^2_{L^2}}.$$

It is apparent now that the duality has decoupled the critical point problem somewhat on $H^1_0(\Omega) \times \mathbb{R}$.

4. A NECESSARY CONDITION

We have

PROPOSITION 3. If v is a critical point of E_1 on H, then $v = \tilde{v} + v_\Gamma\psi$ and

(a) $\tilde{v}$ is a critical point of

$$E_3(u) = \phi^*(-\Delta u) - \frac{1}{2}|\nabla u|^2_{L^2}$$

on the closed affine subspace W of $H^1_0(\Omega)$ defined by

$$0 = I + \int_\Gamma \frac{\partial u}{\partial n}\ .\ \psi(= I + \int_\Omega \Delta u.\psi).$$

(b) v_Γ satisfies

$$\int_\Omega \partial\phi(\tilde{v} + v_\Gamma\psi).\psi \ni I$$

(this being understood as: there exists z in $\partial\phi(\tilde{v} + v_\Gamma\psi)$ such that $\int_\Omega z.\psi = I$).

Remark 4. Since $\phi^*(-\Delta u) = +\infty$ if $-\Delta u$ is not in $L^2(\Omega)$, one can modify the definition of W slightly:

$$W = \{u \in H_0^1(\Omega); -\Delta u \in L^2(\Omega), I + \int_\Omega \Delta u.\psi = 0\}.$$

Proof. We consider a critical point v of E_1 as a critical point of E_2 thanks to Toland's duality and Remark 3.

Consequently, writing v as $\tilde{v} + v_\Gamma \psi$, we first conclude that

$$I + \int_\Omega \Delta u.\psi = 0$$

and that $\tilde{v}$ is a critical point of E_3 on W, so (a) is proved. Part (b) is obvious from (P).

The interesting question is now whether the converse of Proposition 3 holds. At this point, one should notice that E_2 is an 'augmented Lagrangian' for the nonconvex critical point problem defined by E_3, v_Γ being the Lagrange multiplier associated with the affine constraint

$$I + \int_\Omega \Delta u.\psi = 0.$$

Unfortunately E_3 is neither concave nor convex on $H_0^1(\Omega)$ and, through ϕ^*, may have other implicit constraints, so a straightforward answer to this question does not exist. We shall return to this point in Section 6.

From now on we call reduced problem, the problem of finding critical points for E_3 on W.

5. DUALITY FOR THE REDUCED PROBLEM

Upon a change of variable ($\rho = -\Delta u$) the reduced problem for the model case corresponds to the variational approach of Berestycki and Brézis [1]. In this case it was shown in [3] that applying Toland's duality again led to Temam's variational formulation (in [9]). Here we do the same in the general setting.

On $H_0^1(\Omega)$ one writes

$$E_3(u) = \phi^*(-\Delta u) + I_W(u) - \frac{1}{2} |\nabla u|^2_{L^2}$$

where I_W is the so-called indicator function of $W(I_W(u) = 0$ if $u \in W$, $+\infty$ otherwise).

On $H_0^1(\Omega)$ we take $|\nabla u|_{L^2(\Omega)}$ as an equivalent norm and set

$$G_2(u) = \frac{1}{2}|\nabla u|^2_{L^2}, \quad F_2(u) = \phi^*(-\Delta u) + I_W(u).$$

Here again, G_2 is self-conjugate on $H_0^1(\Omega)$ and therefore the solutions of the reduced problem are the critical points on $H_0^1(\Omega)$ of

$$E_4(u) = G_2^*(u) - F_2^*(u) = G_2(u) - F_2^*(u).$$

The determination of F_2^* is straightforward at the beginning since

$$F_2^*(u) = \sup_{\substack{v \in W \\ \Delta v \in L^2}} \{\int_\Omega \nabla u.\nabla v - \phi^*(-\Delta v)\}$$

$$= \sup_{\substack{v \in H_0^1(\Omega) \\ I + \int_\Omega \Delta v.\psi = 0}} \{\int_\Omega -\Delta v.u - \phi^*(-\Delta v)\}.$$

Since $-\Delta v$ can be any θ in $L^2(\Omega)$, we get

$$F_2^*(u) = \sup_{\substack{\theta \in L^2(\Omega) \\ \int_\Omega \theta.\psi = I}} \{\int_\Omega u.\theta - \phi^*(\theta)\}$$

so that F_2^* is the $L^2(\Omega)$ conjugate of

$$\theta \mapsto \phi^*(\theta) + I_{(\int_\Omega \theta\psi = I)}.$$

Classical results on convex conjugate functions (e.g. [5]) imply that F_2^* is the lower-semi-continuous hull of the so-called inf-convolution j of the conjugates of each term in the sum.

Clearly $(\phi^*)^* = \phi$ and the conjugate of

$$I_{\{\int_\Omega \theta\psi = I\}} \text{ is given by } \begin{cases} -cI \text{ if } u = -c\psi \text{ for some c in } \mathbb{R} \\ +\infty \text{ otherwise.} \end{cases}$$

Consequently

$$j(u) = \inf_{c \in \mathbb{R}} \{\phi(u + c\psi) - cI\}.$$

There is no simple condition which guarantees that j is itself l.s.c., except for a sufficient one which also implies that j is exact, i.e., the infimum in the definition of j is actually achieved:

(H1) $$\lim_{|c| \to +\infty} \{\phi(u + c\psi) - cI\}$$

$$= + \infty \text{ uniformly for } u \text{ in compact sets of } L^2(\Omega).$$

Actually, under (H1), the values of c, denoted c(u), for which the infimum is reached, are characterized by

$$\int_\Omega \partial\phi(u + c(u)\psi) \cdot \psi \ni I.$$

Hypothesis (H1) can be strengthened into a more amenable form which only involves the asymptotic behaviour of ϕ in the direction ψ:

(H2) $$\lim_{c \to \pm\infty} \frac{\phi(c\psi)}{|c|} > \pm I.$$

(Note that $\lim_{c \to \pm \infty} \frac{\phi(u+c\psi)}{|c|}$ is independent of u.)

In a sense (H2) is almost necessary for existence of a solution for (P) since for such a solution $u = \tilde{u} + u_\Gamma \psi$ one has

$$\int_\Omega \partial\phi(\tilde{u} + u_\Gamma \psi) \cdot \psi \ni I$$

so that

$$\lim_{c \to \pm \infty} \frac{\phi(\tilde{u} + c\psi)}{|c|} \geqslant I.$$

Under (H1) E_4 can be written as

$$E_4(u) = \frac{1}{2} |\nabla u|^2_{L^2} - \phi(u + c(u)\psi) + c(u)I$$

which, for the model problem, is Temam's formulation:

$$E_4(u) = E_1(u + c(u)\psi),$$

so that the critical points of E_4 are associated with the critical points of the restriction of E_1 to the graph of the 'mapping' $u \mapsto c(u)$ (possibly multi-valued!) also defined by the 'equation'

$$I \in \int_\Omega \partial\phi(v)\cdot\psi.$$

Remark 5. In the case where $\partial\phi$ is the projection P_K on a closed convex subset K of $L^2(\Omega)$ (in the model case K is the positive cone of $L^2(\Omega)$), and provided (H1) is satisfied, j has a simple expression which does not explicitly involve c(u). One can check the following (see [2]):

$$\phi(u) = \int_\Omega P_K u \left(u - \frac{1}{2} P_K u\right)$$

$$\phi^*(u) = \frac{1}{2} |u|^2_{L^2} + I_K(u),$$

so j is the convex conjugate of

$$\phi^* + I_{\tilde{K}}, \text{ where } \tilde{K} = K \cap \{u \in L^2(\Omega), \int_\Omega u\psi = I\},$$

and hypothesis (H1) implies that $\tilde{K}$ is not empty. Therefore

$$j(u) = \int_\Omega \left(u - \frac{1}{2} P_{\tilde{K}} u\right) P_{\tilde{K}} u.$$

6. AN EXISTENCE RESULT

Due to the Lagrangian structure of E_2 with respect to E_3, a critical point of E_3 is one for E_2 provided the Lagrange multiplier exists. It turns out that (H1) is just enough to guarantee

PROPOSITION 4. Under (H1), if u is a critical point for E_3 on $H_0^1(\Omega)$, then $u + c(u)\psi$ is a critical point for E_1 on H, and therefore a solution of (P).

Proof. Since ∂G_2 is the identity on H_0^1, u is a critical point of E_3 if and only if $u \in \partial F_2(u)$ or, equivalently,

$$F_2(u) + F_2^*(u) = |u|^2_{H_0^1} = \int_\Omega |\nabla u|^2.$$

This, explicitly, means:

$$-\Delta u \in L^2(\Omega), \quad I + \int_\Omega \Delta u.\psi = 0$$

and

$$\phi^*(-\Delta u) + \phi(u + c(u)\psi) - c(u)I = \int_\Omega |\nabla u|^2 = \int_\Omega -\Delta u.u.$$

Combining the above equalities one gets

$$\phi^*(-\Delta u) + \phi(u + c(u)\psi) = \int_\Omega -\Delta u\ (u + c(u)\psi),$$

which is another way of saying

$$-\Delta u \in \partial\phi(u + c(u)\psi).$$

Since ψ is harmonic, $-\Delta u = -\Delta(u + c(u)\psi)$ and this completes the proof.

Remark 6. (H1) is a natural condition to introduce since we do not know a priori, which are the critical points u for E_1, but we know that for the values of $\tilde{u}$ associated with such u, f is l.s.c. and exact.

It is easier, it seems to find critical points for E_3 than for E_4 because the constraint is affine. We shall only give a simple sufficient condition:

(H3) There exist two real numbers α and β such that

$$\alpha > \frac{1}{2} \text{ and } \forall u \in W \quad \phi^*(-\Delta u) \geqslant \alpha|\nabla u|^2_{L^2} - \beta.$$

PROPOSITION 5. Under (H1) and (H3), (P) has at least one solution, which corresponds to the absolute minimum of E_3 on W. This solution has been called the variational solution (for the model problem) (cf. [1], [8], [9]).

Proof. By (H3) E_3 is bounded below on W since

$$E_3(u) \geqslant (\alpha - \frac{1}{2})\ |\nabla u|^2_{L^2} - \beta.$$

Let U_n be a minimizing sequence. From the definition of E_3 we get that ∇u_n is bounded in $L^2(\Omega)$ and $\phi^*(-\Delta u_n)$ is bounded also. Since ϕ is bounded on a neighbourhood of the origin in $L^2(\Omega)$, say by C_ε on the ball of radius ε, by duality one gets

$$\phi^*(v) \geqslant \varepsilon|v|_{L^2} - C_\varepsilon.$$

Consequently, $-\Delta u_n$ is bounded in $L^2(\Omega)$, so the u_n are strongly precompact in $H_0^1(\Omega)$ and any weak cluster point of the sequence (u_n) in $H^2(\Omega)$ achieves the minimum of E_3 on W (because ϕ^* is weakly l.s.c. on $L^2(\Omega)$).

We introduce the following hypotheses which will be used to give sufficient conditions implying (H3).

(H4) $\exists t$ and $s \in [1, +\infty]$, $\exists\alpha > 0$:

if $s = +\infty$, $j(u)$ is bounded above on $\{u \in L^2(\Omega) \mid |u|_{L^t(\Omega)} \leqslant \alpha\}$

if $s < +\infty$, $j(u) - \alpha|u|^s_{L^t(\Omega)}$ is bounded above on $L^2(\Omega)$.

(H5) $\exists p,q \in [1, +\infty]$:

$\forall a > 0$, $\exists Ca \in \mathbb{R}$ with

$q = +\infty$ $\phi(v) \leqslant C_a$ on $\{v \in L^2(\Omega);\ |v|_{L^p(\Omega)} \leqslant \frac{1}{a}\}$

$q < +\infty$ $\phi(v) \leqslant a|v|^q_{L^p(\Omega)} + C_a.$

Notice that since $j \leqslant \phi$, (H5) with p,q implies (H4) with $t \geqslant p$, $s \geqslant q$.

PROPOSITION 6. Each of the following conditions implies (H3) (hence together with (H1) or (H2) the existence of a solution for (P)):

(i) (H4) with $t \leqslant \frac{2N}{N-2}$ ($t < +\infty$ if $N = 2$) and $s < 2$.

(ii) (H5) with $p \leqslant \frac{2N}{N-2}$ ($p < +\infty$ if $N = 2$) and $q \leqslant 2$.

(iii) (H4) and (H5) with $p < \frac{2N}{N-2}$, $q > 2$, $t \geqslant \frac{2N}{N-2}$ and

$$\frac{1}{s}\left(\frac{1}{p} + \frac{1}{N} - \frac{1}{2}\right) + \frac{1}{q}\left(\frac{1}{2} - \frac{1}{N} - \frac{1}{t}\right) \geqslant \frac{1}{2}\left(\frac{1}{p} - \frac{1}{t}\right).$$

(iv) (H4) and (H5) with $p > \frac{2N}{N-2}$, $t \leqslant \frac{2N}{N-2}$, $s > 2$ and

$$\frac{1}{s}\left(\frac{1}{2} - \frac{1}{N} - \frac{1}{p}\right) + \frac{1}{q}\left(\frac{1}{t} + \frac{1}{N} - \frac{1}{2}\right) \geqslant \frac{1}{2}\left(\frac{1}{t} - \frac{1}{p}\right).$$

Proof. By duality, (H4) is equivalent to

$(\tilde{H}4)$ $\exists\tilde{\alpha},\tilde{\beta}$ with $|\rho|_{L^{t'}} \leq \tilde{\alpha}([\phi^*(\rho)]^+)^{1/s'} + \tilde{\beta}$

whenever $\int_\Omega \rho\psi = I$ and (H5) is equivalent to

$(\tilde{H}5)$ $\forall\varepsilon > 0$ $\exists C_\varepsilon \in \mathbb{R}$ such that:

$$|\rho|_{L^{p'}(\Omega)} \leq \varepsilon([\phi^*(\rho)]^+)^{1/q'} + C_\varepsilon.$$

Now (i) and (ii) trivially imply (H3).

Let us consider (iii) and (iv). By Sobolev's imbedding theorem one has for u in $H_0^1(\Omega) \cap H^2(\Omega)$:

$$|\nabla u|^2_{L^2(\Omega)} = \int_\Omega -\Delta u.u \leq C\ |-\Delta u|^2_{L^r},$$

where C depends only upon Ω and $r = 2N/(N+2)$. By interpolation one then has

$$|\nabla u|^2_{L^2(\Omega)} \leq C\ |\Delta u|^{2(1-\theta)}_{L^{t'}}\ |\Delta u|^{2\theta}_{L^{p'}}$$

with

$$\frac{\theta}{p'} + \frac{1-\theta}{t'} = \frac{1}{r} = \frac{1}{2} + \frac{1}{N} \quad (\theta \in [0,1]).$$

Making use of $(\tilde{H}4)$ and $(\tilde{H}5)$, one deduces that

$$|\nabla u|^2_{L^2} \leq C'\varepsilon + \varepsilon^{2\theta}\ C_2\ ([\phi^*(-\Delta u)]^+)^{2(\frac{1-\theta}{s'} + \frac{\theta}{q'})}.$$

Both conditions (iii) and (iv) guarantee that $\theta > 0$ and $2(\frac{1-\theta}{s'} + \frac{\theta}{q'}) \leq 1$.

7. FINAL REMARKS

Hypothesis (H4) seems hard to check except sometimes in its dual form $(\tilde{H}4)$.

This is definitely the case if $\phi^*(\rho) < +\infty$ implies $\rho \geq 0$ a.e. in Ω and if ψ is bounded below away from zero in Ω. Then

$$\int_\Omega \rho\psi = I \quad \text{implies} \quad |\rho|_{L^1(\Omega)} \text{ bounded},$$

so (H4) is satisfied with $t = +\infty$ and $s = 1$.

• In the case (iii), for $t = +\infty$, and $s = 1$, the condition becomes

$$\frac{1}{p} + \frac{1}{q}\left(1 - \frac{2}{N}\right) \geqslant 1 - \frac{2}{N}.$$

If in particular $p = q$ (which happens if ϕ is local, i.e. derives from a convex function of the real variable), this condition reduces to

$$p = q \leqslant \frac{2(N-1)}{N-2} ,$$

which is the growth condition obtained in [1].

If, on the contrary, ϕ is not local but can be compared to a power of the L^2 norm (i.e. $p = 2$), then the condition on q becomes

$$\frac{1}{q}\left(1 - \frac{2}{N}\right) \geqslant \frac{1}{2} - \frac{2}{N}$$

i.e.

$N \leqslant 3$ no condition on q

$N = 4$ $q < +\infty$

$N \geqslant 5$ $q \leqslant 2 + \frac{4}{N-4}.$

For example $q = 2 + \varepsilon$ is always acceptable, and this corresponds to the result in [2], where $\partial\phi$ is the projection P_K on the convex set K, in which case $p = 2$, $q > 2$ guarantee (H5).

REFERENCES

[1] H. Berestycki and H. Brézis, On a free boundary problem arising in plasma physics. Non-linear Analysis, T.M.A. 4 (3), 415-436 (1980).

[2] R. Cipolatti, Sur un problème de valeur propre non linéaire, C.R.A.S. Paris 293 (I), 455-458 (1981).

[3] A. Damlamian, Application de la dualité non convexe à un problème non linéaire à frontière libre. C.R.A.S. Paris 286(A), 153-155 (1978).

[4] B. Heron and M. Sermange, Non convex methods for computing free boundary equilibria of axially symmetric plasmas, Rapport INRIA No. 108, Rocquencourt, France (1981).

[5] J.J. Moreau, Fonctionnelles convexes; Séminaire J. Leray, Collège de France (1966).

[6] J.P. Puel, A free boundary non-linear eigenvalue problem, in Proc. of the Int. Sympos. on continuous mechanics and P.D.E., G. M. de la Penha and L.A. Medeiros (eds.), Rio de Janeiro (1978).

[7] D.G. Schaeffer, Non uniqueness in the equilibrium shape of a confined plasma. Com. Part. Diff. Equ. 2(6), 587-600 (1977).

[8] R. Temam, A non linear eigenvalue problem : the shape at equilibrium of a confined plasma. Arch. Rat. Mec. Anal. 60, 51 (1976).

[9] R. Temam, Remarks on a free boundary value problem arising in plasma physics. Comm. in P.D.E. 2(6), 563 (1977).

[10] J.F. Toland, Duality in non convex optimization. J. of Math. and Appl. 66, 339 (1978).

[11] J.F. Toland, A duality principle for non convex optimization and the calculus of variations. Arch. Rat. Mec. Anal. 71, 41 (1979).

R. Cipolatti
Département de Mathématiques,
Université Paris-Sud F 91405,
Orsay Cedex
France

and

Instituto de Matematica,
UFRJ,
Rio de Janeiro,
Brazil

A. Damlamian
Centre de Mathématiques,
Ecole Polytechnique F 91128,
Palaiseau Cedex
France

J-M CORON

Nontrivial periodic solutions of a nonlinear wave equation

0. INTRODUCTION

Let $g:R \to R$ be a continuous function such that $g(0) = 0$. We seek nontrivial T-periodic (in t) solutions of the following nonlinear wave equation

$$u_{tt} - u_{xx} + g(u) = 0 \tag{1}$$

under Dirichlet boundary conditions. By nontrivial we mean that $g(u(x,t)) \neq 0$ on a set (x,t) of positive measure.

In Section 1, we assume that u tends to zero as x tends to $+\infty$; we prove that if $g'(0) < (2\pi/T)^2$, u is independent of t.

In Section 2, we seek u such that $u(\pi,t) = u(0,t) = 0$.

1. u TENDS TO ZERO AS x TENDS TO + ∞

We assume that $g \in C^2(R,R)$ and that $u \in C^2(R^2,R)$ is a T-periodic solution of (1) which satisfies the following conditions:

$$\int_0^{+\infty} dx \int_0^T |u(x,t)|\, dt < +\infty \tag{2}$$

$$\lim_{x\to+\infty} \int_0^T (u_t^2(x,t) + u_x^2(x,t))\, dt = 0. \tag{3}$$

We prove

THEOREM 1 (see [9])

If

$$g'(0) < \left(\frac{2\pi}{T}\right)^2, \tag{4}$$

then u is independent of t.

From Theorem 1 one can easily deduce that if u satisfies (1), (2) and (3) we have:

COROLLARY 1. If $g(u) = u^3$ then $u \equiv 0$.

COROLLARY 2. If $g(u) = \sin u$, $T < 2\pi$ and $\lim_{x\to-\infty} \max_{t\in[0,T]} |u(x,t)| = 0$, then $u \equiv 0$.

REMARK. When $g(u) = \sin u$ and $T > 2\pi$ there exists a solution u of (1), T periodic in t, dependent on t, such that (2) and (3) are satisfied (see [12]):

$$u(x,t) = 4 \arctan \frac{\varepsilon \sin \frac{2\pi t}{T}}{\cosh \frac{2\pi\varepsilon x}{T}}$$

where $\varepsilon = \left[\left(\frac{T}{2\pi}\right)^2 - 1\right]^{1/2}$.

PROOF OF THEOREM 1. First we remark that it follows easily from (2) and (3) that:

$$\lim_{x\to+\infty} \max_{t\in[0,T]} |u(x,t)| = 0. \tag{5}$$

Let

$$c(x) = \frac{1}{T}\int_0^T u(x,t)\, dt$$

and

$$v(x,t) = u(x,t) - c(x);$$

we have

$$\int_0^T v(x,t)\, dt = 0. \tag{6}$$

It is easy to see that there exists a continuous function h from R^2 into R such that

$$g(v+c) = g(v) + g(c) + cvh(v,c). \tag{7}$$

Let

$$G(x) = \int_0^x g(s)\, ds.$$

We multiply (1) by v_x and integrate from 0 to T (x is fixed); we have

$$\frac{d}{dx}\{\int_0^T\left(\frac{v_t^2}{2}+\frac{v_x^2}{2}-G(v)\right)dt\} = c(x)\int_0^T vv_x h(c,v)\,dt. \tag{8}$$

Let

$$m(x) = \frac{1}{2}\int_0^T (v^2(x,t)+v_x^2(x,t))\,dt.$$

Using (6), we have

$$\int_0^T v_t^2(x,\tau)\,d\tau \geqslant \left(\frac{2\pi}{T}\right)^2\int_0^T v^2(x,\tau)\,d\tau. \tag{9}$$

We integrate (8) from x to $+\infty$ and use (3), (4), (5) and (9); there exist $r > 0$ and $x_0 > 0$ such that

$$m(x) \leqslant r\int_x^{+\infty} m(y)|c(y)|\,dy \quad \forall x > x_0. \tag{10}$$

Using (2) we have

$$\int_0^{+\infty} |c(y)|dy < +\infty. \tag{11}$$

It follows easily from (10) and (11) that

$$m(x) = 0 \text{ for } x \text{ large enough}$$

and then u is independent of t.

2. $\underline{u(0,t) = u(\pi,t) = 0}$

This section is divided into two parts:

(a) We assume g is nondecreasing.
(b) We make no assumption of monotony about g.

(a) g is nondecreasing

Let H be $L^2((0,\pi)\times(0,T))$. Let A be the linear operator defined by

$$D(A) = \{u \in C^2([0,\pi]\times[0,T])/u(0,t) = u(\pi,t) = 0 \quad \forall t \in [0,T]$$
$$u(x,0) = u(x,T) \text{ and } u_t(x,0) = u_t(x,T) \quad \forall x \in [0,\pi]\}$$

$$Au = u_{tt} - u_{xx}.$$

Let A be the adjoint of A in H. We seek u in H such that

$$Au + g(u) = 0 \quad \text{and} \quad g(u) \not\equiv 0. \tag{12}$$

Let us recall some generalities about A:

A is a self-adjoint operator; the eigenvalues of A are $j^2 - ((2\pi/T)k)^2$, $j = 1,2,3,\ldots$ and $k = 0,1,2,\ldots$, and the corresponding eigenfunctions are

$$\sin jx \sin\left(\frac{2\pi kt}{T}\right) \text{ and } \sin jx \cos\left(\frac{2\pi kt}{T}\right).$$

Moreover if (T/π) is a rational number, R(A) is closed and A^{-1} (from R(A) into itself) is compact.

In all that follows we shall assume that T/π is a rational number. We denote by $\lambda_{-1}(T)$ the first negative eigenvalue.

We have the following theorems:

THEOREM 2. If

$$\lim_{u\to 0} \frac{g(u)}{u} > |\lambda_{-1}(T)|$$

and

$$\overline{\lim_{|u|\to+\infty}} \frac{g(u)}{u} < |\lambda_{-1}(T)|$$

then (12) has at least one solution.

(For a proof of Theorem 2 see [7]. For theorems about multiple solutions of (12) see Amann and Zehnder [1].)

THEOREM 3. (Brézis and Coron [4]): If

$$\lim_{|u|\to+\infty} \frac{g(u)}{u} = 0 \quad \text{and} \quad g \not\equiv 0, \tag{13}$$

then for T large enough (and rational multiple of π), (12) has at least a solution.

We just sketch a proof of Theorem 3 (for a complete proof see [4]).

Instead of (12) we solve:

$$A^{-1}v + (g+\varepsilon)^{-1}(v) \in N(A) \tag{14}$$

where $N(A)$ is the kernel of A and $\varepsilon > 0$. To solve (13) we minimize $F(v)$ where

$$F_\varepsilon(v) = \tfrac{1}{2}(A^{-1}v,v) + \int_0^\pi \int_0^T (G_\varepsilon)^*(v)\, dx\, dt,\ v \in R(A)$$

($(G_\varepsilon)^*$ is the conjugate of G_ε, where $G_\varepsilon(x) = \frac{\varepsilon}{2}|x|^2 + \int_0^x g(s)\, ds$).

One can prove that for T large enough

$$\min_{v \in R(A)} F_\varepsilon(v) < 0 \qquad \forall \varepsilon > 0.$$

With this minimization we find a solution of

$$Au_\varepsilon + \varepsilon u_\varepsilon + g(u_\varepsilon) = 0, \qquad g(u_\varepsilon) \not\equiv 0$$

after letting ε go to zero.

THEOREM 4. If

$$\lim_{u \to +\infty} \frac{g(u)}{u} = +\infty$$

and if there exists constants $\alpha > 0$ and c such that

$$\tfrac{1}{2}\, tg(t) - G(t) \geqslant \alpha|g(t)| - c \text{ for all } t$$

then for all rational multiples T of π, (12) has at least one solution.

This theorem, with slightly stronger assumptions, is due to Rabinowitz [13]. For the assumptions of Theorem 4 and a different proof see Brézis et al. [5]. See also Benci and Fortunato [3] for many complements.

THEOREM 5. If

$$\lim_{u \to 0} \frac{g(u)}{u} = 0 \qquad \lim_{|u| \to +\infty} \frac{g(u)}{u} = 0 \qquad g \not\equiv 0,$$

then, for T large enough (and a rational multiple of π), (12) has at least two solutions u and v such that

$$\int_0^\pi \int_0^T (\tfrac{1}{2} Au.u + G(u))\, dx\, dt \leqslant 0 < \int_0^\pi \int_0^T (\tfrac{1}{2} Av.v + G(v))\, dx\, dt.$$

One can find a proof of Theorem 5 in [8]: u is obtained by use of the 'mountain pass' theorem of Ambrosetti and Rabinowitz [2] and a duality argument; v is obtained by minimization and a duality argument. This approach has been stimulated by Clarke and Ekeland [6] and Ekeland [11].

(b) No assumption of monotony about g

We begin by a remark to show the use of monotony. If

$$Au_n + g(u_n) \to 0 \text{ in } H \ (n \to +\infty) \tag{15}$$

and

$$u_n \rightharpoonup u \qquad \text{in } H \ (n \to +\infty) \text{ (weakly)} \tag{16}$$

then if g is monotone one can prove, using Minty's trick, that

$$Au + g(u) = 0.$$

But there are functions g and u_n such that (15) and (16) hold and

$$Au + g(u) \neq 0;$$

however, if we assume that $u_n \in R(A)$ and (15) and (16) hold, then

$$Au + g(u) = 0$$

without assumption of monotony about g.

Then a natural idea to avoid this assumption is to find a subspace invariant by A and g included in R(A). With the help of this subspace the proof is identical to the already known proofs. Let us give an example of such a subspace: if

$$T = 2\pi,$$

let

$$E = \{u \in H \mid u(\pi-x,\pi+t) = u(x,t) \quad \text{a.e. } x \in (0,\pi),\ t \in R\}$$

E is invariant by A and g and is included in R(A).

One can now improve Theorem 4 (for example):

THEOREM 6. If there exists constants r, $\theta \in (0,\frac{1}{2})$, a, b, p such that

$$|s| \leqslant r \Rightarrow 0 < \int_0^s g(\tau)d\tau \leqslant \theta s g(s),$$

$$|g(s)| \leqslant \alpha |s|^p + \beta,$$

then for all rational multiples T of π, and for all real numbers R, there exists a continuous solution of (11) such that

$$\max_{\substack{x \in [0,\pi] \\ t \in [0,T]}} |u(x,t)| \geqslant R.$$

For a proof of Theorem 6 and other theorems without assumption of monotony see [10].

REFERENCES

[1] H. Amann and E. Zehnder, Multiple periodic solutions for a class of nonlinear autonomous wave equations, Houston J. Math. 7(2), 147-174 (1981).

[2] A. Ambrosetti and P. Rabinowitz, Dual variational methods in critical point theory and applications, J. Funct. Anal. 349-381 (1973).

[3] V. Benci and D. Fortunato, The dual method in critical point theory. Multiplicity results for indefinite-functionals (to appear).

[4] H. Brézis and J.M. Coron, Periodic solutions of nonlinear wave equations and Hamiltonian systems, Amer. J. Math. 103 (3), 559-570 (1981).

[5] H. Brézis, J.M. Coron and L. Nirenberg, Free vibrations for a non linear wave equation and a theorem of P. Rabinowitz, Comm. Pure Appl. Math. 33, 667-689 (1980).

[6] F. Clarke and I. Ekeland, Hamiltonian trajectories having prescribed minimal period, Comm. Pure Appl. Math. 33, 103-116 (1980).

[7] J.M. Coron, Résolution de l'équation Au+Bu = f où A est linéaire et B dérive d'un potentiel convexe, Ann. Fac. Sc. Toulouse 1, 215-234 (1979).

[8] J.M. Coron, Solutions périodiques non-triviales d'une équation des ondes, Comm. in P.D.E., 6, 7, 829-848 (1981).

[9] J.M. Coron, Période minimale pour une corde vibrante de longueur infinie, C.R. Acad. Sci. Paris 294, 127-129 (1982).

[10] J.M. Coron, Periodic solutions of a nonlinear wave equation witnout assumption of monotonicity, Math.Ann.262, 273-285 (1983).

[11] I. Ekeland, Periodic solutions of Hamiltonian equations and a theorem of P. Rabinowitz, J. Diff. Eq. 34, 523-534 (1979).

[12] G.L. Lamb, Elements of Soliton Theory, Wiley-Interscience (1980).

[13] P. Rabinowitz, Free vibrations for a semilinear wave equation, Comm. Pure Appl. Math. 31, 31-68 (1978).

Jean-Michel Coron
ENSMP, CAI,
35, Rue Saint-Honoré,
77305 Fontainebleau Cedex,
France

G DÍAZ

On Hamilton–Jacobi inequalities with obstacle

1. INTRODUCTION

$\Omega \subset \mathbb{R}^N$ being a bounded open satisfying the uniform exterior sphere condition, H a continuous function $H:\mathbb{R}^N \to \mathbb{R}$, and f and ψ two given functions we are interested in studying the next first order problem

$$(P) \begin{cases} \max \{u(x) + H(Du(x)) - f(x), u(x) - \psi(x)\} = 0 & x \in \Omega \\ u(x) = 0 & x \in \Gamma = \partial\Omega \end{cases}$$

called the Hamilton-Jacobi inequalities with obstacle ψ.

The main interest in this paper is to prove

THEOREM 1.1. Let $f \in W^{1,\infty}(\Omega)$ and $\psi \in W^{1,\infty}(\Omega)$ such that $\psi|_\Gamma \geq 0$. Assume

there exist two positive constants L and γ such that $H(q) \geq \alpha|q| - \gamma$, for all $|q| > L$, being α some constant verifying $\alpha > \frac{N-1}{R}$ and R the radius of the exterior spheres; (1.1)

there exists a function $v \in W^{2,p}(\Omega) \cap W_0^{1,p}(\Omega)$ $(p > N)$ such that $\Delta v \geq 0$ a.e. in Ω, $v \leq \psi$ in $\bar{\Omega}$ and $v + H(Dv) \leq f$ in Ω. (1.2)

Then there exists a unique function $u \in W_0^{1,\infty}(\Omega)$ verifying (P).

We note that for $\psi \geq 0$ in $\bar{\Omega}$ and $f \geq H(0)$ in Ω the zero function can be considered in place of v.

It would be absurd to claim that our theoretical or practical knowledge of the Hamilton-Jacobi inequality is anywhere near complete. However a good introduction to the Hamilton-Jacobi equation can be found in Benton [1].

In Section 2 we study problems approximating to (P) by means of 'vanishing viscosity' and 'penalization' methods. For each $0 < \nu < 1$ and $0 < \varepsilon < 1$ we consider the problem

$$(P_\varepsilon^\nu)\begin{cases} -\nu\Delta u_\varepsilon^\nu(x) + u_\varepsilon^\nu(x) + H(Du_\varepsilon^\nu(x)) + \beta_\varepsilon(u_\varepsilon^\nu - \hat{\psi}) = f(x), \ x \in \Omega \\ u_\varepsilon^\nu(x) = 0, \quad x \in \Gamma \end{cases}$$

where $\hat{\psi}$ is a smooth approximation of ψ such that $\psi \leqslant \hat{\psi}$, and $\beta_\varepsilon(t) = (1/\varepsilon)\beta(t)$ for a smooth function β such that $\beta(t) = 0$ if $t \leqslant 0$ and $0 < \beta'(t) \leqslant 1$ if $t > 0$.

The idea is to obtain precompactness of the family

$$\{u_\varepsilon^\nu \mid 0 < \nu < 1, 0 < \varepsilon\}$$

of the solutions of (P_ε^ν), and to construct a function u which is a candidate solution of (P).

In Section 3 we prove that u is the unique function satisfying (P).

In order to do it we adapt to (P) a notion of solution - called the 'viscosity' solution - recently proposed by Crandall and Lions [3] for which one has strong uniqueness results and stability theorems. We note that if $u \in W_0^{1,\infty}(\Omega)$ is a solution of (P) one has

$$u(x) + H(Du(x)) = f(x) \text{ a.e. } x \in [u < \psi].$$

Then

<u>DEFINITION</u>. Given $f \in C(\Omega)$, $\psi \in C(\bar{\Omega})$ such that $\psi_{|\Gamma} \geqslant 0$ and $u \in C_0(\bar{\Omega})$ such that $u \leqslant \psi$ in $\bar{\Omega}$. We will say u is a viscosity subsolution of (P) if for every $\phi \in \mathcal{D}(\Omega)^+$ and $k \in \mathbb{R}$ one has

$$\left.\begin{aligned} &E_+(\phi(u-k),\Omega) \neq \emptyset \Rightarrow \exists\, x_0 \in E_+(\phi(u-k),\Omega) \text{ such that} \\ &u(x_0) + H\left(-\frac{(u(x_0)-k)}{\phi(x_0)}\ D\phi(x_0)\right) \leqslant f(x_0). \end{aligned}\right\} \quad (1.3)$$

u is a viscosity supersolution of (P) if for every $\phi \in \mathcal{D}([u < \psi])^+$ and $k \in \mathbb{R}$ one has

$$\left.\begin{aligned} &E_-(\phi(u-k), [u < \psi]) \neq \emptyset \Rightarrow \exists\, x_0 \in E_-(\phi(u-k), [u < \psi]) \text{ such that} \\ &u(x_0) + H\left(-\frac{(u(x_0)-k)}{\phi(x_0)}\ D\phi(x_0)\right) \geqslant f(x_0). \end{aligned}\right\} \quad (1.4)$$

Clearly u is a viscosity solution of (P) if (1.3) and (1.4) hold.

(We denote by $E_+(g,A)$ (respectively $E_-(g,A)$) the interior positive (respectively, negative) extreme set of g in A.)

We note that if $x_0 \in \Omega$ satisfies

$$\phi(u-k)(x_0) = \max_{\Omega} (\phi(u-k))(x) > 0$$

and $\exists D(\phi(u-k))$ at x_0, then

$$H(Du(x_0)) = H\left(- \frac{(u(x_0)-k)}{\phi(x_0)} D\phi(x_0)\right).$$

We refer to Crandall and Lions [3] for general commentaries and consistency results between the viscosity solution and the $W^{1,\infty}$ solution.

The approximation and uniqueness permit us to derive certain properties about the common boundary - the free boundary - to $[u = \psi]$ and $[u < \psi]$, following results analogous to those obtained in Díaz [4].

We remark that the problem has been studied with respect to a conservation law, i.e.

$$H(Du) = \sum_{i=1}^{N} D_i(\phi_i(u)),$$

by Díaz and Veron [5].

2. THE SETTING

Assume Ω with smooth boundary

PROPOSITION 2.1. Let $H : \mathbb{R}^N \to \mathbb{R}$ a continuous differentiable function with bounded gradient and $f \in W^{1,\infty}(\Omega)$. Assume that

there exists a function $v \in W^{2,p}(\Omega) \cap W_0^{1,p}(\Omega)$ (for $p > N$) such that $\Delta v \geqslant 0$ a.e. in Ω, $v \leqslant \psi$ in $\bar{\Omega}$ and $v + H(Dv) - f \leqslant 0$ in Ω (2.1)

there exists two positive constants L and γ such that $H(q) \geqslant \alpha|q| - \gamma$ for $|q| > L$. (2.2)

Then, for each $\nu > 0$, $\varepsilon > 0$ there exists a unique solution $u_\varepsilon^\nu \in W^{3,p}(\Omega)$ of (P_ε^ν) such that

$$\|(u_\varepsilon^\nu - \hat{\psi})^+\|_\infty \leq \beta^{-1}(\varepsilon\|(\nu\Delta\hat{\psi} - \hat{\psi} - H(D\hat{\psi}) + f)^+\|_\infty). \tag{2.3}$$

Proof.

The proof is based on the 'sub- and supersolution' method. By (2.2) and the continuity of H there exists a positive constant γ_1 depending on α, L and γ such that

$$H(q) \geq \alpha|q| - \gamma_1 \text{ for all } q \in \mathbb{R}^N. \tag{2.4}$$

Then, from Lions [6], we derive the existence of a solution $u_\varepsilon^\nu \in W^{3,p}(\Omega)$ of (P_ε^ν) such that $v \leq u_\varepsilon^\nu \leq \bar{v}$ in $\bar{\Omega}$, where $\bar{v} \in W^{3,p}(\Omega)$ is the unique solution of

$$\begin{aligned} -\nu\Delta\bar{v} + \alpha|D\bar{v}| + \bar{v} &= \gamma_1 + f(x), \quad x \in \Omega \\ \bar{v}|_\Gamma &= 0 \end{aligned} \tag{2.5}$$

(since $\alpha|q|$ is a convex function in $q \in \mathbb{R}^N$ and $v + \alpha|Dv| \leq \gamma_1 + f$, the existence of $\bar{v}$ follows from [6]).

The uniqueness is obtained by the maximum principle.

Finally, let $x_0 \in \bar{\Omega}$ such that $(u_\varepsilon^\nu - \hat{\psi})^+(x_0) = \|(u_\varepsilon^\nu - \hat{\psi})^+\|_\infty$. With no loss of generality, we may assume $x_0 \in \Omega$ and $(u_\varepsilon^\nu - \hat{\psi})^+(x_0) > 0$. The maximum principle implies

$$\begin{aligned} \beta_\varepsilon(u_\varepsilon^\nu - \hat{\psi})(x_0) &\leq -\nu\Delta(u_\varepsilon^\nu - \hat{\psi})(x_0) + u_\varepsilon^\nu(x_0) + H(Du_\varepsilon^\nu(x_0)) + \beta_\varepsilon(u_\varepsilon^\nu - \hat{\psi})(x_0) \\ -\hat{\psi}(x_0) - H(D\hat{\psi}(x_0)) &= \nu\Delta\hat{\psi}(x_0) - \hat{\psi}(x_0) - H(D\hat{\psi}(x_0)) + f(x_0) \end{aligned} \tag{2.6}$$

and the monotonicity of β_ε concludes (2.3).

PROPOSITION 2.2. Let $H:\mathbb{R}^N \to \mathbb{R}$ be a continuous function verifying (2.1) and (2.2), and $f \in W^{1,\infty}(\Omega)$; then there exists a unique solution $u_\varepsilon^\nu \in W^{3,p}(\Omega)$ of (P_ε^ν) such that

$$\|(u_\varepsilon^\nu - \hat{\psi})^+\|_\infty \leq \beta^{-1}(\varepsilon\|(\nu\Delta\hat{\psi} - \hat{\psi} - H(D\hat{\psi}) + f)^+\|_\infty) \tag{2.7}$$

$$\|u_\varepsilon^\nu\|_{W^{1,\infty}(\Omega)} \leq \max\{\|v\|_{W^{1,\infty}}, \|\bar{v}\|_{W^{1,\infty}}, \|\hat{\psi}\|_{W^{1,\infty}}, \|f\|_{W^{1,\infty}}\}. \quad (2.8)$$

Proof. Indeed, assume for the moment H as in Proposition 2.1. Then it is clear that from $v \leq u_\varepsilon^\nu \leq \bar{v}$ in $\bar{\Omega}$ we obtain

$$\frac{\partial \bar{v}}{\partial n} \leq \frac{\partial u_\varepsilon^\nu}{\partial n} \leq \frac{\partial v}{\partial n} \text{ on } \Gamma, \quad (2.9)$$

where n is the exterior unitary normal vector to Γ.

With no loss of generality we may assume that there exists i_0 $(1 \leq i_0 \leq N)$ and $x_0 \in \bar{\Omega}$ such that

$$0 < |D_{i_0} u_\varepsilon^\nu (x_0)| = \|D_{i_0} u_\varepsilon^\nu\|_\infty = \max_{1\leq k\leq N} \{\|D_k u_\varepsilon^\nu\|_\infty, \|D_{i_0} \hat{\psi}\|_\infty\}. \quad (2.10)$$

We also assume $x_0 \in \Omega$ because otherwise (2.8) on follows from (2.9). Clearly, one has

$$|D_{i_0} u_\varepsilon^\nu(x_0)|^2 = \|(D_{i_0} u_\varepsilon^\nu)^2\|_\infty = \max_{1\leq k\leq N}\{\|(D_k u_\varepsilon^\nu)^2\|_\infty, \|(D_{i_0}\hat{\psi})^2\|_\infty\}. \quad (2.11)$$

Consider $w = (D_{i_0} u_\varepsilon^\nu)^2$, $D_i w = 2D_{i_0} u_\varepsilon^\nu \cdot D_{i_0 i} u_\varepsilon^\nu$

$$D_{ii} w = 2(D_{i_0 i} u_\varepsilon^\nu)^2 + 2D_{i_0} u_\varepsilon^\nu \cdot D_{i_0 ii} u_\varepsilon^\nu.$$

Then, by Bony's principle [2] and (P_ε^ν) one has

$$0 \leq \lim_{x\to x_0} \text{ess sup}(-\nu\Delta w(x) + \frac{\partial H}{\partial q_i}(Du_\varepsilon^\nu)D_i w(x)) = \lim_{x\to x_0} \text{ess sup}(-2\nu(D_{i_0 i} u_\varepsilon^\nu(x))^2$$

$$- 2\nu D_{i_0} u_\varepsilon^\nu(x) D_{i_0 ii} u_\varepsilon^\nu(x) + 2\frac{\partial H}{\partial q_i}(-) D_{i_0} u_\varepsilon^\nu D_{i_0 i} u_\varepsilon^\nu)$$

$$\leq -2(D_{i_0} u_\varepsilon^\nu(x_0))^2 + 2|D_{i_0} u_\varepsilon^\nu(x_0)|\cdot\|D_{i_0} f\|_\infty$$

$$-2 D_{i_0} u_\varepsilon^\nu(x_0)\ \beta_\varepsilon'(u_\varepsilon^\nu - \psi)(x_0) D_{i_0}(u_\varepsilon^\nu - \hat{\psi})(x_0) \quad (2.12)$$

but

$$2 D_{i_0} u_\varepsilon^\nu(x_0)(D_{i_0} u_\varepsilon^\nu(x_0) - D_{i_0}\hat{\psi}(x_0)) \geq (D_{i_0} u_\varepsilon^\nu(x_0))^2 - (D_{i_0}\hat{\psi}(x_0))^2 \geq 0$$

hence

$$|D_{i_0} u_\varepsilon^\nu (x_0)| \leq \|D_{i_0} f\|_\infty. \quad (2.13)$$

For the general case we may repeat the above argument for a smooth approximation, $\hat{H}$, of H such that $\hat{H} \leq H$. (We note that the inequality (2.8) is uniform in $\hat{H}$.)

COROLLARY 2.3. Under assumptions of Proposition 2.2, for each $\varepsilon > 0$ there exists a sequence of ν (we again denote the sequence by ν) and a function $u_\varepsilon \in W^{1,\infty}(\Omega)$ such that

$$u_\varepsilon^\nu \to u_\varepsilon \text{ uniformly in } \bar{\Omega}$$
$$u_\varepsilon^\nu \rightharpoonup u_\varepsilon \text{ weakly in } W^{1,\infty}(\Omega). \tag{2.14}$$

There exists a sequence of ε (we again denote the sequence by ε) and a function $u \in W^{1,\infty}(\Omega)$ such that

$$u_\varepsilon \to u \text{ uniformly on } \bar{\Omega}$$
$$u_\varepsilon \rightharpoonup u \text{ weakly in } W^{1,\infty}(\Omega) \tag{2.15}$$

and

$$u(x) \leq \psi(x), \quad \forall x \in \bar{\Omega}. \tag{2.16}$$

Proof. If we prove that $\max\{\|\bar{v}\|_\infty, \|D\bar{v}\|_{\infty,\Gamma}\} \leq c$, for some constant independent of ν, an argument as in (2.10)-(2.12) yields $\|\bar{v}\|_{W^{1,\infty}(\Omega)} \leq c$, for some constant independent of ν, and then the corollary follows easily from (2.7) and (2.8).

By using the maximum principle on (2.5) we obtain $\|\bar{v}\|_\infty \leq \|f\|_\infty + \gamma_1$. On the other hand, for any, and fixed, $x_0 \in \Gamma$ there exists a ball $B = B(y_0,R)$ such that $\{x_0\} = \bar{B} \cap \bar{\Omega}$. (We recall that Ω satisfies a uniform exterior sphere condition.) Then the function

$$\tilde{\psi}(r) = \frac{\gamma_1 + \|f\|_\infty}{\alpha - \frac{N-1}{R}} r, \quad r = |x - y_0| - R, \qquad x \in \bar{\Omega} \tag{2.17}$$

verifies $\alpha\tilde{\psi}'(r) - \gamma_1 - \|f\|_\infty = \frac{N-1}{R}\tilde{\psi}'(r) \geq \nu\frac{N-1}{R}\tilde{\psi}'(r)$, for any $0 < \nu < 1$ and hence

$$-\nu\tilde{\psi}''(r) - \nu\frac{N-1}{R}\tilde{\psi}'(r) + \alpha\tilde{\psi}'(r) + \tilde{\psi}(r) \geq \gamma_1 + f.$$

Hence one has

$$-\nu\Delta\tilde{\psi}(r) + \alpha|D\tilde{\psi}(r)| + \tilde{\psi}(r) \geqslant \gamma_1 + f \text{ in } \Omega$$
$$\tilde{\psi}(r) \geqslant 0 \qquad \text{in } \bar{\Omega} \tag{2.18}$$

and by comparing with (2.5) we conclude

$$|D\bar{v}(x_0)| \leqslant \max\left\{|Dv(x_0)|, \ \frac{\gamma_1 + ||f||_\infty}{\alpha - \frac{N-1}{R}}\right\}.$$

Remark 2.1. By the maximum principle we deduce that $0 < \varepsilon < \varepsilon'$ implies $u_\varepsilon^\nu \leqslant u_{\varepsilon'}^\nu$ on $\bar{\Omega}$, and therefore $u_\varepsilon \leqslant u_{\varepsilon'}$ on $\bar{\Omega}$.

3. THE VISCOSITY SOLUTION OF (P)

By means of the viscosity solution notion we will prove, in this section, that u is the unique $W_0^{1,\infty}$ solution of (P).

LEMMA 3.1. For each $\varepsilon > 0$, the function u_ε constructed in Corollary 2.3 verifies for every $\phi \in \mathcal{D}(\Omega)^+$ and $k \in \mathbb{R}$

$$E_+(\phi(u_\varepsilon-k),\Omega) \neq \emptyset \Rightarrow \exists\, x_0 \in E_+(\phi(u_\varepsilon-k),\Omega) \text{ such that}$$
$$u_\varepsilon(x_0) + H\left(-\frac{u_\varepsilon(x_0)-k)}{\phi(x_0)} D\phi(x_0)\right) \leqslant f(x_0), \tag{3.1}$$

and

$$E_-(\phi(u_\varepsilon-k),\Omega) \neq \emptyset \Rightarrow \exists x_0 \in E_-(\phi(u_\varepsilon-k),\Omega) \text{ such that}$$
$$u_\varepsilon(x_0) + H\left(-\frac{(u_\varepsilon(x_0)-k)}{\phi(x_0)} D\phi(x_0)\right) + \beta_\varepsilon(u_\varepsilon(x_0) - \hat{\psi}(x_0)) \geqslant f(x_0) \tag{3.2}$$

Furthermore, if ε is small enough, then for every $\phi \in \mathcal{D}([u < \psi])^+$ and $k \in \mathbb{R}$

$$E_-(\phi(u_\varepsilon-k), [u < \psi]) \neq \emptyset \Rightarrow \exists x_0 \in E_-(\phi(u_\varepsilon-k), [u<\psi]) \text{ such that}$$
$$u_\varepsilon(x_0) + H\left(-\frac{(u_\varepsilon(x_0)-k)}{\phi(x_0)} D\phi(x_0)\right) \geqslant f(x_0), \tag{3.3}$$

where u is the function constructed in Corollary 2.3.

Proof. Indeed, fix $\phi \in \mathcal{D}(\Omega)^+$, $k \in \mathbb{R}$ and assume $E_+(\phi(u_\varepsilon - k),\Omega) \neq \emptyset$. Then for large n there exists $x_n \in E_+(\phi(u_\varepsilon^{\nu_n} - k),\Omega)$ (being $\{\nu_n\}_n \downarrow 0$), and we may assume, passing to a subsequence if necessary $x_n \to x_0 \in E_+(\phi(u_\varepsilon - k),\Omega)$.

On the other hand, a simple computation on supp ϕ in (P_ε^ν) yields

$$
\begin{aligned}
0 &= \frac{1}{\phi(x)}\,\phi(x)(-\nu\Delta u_\varepsilon^\nu(x)) + u_\varepsilon^\nu(x) + H(Du_\varepsilon^\nu(x)) - f(x) + \beta_\varepsilon(u_\varepsilon^\nu(x) - \hat{\psi}(x)) \\
&= -\nu\,\frac{1}{\phi(x)}\,\Delta(\phi(x)(u_\varepsilon^\nu(x)-k)) + \nu(u_\varepsilon^\nu(x)-k)\,\frac{\Delta\phi(x)}{\phi(x)} + 2\,\frac{D\phi(x)D(\phi(x)(u_\varepsilon^\nu(x)-k))}{(\phi(x))^2} \\
&\quad - 2\nu\,\frac{(u_\varepsilon^\nu(x)-k)}{(\phi(x))^2}\,|D\phi(x)|^2 + u_\varepsilon^\nu(x) + H(Du_\varepsilon^\nu(x)) - f(x) + \beta_\varepsilon(u_\varepsilon^\nu(x) - \hat{\psi}(x)).
\end{aligned}
\tag{3.4}
$$

But $x_n \in E_+(\phi(u_\varepsilon^{\nu_n}-k),\Omega)$ implies $D(\phi(x_n)(u_\varepsilon^{\nu_n}(x_n)-k)) = 0$ and $\Delta\phi(x_n)(u_\varepsilon^{\nu_n}(x_n) - k) \leqslant 0$, so letting $n \to \infty$ in (3.4) evaluated at $\nu = \nu_n$ and $x = x_n$ we conclude (3.1). (We recall that $\beta_\varepsilon \geqslant 0$.) By an analogous argument we obtain (3.2) and (3.3). (We recall from Remark 2.1 that if $\bar{x} \in [u < \psi]$, then for small ε one has $u(\bar{x}) \leqslant u_\varepsilon(\bar{x}) < \psi(\bar{x}) \leqslant \hat{\psi}(\bar{x})$.)

Stability results obtained in [3], together with Lemma 3.1 ((3.1) and (3.2)), imply.

THEOREM 3.2. The function u, constructed in Corollary 2.3, is a viscosity solution of (P).

In order to obtain uniqueness we need a previous result.

LEMMA 3.3. For each $\varepsilon > 0$, the function u_ε verifies for every $\phi \in C(\Omega)^+$ and $k \in \mathbb{R}$

$$
u_\varepsilon(x) + H\left(-\frac{(u_\varepsilon(x)-k)}{\phi(x)}\,D\phi(x)\right) + \beta_\varepsilon(u_\varepsilon(x) - \hat{\psi}(x)) \geqslant f(x) \tag{3.5}
$$

for all $x \in E_-(\phi(u_\varepsilon - k),\Omega) \cap d(\phi)$. (Here $d(\phi)$ denotes the differentiable point set of ϕ.)

Proof. Indeed, if $x_0 \in d(\phi)$, then by Lemma 1.4 of [3] there exists $\phi_+ \in C_0^1(\bar{\Omega})$ such that $\phi_+(x_0) = \phi(x_0)$, $D\phi_+(x_0) = D\phi(x_0)$, $\phi_+(x_0) > \phi(x)$, $\forall x \in \Omega$, $x \neq x$. Then, if

$$x_0 \in d(\phi) \cap E_-(\phi(u_\varepsilon - k), \Omega)$$

one has $\{x_0\} = E_-(\phi_+(u_\varepsilon - k), \Omega)$. Choose $\{\phi_n\} \subset \mathcal{D}(\Omega)^+$ such that $\{\phi_n\}_n \to \phi_+$, $\{D\phi_n\}_n \to D\phi_+$ both uniformly in $\bar{\Omega}$. For large n $\phi_n(x_0)\,(u_\varepsilon(x_0) - k) < 0$ so $E_-(\phi_n(u_\varepsilon - k), \Omega) \neq \emptyset$. Then by (3.2) there exists $x_n \in E_-(\phi_n(u_\varepsilon - k), \Omega)$ such that

$$u_\varepsilon(x_n) + H\left(-\frac{u_\varepsilon(x_n) - k}{\phi_n(x_n)} D\phi_n(x_n)\right) + \beta_\varepsilon(u_\varepsilon(x_n) - \hat{\psi}(x_n)) \geqslant f(x_n). \quad (3.6)$$

Passing to a subsequence if necessary we may assume $x_n \to \bar{x} \in E_-(\phi_+(u_\varepsilon - k), \Omega)$ and a fortiori $\bar{x} = x_0$. The proof is concluded by letting $n \to \infty$ in (3.6).

<u>THEOREM 3.3</u>. For any viscosity subsolution $\bar{u}$ of (P) one has $\bar{u} \leqslant u$ on $\bar{\Omega}$.

<u>Proof</u>. The idea is to show that the maximum of $\bar{u} - u_\varepsilon$ is attained on the boundary, where u_ε is the approximation constructed in Corollary 2.3. Let $\phi \in \mathcal{D}(\Omega)^+$, $0 \leqslant \phi \leqslant 1$ and $\phi(0) = 1$, and

$$M_\varepsilon = \max_{x,y \in \Omega} \phi(x-y)(\bar{u}(x) - u_\varepsilon(y)) \equiv \phi(x_0 - y_0)(\bar{u}(x_0) - u_\varepsilon(y_0)).$$

For any $x \in \Omega$ $\bar{u}(x) - u_\varepsilon(x) = \phi(x-x)(\bar{u}(x) - u_\varepsilon(x)) \leqslant M_\varepsilon$ so $\|(\bar{u} - u_\varepsilon)^+\|_\infty \leqslant M_\varepsilon$ and with no loss of generality we may assume $M_\varepsilon > 0$, $x_0, y_0 \in \Omega$, because otherwise the proof is obvious. Then

$$x_0 \in E_+(\phi(\cdot - y_0)(\bar{u}(\cdot) - u_\varepsilon(y_0)), \Omega)$$

and

$$y_0 \in E_-(\phi(x_0 - .)(\bar{u}(x_0) - u_\varepsilon(\cdot)), \Omega).$$

We note that $\bar{u}$ is also a viscosity subsolution of the equation involved to (P) On the other hand, in [3] analogous results to (3.5) have been obtained for viscosity subsolution of the equation (clearly without the terms β_ε). Hence by (3.5) one has

$$\bar{u}(x_0) + H\left(-\frac{(\bar{u}(x_0) - u_\varepsilon(y_0))}{\phi(x_0 - y_0)} D\phi(x_0 - y_0)\right) - f(x_0) \leqslant 0$$
$$\leqslant u_\varepsilon(y_0) + H\left(-\frac{(\bar{u}(x_0) - u_\varepsilon(y_0))}{\phi(x_0 - y_0)} D\phi(x_0 - y_0)\right) + \beta_\varepsilon(u_\varepsilon(y_0) - \hat{\psi}(y_0)) - f(y_0). \quad (3.7)$$

The assumption $M_\varepsilon > 0$ implies $u_\varepsilon(y_0) < \bar{u}(x_0)$, so

$$0 \leqslant \beta_\varepsilon(u_\varepsilon(y_0) - \hat{\psi}(y_0)) + f(x_0) - f(y_0). \tag{3.8}$$

Choosing ϕ to be supported in $B(0,\alpha)$ (so $|x_0 - y_0| \leqslant \alpha$), (3.8) implies

$$0 \leqslant \beta_\varepsilon(u_\varepsilon(y_0) - \hat{\psi}(y_0)) + \rho_f(\alpha) \tag{3.9}$$

where $\rho_f(\alpha) = \sup\{|f(x) - f(y)| : |x - y| \leqslant \alpha\}$ is the modulus of continuity of f.

Letting $\alpha \to 0$ one has $0 \leqslant \beta_\varepsilon(u_\varepsilon(\bar{x}) - \hat{\psi}(\bar{x}))$, being $\bar{x} \in \Omega$ such that $\|(\bar{u} - u_\varepsilon)^+\|_\infty = \bar{u}(\bar{x}) - u_\varepsilon(\bar{x})$. Finally the monotonicity of β_ε and $\bar{u}(\bar{x}) \leqslant \psi(\bar{x}) \leqslant \hat{\psi}(\bar{x})$ imply

$$\|(\bar{u} - u_\varepsilon)^+\|_\infty = \bar{u}(\bar{x}) - u_\varepsilon(\bar{x}) \leqslant \hat{\psi}(\bar{x}) - \hat{\psi}(\bar{x}) = 0,$$

and the proof is completed by letting $\varepsilon \downarrow 0$.

Reasoning as in Lemma 3.3 we obtain from (3.1)

LEMMA 3.4. For each $\varepsilon > 0$ the function u_ε verifies for every $\phi \in C(\Omega)^+$ and $k \in \mathbb{R}$

$$u_\varepsilon(x) + H\left(- \frac{(u_\varepsilon(x)-k)}{\phi(x)} D\phi(x)\right) \leqslant f(x) \tag{3.10}$$

for all $x \in E_+(\phi(u_\varepsilon - k),\Omega) \cap d(\phi)$.

THEOREM 3.5. For any viscosity supersolution $\hat{u}$ of (P) one has $u \leqslant \hat{u}$ on $\bar{\Omega}$.

Proof. Obviously, $\bar{\Omega} = [\hat{u} = \psi] \cup [\hat{u} < \psi]$.

On $[\hat{u} = \psi]$ one has trivially $u(x) \leqslant \psi(x) = \hat{u}(x)$.

On the other hand, let $\phi \in \mathcal{D}([\hat{u} < \psi])^+$ such that $0 \leqslant \phi \leqslant 1$ and $\phi(0) = 1$; then by extending ϕ to a $\tilde{\phi} \in \mathcal{D}(\Omega)^+$ by 0 we obtain by Lemma 3.4 and Theorem III.1 of [3]

$$u_\varepsilon(x) \leqslant \hat{u}(x), \quad \forall x \in [\hat{u} < \psi] \tag{3.11}$$

where u_ε is the approximation of u. Finally letting $\varepsilon \downarrow 0$ we conclude the proof.

Remark 3.1. By repeating the relative results of Crandall and Lions [3] one obtains the consistency between the viscosity solution notion and the $W^{1,\infty}$ (or $W^{1,\infty}_{loc}$) solution of (P). More precisely: let u be a subsolution (respectively, supersolution) of (P). Then

$$u(x) + H(Du(x)) \leqslant f(x) \text{ on } d(u); \tag{3.12}$$

respectively

$$u(x) + H(Du(x)) \geqslant f(x) \text{ on } d(u) \cap [u < \psi]. \tag{3.13}$$

Remark 3.2. From the uniqueness proof we may deduce that

$$\|u(\psi_1) - u(\psi_2)\|_\infty \leqslant \|\psi_1 - \psi_2\|_\infty, \tag{3.14}$$

$$\|u(\psi_1) - u(\psi_2))^+\|_\infty \leqslant \|(\psi_1 - \psi_2)^+\|_\infty \tag{3.15}$$

being $u(\psi_i)$ (i = 1,2) the solution of (P) relative to ψ_i.

Remark 3.3. Assume that Γ is connected and $\psi + H(D\psi) - f \leqslant 0$ a.e. in Ω, then, by using comparison results, we may derive from $\psi|_\Gamma > 0$ that $[u < \psi]$ is connected.

REFERENCES

[1] S. Benton, The Hamilton-Jacobi Equation. A Global Approach, Academic Press, New York (1977).

[2] J.M. Bony, Principe du maximum dans les espaces de Sobolev, C.R.A.S. Paris 265, 333 (1967).

[3] M.G. Crandall and P.L. Lions, Viscosity solutions of Hamilton-Jacobi equations, MRC. Univ. of Wisconsin, Madison (1981).

[4] G. Díaz, Fully nonlinear inequalities and certain questions about their free boundary (to appear).

[5] J.I. Díaz and L. Veron, Existence theory and qualitative properties of the solutions of some first order quasilinear variational inequalities (to appear).

[6] P.L. Lions, Resolution des problèmes élliptiques quasilinéaires, Arch. Rat. Mech. Anal. 74, 335 (1980).

G. Díaz
Departamento Ecuaciones Funcionales,
Facultad de Matemáticas,
Universidad Complutense,
Madrid-3
Spain

J I DÍAZ

On two nonlinear parabolic equations in duality arising in thermal control: the L^∞ and L^1 semigroup approach and the asymptotic behaviour

1. INTRODUCTION

Let Ω be a smooth bounded domain in $\mathbb{R}^N$. Given $\psi \in L^2(\Omega)$ with $\psi \geq 0$ a.e. and $u_0 \in H_0^1(\Omega)$, we consider the problem

$$\left.\begin{array}{ll} u_t = \min\{\psi, \Delta u\} & \text{on } (0,\infty) \times \Omega \\ u = 0 & \text{on } (0,\infty) \times \partial\Omega \\ u(0,x) = u_0(x) & \text{on } \Omega. \end{array}\right\} \quad (1)$$

Such problems arise in heat control theory when the temporal temperature variation of a body or fluid Ω is not allowed to be greater than a given positive function ψ (called the 'obstacle'). See [10, Chapter 2].

Problem (1) can be expressed in a weak form by means of the following evolution variational inequality:

$$\left.\begin{array}{l} u_t \in K, \quad K = \{v \in H_0^1(\Omega) \mid v \leq \psi \text{ a.e. on } \Omega\} \\ \displaystyle\int_\Omega u_t(v-u_t)dx + \int_\Omega \text{grad } u \cdot \text{grad } (v-u_t)\, dx \geq 0 \quad \forall v \in K \text{ and } t > 0. \end{array}\right\} \quad (2)$$

The existence and uniqueness of a solution of (2), for each $u_0 \in H_0^1(\Omega)$, was proved by Brézis in [5] (see also [3]). Also the asymptotic behaviour is considered in [5] by means of the abstract result on asymptotic behaviour of solutions of evolution equations. It is shown there that $u(t,x)$ converges weakly in $H_0^1(\Omega)$, when $t \to \infty$, to a function $u_\infty(x) \in H_0^1(\Omega)$ satisfying

$$\min\{\Delta u_\infty, \psi\} = 0 \text{ on } \Omega \quad (3)$$

in the sense that

$$\int_\Omega \text{grad } u_\infty \cdot \text{grad } v \;\; dx \geq 0 \quad \forall v \in K.$$

Nevertheless it is neither known how the solution selects an equilibrium point among all of them nor if the convergence also holds in the strong topology of $H_0^1(\Omega)$. Both questions were proposed in [5] and they are, essentially, the main aims of this work.

Our methods for the study of the asymptotic behaviour are based on considerations made in terms of strong solutions, i.e. solutions which satisfy (1) a.e. Because of this we will first consider some regularity results. It is not difficult to see that if the solution u of (2) is such that $\Delta u(t,\cdot) \in L^1(\Omega)$, for $t > 0$, then u is a strong solution. Nevertheless not every solution of (2) is a strong solution (for instance, if $\psi \equiv 0$ and $\Delta u_0 \geqslant 0$ in $\mathcal{D}'(\Omega)$, then u is a strong solution iff $\Delta u_0 \in L^1(\Omega)$).

For the strong solutions, problem (1) can be equivalently formulated by

$$P \begin{cases} u_t(t,x) + \beta(x,-\Delta u(t,x)) = 0 & \text{on } (0,\infty) \times \Omega \\ u(t,x) = 0 & \text{on } (0,\infty) \times \partial\Omega \\ u(0,x) = u_0(x) & \text{on } \Omega \end{cases}$$

where

$$\beta(x,r) = -\min\{\psi(x),-r\} \quad \text{a.e. } x \in \Omega, \ \forall r \in \mathbb{R} \tag{4}$$

To prove the existence of strong solutions we shall consider the 'dual' problem

$$P^* \begin{cases} v_t(t,x) - \Delta\beta(x,v(t,x)) = 0 & \text{on } (0,\infty) \times \Omega \\ \beta(x,v(t,x)) = 0 & \text{on } (0,\infty) \times \partial\Omega \\ v(0,x) = v_0(x) & \text{on } \Omega. \end{cases}$$

It is clear, at least formally, that the existence of solutions v of P* in $L^1(\Omega)$ (i.e. such that $v(t,\cdot) \in L^1(\Omega)$ for $t > 0$) implies the existence of strong solutions of (1) by using the relation $v = -\Delta u$.

The existence of solutions of P* has been very much studied recently but the term $\beta(x,r)$ (a maximal monotone graph of $\mathbb{R}^2$ a.e. $x \in \Omega$) is always taken in the following two cases:

(a) $\beta(x,r)$ is independent of x

(b) $\beta(x,r)$ is onto a.e. $x \in \Omega$

(see [6]). Notice that the $\beta(x,r)$, given in (4), is neither in case (a) nor in case (b). Nevertheless, we shall show that P* is a 'well-posed' problem in $L^1(\Omega)$ (in the semigroup sense) when $\psi \in H^1(\Omega)$ and $(-\Delta\psi)^- \in L^2(\Omega)$.

If the obstacle ψ is assumed such that $\psi \in C^2(\bar{\Omega})$, then we shall prove that P is 'well posed' on $L^\infty(\Omega)$ and then the solution u of (2) satisfies $\Delta u(t,\cdot) \in L^\infty(\Omega)$ for $t > 0$.

Finally, using an abstract comparison result given in [1] we obtain some useful estimates allowing us to prove our main result on the asymptotic behaviour of the solutions of (1): if $\psi > 0$ and $\Delta\psi \geqslant 0$ a.e. on Ω, then $u(t,\cdot)$ converges strongly in $H_0^1(\Omega)$ to the equilibrium point zero. If, in addition, $\psi(x) \geqslant \delta > 0$ a.e. $x \in \Omega$ (for some δ) then the asymptotic behaviour is completely described in the sense that we show the solution verifies the <u>linear</u> heat equation $u_t = \Delta u$ on $(T_0,\infty) \times \Omega$ for an adequate finite time T_0. For other answers on the strong convergence and the selection of the equilibrium point we refer to [9].

2. THE SEMIGROUP APPROACH TO P* AND P

Problem P* can be formulated as an abstract Cauchy problem on the $L^1(\Omega)$ space

$$\left.\begin{aligned} &\frac{dv}{dt} + Av \ni 0 \text{ in } L^1(\Omega), \text{ on } (0,\infty) \\ &v(0) = v_0, \end{aligned}\right\} \qquad (5)$$

A being the operator in $L^1(\Omega)$ given by

$$\left.\begin{aligned} &D(A) = \{w \in L^1(\Omega): \beta(x,w(x)) \in W_0^{1,1}(\Omega) \text{ and } \Delta\beta(x,w(x)) \in L^1(\Omega)\} \\ &Aw = -\Delta\beta(\cdot,w(\cdot)) \text{ if } w \in D(A). \quad (\beta \text{ given in (4)}). \end{aligned}\right\} \quad (6)$$

In order to prove that (5) is 'well posed' on $L^1(\Omega)$ we shall apply the results on evolution equations governed by accretive operators (see e.g. [7]). We have

<u>THEOREM 1</u>. (a) The operator A is T accretive in $L^1(\Omega)$. (b) Let $\psi \in H^1(\Omega)$ be such that $\psi \geqslant 0$ a.e. and $\Delta\psi$ is a measure with $(-\Delta\psi)^- \in L^2(\Omega)$. Then the operator A is m-accretive in $L^1(\Omega)$. (c) Assume ψ as in part (b). Then $\overline{D(A)}^{L^1(\Omega)} = L^1(\Omega)$.

<u>Proof</u>. (a) Let $[u,v], [\hat{u},\hat{v}] \in A$ and consider the operator $-\Delta$ defined in

$L^1(\Omega)$ by $D(-\Delta) = \{w \in W_0^{1,1}(\Omega) \mid \Delta w \in L^1(\Omega)\}$. Then $u^* = \beta(\cdot,u) - \beta(\cdot,\hat{u})$ belongs to $D(-\Delta)$, and taking

$$\alpha^*(x) = \begin{cases} 1 \text{ if } (u-\hat{u})(x) \geqslant 0 \text{ and } u^*(x) \geqslant 0 \\ \\ 0 \text{ if } (u-\hat{u})(x) \leqslant 0 \text{ and } u^*(x) \leqslant 0 \\ \quad \text{or } (u-\hat{u})(x) < 0 \text{ and } u^*(x) = 0, \end{cases}$$

we have that $\alpha^*(x) \in \text{sign}^+(u(x)-u(x)) \cap \text{sign}^+ u^*(x)$ and that

$$\int_\Omega (Au-A\hat{u})\alpha^* \, dx \geqslant 0$$

because $-\Delta$ is a strongly T-accretive operator in $L^1(\Omega)$ (see [7]). (b) Given $f \in L^1(\Omega)$ it is easy to see that u satisfies $u + \lambda Au = f$ if and only if the function $h(x) = \beta(x,u(x))$ satisfies $h \in W_0^{1,1}(\Omega)$, $\Delta h \in L^1(\Omega)$ and

$$\left.\begin{array}{ll} -\lambda\Delta h(x) + h(x) + \gamma(h(x) + \psi(x)) \ni f(x) & \text{a.e. } x \in \Omega \\ h = 0 & \text{on } \partial\Omega \end{array}\right\} \tag{7}$$

being $\gamma(r)$ the maximal monotone graph of $\mathbb{R}^2$ defined by

$$\begin{array}{l} \gamma(r) = 0 \text{ if } r > 0,\ \gamma(0) = (-\infty,0] \text{ and} \\ \gamma(r) = \emptyset \text{ (the empty set) if } r < 0. \end{array} \tag{8}$$

Arguing as in [6], to prove the existence of solutions of (7) it suffices to consider f in a dense set of $L^1(\Omega)$. Actually, when $f \in L^2(\Omega)$, we can choose h as the unique solution of the variational inequality

$$\left.\begin{array}{l} h(x) \geqslant -\psi(x) \quad \text{a.e. } x \in \Omega \\ -\lambda\Delta h + h \geqslant f \text{ a.e. on } \Omega \\ (h+\psi)(-\lambda\Delta h+h-f)=0 \text{ a.e.on } \Omega \\ h = 0 \text{ on } \partial\Omega \end{array}\right\} \tag{9}$$

and by the regularity result of [3] we know that $h \in H^2(\Omega) \cap H_0^1(\Omega)$. Then h solves (7) and equivalently $R(I + \lambda A) = L^1(\Omega)$ for every $\lambda > 0$. (c) Take $f \in L^\infty(\Omega)$ and for each $\lambda > 0$ let $z_\lambda \in H_0^1(\Omega) \cap H^2(\Omega)$ be the solution of (7). By Theorem I.1 of [3] we get

$$\|\lambda\Delta z_\lambda\|_{L^2(\Omega)} \leqslant \|f\|_{L^2(\Omega)} + C\,\|(-\lambda\Delta\psi)^-\|_{L^2(\Omega)} \quad (C \text{ indep. of } \lambda). \tag{10}$$

Then $\{\lambda z_\lambda\}$ converges weakly in $H^2(\Omega)$ and by the comparison results it is shown that $\{\lambda z_\lambda\} \to 0$ in $L^2(\Omega)$. Setting $y_\lambda(x) = f(x) + \lambda\Delta z_\lambda(x)$ it is clear that $y_\lambda \in D(A)$ and y_λ converges (weakly) to f in $L^2(\Omega)$. Finally from (10) we deduce that

$$\lim_{\lambda\to 0} \|y_\lambda\|_{L^2(\Omega)} = \|f\|_{L^2(\Omega)}$$

and then y_λ converges strongly in $L^2(\Omega)$.

Remark. Problem P* is also well posed on the space $H^{-1}(\Omega)$. Indeed, from the result of [5] it is easy to see that for every $\psi \in L^2(\Omega)$, $\psi \geqslant 0$ a.e. on Ω, P* is governed by a maximal monotone operator on the Hilbert space $H^{-1}(\Omega)$. Such an operator can be characterized as the subdifferential of an adequate convex, l.s.c., functional on $H^{-1}(\Omega)$ in the following cases: (a) $\psi(x) \equiv \delta$ (see [4]), (b) $\psi \in H^1(\Omega)$ (unpublished result of A. Damlamian). Finally, we refer to the lecture of M.F. Bidaut-Véron for a very complete discussion about nonlinear equations with terms depending on x (such as (7)).

The existence of strong solutions of (1) is now a consequence of Theorem 1.

PROPOSITION 1. Assume $\psi \in H^1(\Omega)$ such that $\psi \geqslant 0$ on Ω and $(-\Delta\psi)^- \in L^2(\Omega)$. Let $u_0 \in H_0^1(\Omega)$ with $\Delta u_0 \in L^1(\Omega)$. Then the weak solution of (1) satisfies $\Delta u \in C([0,\infty) : L^1(\Omega))$.

Proof. From Theorem 1 it is enough to show that $-\Delta u(t,\cdot)$ coincides with $v(t)$, the unique $L^1(\Omega)$-semigroup solution of P* corresponding to the initial datum $v_0 = -\Delta u_0$. Using the continuity in $L^1(\Omega)$ of the semigroup generated by A, we can suppose $u_0 \in H^1(\Omega) \cap H^2(\Omega)$. On the other hand, we recall that u is given by the solution of

$$\left.\begin{aligned} &\frac{du}{dt} + Bu \ni 0 \text{ in } H_0^1(\Omega), \text{ on } (0,\infty) \\ &u(0) = u_0, \end{aligned}\right\} \tag{11}$$

where B is the maximal monotone operator on $H_0^1(\Omega)$ defined by

$$Bu = -\partial\phi^*(-u) \tag{12}$$

where ϕ^* is the conjugate convex function of

$$\phi(z) = \begin{cases} \frac{1}{2}\int_\Omega |z|^2\, dx & \text{if } z \in K \\ +\infty & \text{if } z \notin K \end{cases} \tag{13}$$

(see [5]). Then if $v(t) = \lim_{n\to\infty} v_n(t)$ with $v_n(t) = a_k^n$ for $k\lambda_n \leq t < (k+1)\lambda_n$ where $a_k^n = (I + \lambda_n A)^{-K}(-\Delta u_0)$ and $\lambda_n \to 0$, it is easy to see that $a_k^n \in L^2(\Omega)$ and that the functions $b_k^n = (-\Delta)^{-1}\ a_k^n$ satisfy $b_k^n = (I + \lambda_n B)^{-K}u_0$. Then

$$-\Delta u(t,\cdot) = \lim v_n(t) = v(t).$$

(The details can be found in [9]).

To prove a further regularity result to (1), consider P formulated as the following abstract Cauchy problem:

$$\left.\begin{aligned} &\frac{d\bar{u}}{dt} + C\bar{u} \ni 0 \text{ in } L^\infty(\Omega), \text{ on } (0,\infty) \\ &\bar{u}(0) = \bar{u}_0, \end{aligned}\right\} \tag{14}$$

C being the operator on $L^\infty(\Omega)$ given by

$$\left.\begin{aligned} &D(C) = \{w \in L^\infty(\Omega) \cap H_0^1(\Omega): \Delta w \in L^\infty(\Omega),\ \min(\psi,\Delta w) \in L^\infty(\Omega)\} \\ &Cw = -\min\{\psi,\Delta w\} \quad \text{if } w \in D(C). \end{aligned}\right\} \tag{15}$$

THEOREM 2. (i) The operator C is T-accretive in $L^\infty(\Omega)$. (ii) If $\psi \in C^2(\bar{\Omega})$ with $\psi \geq 0$ on $\bar{\Omega}$, C satisfies the range condition $R(I + \lambda C) \supseteq D(C)$, $\forall\lambda > 0$.

Proof. (i) Follows from the maximum principle and (ii) is shown by means of the Brézis-Kinderlehrer regularity result for stationary variational inequalities (see [9]).

PROPOSITION 2. Assume $\psi \in C^2(\bar{\Omega})$, $\psi \geq 0$ on Ω. Let $u_0 \in H_0^1(\Omega)$ be such that $\Delta u_0 \in L^\infty(\Omega)$. Then the weak solution u of (1) satisfies

$$u \in W^{1,\infty}((0,\infty) \times \Omega) \cap L^{\infty}(0,\infty: H^2(\Omega))$$

and $\Delta u(t,\cdot) \in L^{\infty}(\Omega)$ a.e. $t > 0$.

Proof. It is easy to show that $u_0 \in D(C)$. Then the $L^{\infty}(\Omega)$-semigroup solution $\bar{u}$ satisfies

$$\bar{u}(t) \in D(C)$$

and

$$\bar{u} \in W^{1,\infty}([0,\infty) \times \Omega) \cap L^{\infty}(0,\infty:H^2(\Omega))$$

(see [2]). Finally if $b \in D(B) \cap D(C)$, then $(I + \lambda C)^{-1}b = (I + \lambda B)^{-1}b$ for any $\lambda > 0$ and in consequence $u = \bar{u}$.

Remark. The equation of (1) can obviously be written as

$$u_t + \max \{-\Delta u, -\psi\} = 0 \tag{16}$$

and then it is similar to the so-called Bellman's equation of dynamic programming

$$u_t + \max \{L^1u-f^1, L^2u-f^2\} = 0,$$

where L^k is a second order, uniformly elliptic operator ($k = 1,2$). Interesting regularity results for (17) can be found in [11], [12] and [13]. Nevertheless, in (16) $L^2 \equiv 0$ and the above works do not apply.

3. THE ASYMPTOTIC BEHAVIOUR

We fix our attention on the convergence of the weak solution to an equilibrium point of (1). We shall limit our attention to the case $\psi > 0$ a.e. on Ω (then it is obvious that $u_{\infty} = \lim_{t\to\infty} u(t,\cdot)$ is $u_{\infty} \equiv 0$). Other cases are discussed in [9].

If $\psi(x) \equiv \delta > 0$, by using the regularizing effects for the dual equation P* (results of Bênilan, Veron, Evans,...) it is easy to see that the weak solution u satisfies the linear heat equation after a finite time (see [8]). This kind of result is far from being so easy when ψ is a non-constant function. Indeed, due to the non-surjectivity of the graph $\beta(\cdot,r)$ given in (4), no regularizing effect is known for the problem P*. Instead we have the following result which is, with slight modifications, a particular application of the abstract result of [1]:

LEMMA 1. Let $\psi \in H^1(\Omega)$ with $\psi \geq 0$ a.e. on Ω and $(-\Delta\psi)^- \in L^2(\Omega)$. Assume $u_0 \in H_0^1(\Omega)$ such that

$$-\Delta u_0 \in \overline{D^+(A)}^{L^1}, \quad (D^+(A) = \{w \in D(A) \mid Aw \geq 0\}).$$

Then

$$h(t,x; \tilde{v}_0) \leq -\min \{\psi(x), \Delta u(t,x)\} \text{ a.e. } (t,x) \in (0,\infty) \times \Omega$$

where $\tilde{v}_0 = -\min \{\psi, \Delta u_0\}$ and $h(t,x:z)$ denotes the solution of the heat equation

$$\begin{cases} h_t = \Delta h & \text{on } (0,\infty) \times \Omega \\ h = 0 & \text{on } (0,\infty) \times \partial\Omega \\ h(0,x) = z(x) & \text{on } \Omega. \end{cases}$$

THEOREM 3. Assume $\psi \in H^2(\Omega)$ with $\psi \geq 0, \Delta\psi \geq 0$ a.e. on Ω and let $u_0 \in H_0^1(\Omega)$. Then if $\psi(x) > 0$ a.e. $x \in \Omega$, $u(t) \to 0$ (strongly) in $H_0^1(\Omega)$ when $t \to +\infty$. If in addition $\psi(x) \geq \delta$ for some $\delta > 0$, then $u_t = \Delta u$ on $(T_0,\infty) \times \Omega$ where $T_0 = (C/\delta \|\psi\|_{L^1(\Omega)})^{2/N}$ and C is a positive constant depending only on $|\Omega|$.

Proof. Step 1. Assume $\psi \in C^2(\bar{\Omega})$, $\psi \geq 0$, $\Delta\psi \geq 0$ and $u_0 \in H_0^1(\Omega)$ such that $h = -\Delta u_0 \in L^\infty(\Omega)$. Set $u_{0,+}$ and $u_{0,-}$ belonging to $H_0^1(\Omega)$ such that $-\Delta u_{0,+} = h^+$ and $-\Delta u_{0,-} = -h^-$. Let u_+ and u_- be the weak solutions of (1) corresponding to the initial data $u_{0,+}$ and $u_{0,-}$ respectively. By Proposition 2 and the T-accretiveness of A we know that

$$-\Delta u_-(t) \leq -\Delta u(t) \leq -\Delta u_+(t) \text{ in } L^\infty(\Omega), \quad \text{a.e. } t > 0.$$

It is easy to see that $-\Delta u_+(t) \geq 0$ a.e. Then u_+ satisfies the linear heat equation, and so $-\Delta u_+(t) \to 0$ in $L^\infty(\Omega)$ when $t \to +\infty$. On the other hand, it is possible to find a $\hat{u}_0 \in H_0^1(\Omega)$ with $\Delta\hat{u}_0 \in L^\infty(\Omega)$ and such that $-\Delta\hat{u}_0 \leq -\Delta u_{0,-}$ a.e. on Ω as well as $-\Delta\hat{u}_0 \in D^+(A)L^{1(\Omega)}$. Indeed, we can choose $\hat{v}_0 \in L^\infty(\Omega)$ with $\hat{v}_0 \leq \min \{-\psi, -\Delta u_{0,-}\}$ and then $\hat{u}_0 = (-\Delta)^{-1}\hat{v}_0$. (We remark in this case min $\{\psi,\Delta\hat{u}_0\} = \psi$, so $A(-\Delta\hat{u}_0) = \Delta\psi \geq 0)$. Therefore Lemma 1 shows that

$$h(t,x:-\psi(x)) \leq -\min\{\psi(x),\Delta\hat{u}(t,x)\} \leq -\min\{\psi(x),\Delta u(t,x)\} \leq 0,$$

where $\hat{u}$ is the weak solution of (1) corresponding to the initial datum $\hat{u}_0$. From the results on the asymptotic behaviour for the linear heat equation it is well known that there exists a positive constant C (only depending on $|\Omega|$) such that

$$\frac{-C}{t^{N/2}} \|\psi\|_{L^1(\Omega)} \leq h(t,x; -\psi(x)) \leq 0 \text{ a.e. } (t,x) \in (0,\infty) \times \Omega.$$

Thus, the conclusion follows easily.

Step 2. Take $\psi \in C^2(\Omega)$ with $\psi \geq 0$ and $\Delta\psi \geq 0$ a.e. on Ω. Let $u_0 \in H_0^1(\Omega)$. Consider $u_{0,n} \in H_0^1(\Omega)$ with $-\Delta u_{0,n} \in L^\infty(\Omega)$ and $u_{0,n} \to u_0$ in H_0^1 when $n \to \infty$. Then, if $u_n(t)$ is the weak solution of (1) of initial datum $u_{0,n}$, $u_n(t) \to u(t)$ in $H_0^1(\Omega)$ when $n \to \infty$ and so the first assertion follows from the first step. The second assertion can be shown by using the 'exponential formula' and the first step (see [9]).

Step 3. Let $\psi \in H_0^1(\Omega)$ with $\psi \geq 0$, $\Delta\psi \geq 0$ a.e. on Ω and $u_0 \in H_0^1(\Omega)$. Consider $\psi_n \in C^2(\bar{\Omega})$ with $\psi_n \geq 0$, $\Delta\psi_n \geq 0$, $\|\psi_n\|_{L^1(\Omega)} \leq \|\psi\|_{L^1(\Omega)}$, and such that $\psi_n \to \psi$ in $H^2(\Omega)$ when $n \to \infty$. Thanks to some convergence results for variational inequalities it can be shown that

$$(I + \lambda B_n)^{-1}z \to (I + \lambda B)^{-1}z \text{ when } n \to \infty,\ \lambda > 0,\ z \in D(B) \cap D(B_n)$$

(B_n designates the operator B corresponding to the obstacle ψ_n). The conclusion follows from the abstract results on convergence of maximal monotone operators.

REFERENCES

[1] Ph. Bênilan and J.I. Díaz, Comparison of solutions of nonlinear evolution problems with different nonlinear terms, Israel J. Math. 42(3), (1982).

[2] Ph. Bênilan and K. Ha, Equations d'évolution du type $du/dt + \beta(\partial\phi(u)) \ni 0$ dans $L^\infty(\Omega)$, C.R. Acad. Sci. Paris 281, 947-950 (1975).

[3] H. Brézis, Monotonicity methods in Hilbert spaces and some applications to nonlinear partial differential equations, in Contributions to Nonlinear Functional Analysis (E. Zarantonello, ed.), Academic Press, New York (1971).

[4] H. Brézis, Problèmes Unilateraux, J. Math. Pures Appl. 51, 1-164 (1972).

[5] H. Brézis, Asymptotic behaviour of some evolution system, in Nonlinear Evolution Equations (M. G. Crandall, ed.) Academic Press, New York (1978).

[6] H. Brézis and W. Strauss, Semilinear elliptic equations in L^1, J. Math. Soc. Japan 25, 565-590 (1973).

[7] M.G. Crandall, An introduction to evolution governed by accretive operators, in Dynamical Systems. Vol. I. Academic Press, New York (1976).

[8] J.I. Díaz, Propiedades cualitativas de ciertos problemas parabolicos no lineales: Una clasificacion para los modelos de difusion del calor, Memory no XIV of the Real Acad. Ciencias. Madrid (1980).

[9] J.I. Díaz, On a fully nonlinear parabolic equation and the asymptotic behaviour of its solutions, Mathematics Research Center Technical Summary Report #2291, University of Wisconsin-Madison (October 1981). To appear in J. Math. Anal. Appl.

[10] G. Duvaut and J.L. Lions, Les inéquations en mécanique et en Physique, Dunod, Paris (1972).

[11] L.C. Evans and S. Lenhart, The parabolic Bellman equation, Nonlinear Anal. Theory, Meth and Appl. 5, 765-773 (1981).

[12] P.L. Lions, Le problème de Cauchy pour les équations de Hamilton-Jacobi-Bellman (to appear).

[13] M. Schoenauer, Some regularizing effects for the Hamilton-Jacobi-Bellman equation for two parabolic operators, Comm. Partial Differential Equations 6 (8), 929-949 (1981).

J.I. Díaz
Dept de Ecuaciones Funcionales,
Facultad de Matematicas
Universidad Complutense de Madrid,
Madrid-3,
Spain

J I DÍAZ & J HERNÁNDEZ

Free boundary problems for some stationary reaction–diffusion systems

1. INTRODUCTION

We give here a brief survey of some recent results concerning the existence of free boundaries for a class of reaction-diffusion systems arising in combustion theory. Complementary results and complete proofs can be found in [11].

Here we consider a model system describing a single, irreversible non-isothermic stationary reaction

$$\left.\begin{aligned} -\Delta u + \mu^2 F(u) e^{\frac{\gamma(v-1)}{v}} &= 0 \quad \text{in } \Omega \\ -\Delta v - \nu\mu^2 F(u) e^{\frac{\gamma(v-1)}{v}} &= 0 \quad \text{in } \Omega \end{aligned}\right\} \tag{1.1}$$

$$u = v = 1 \quad \text{on } \partial\Omega \tag{1.2}$$

where Ω is a bounded open subset of $\mathbb{R}^n$ with smooth boundary $\partial\Omega$, μ^2 is the Thiel number, ν is the Prater temperature and γ is the Arrhenius number (cf. [2]). The function F is assumed to be increasing and such that $F(0) = 0$, $F(1) = 1$ and $F(s) > 0$ for $s > 0$. The unknowns u and v are non-negative and they represent, respectively, the concentration and temperature of the reactant. Often F takes the form $F(u) = u^p$, where $p \geqslant 0$ is the reaction order (cf. [2]). In the case $p = 0$, F is given by $F(0) = 0$ and $F(s) = 1$ if $s > 0$. (Thus, F is discontinuous.)

Existence and uniqueness results for the parabolic problem associated with (1.1) and (1.2) were given in [1] and [3] for $p \geqslant 1$. Existence and, in some particular cases, uniqueness for the elliptic problem can be found in [1] or [12], again for $p \geqslant 1$. The case $0 \leqslant p < 1$ is considered in [2], p. 311 (cf. also [14]) but existence theorems are not given. It is shown in [2] and [14] that for $p = 0$ and μ large enough, strictly positive solutions cannot exist. It is also shown, in particular examples, that the set $\Omega_0 = \{x \in \Omega | \ u(x) = 0\}$, which is called the *dead core*, has positive measure for $0 \leqslant p < 1$.

The main idea used in [14] and other papers is to reduce (1.1) and (1.2) to a nonlinear elliptic equation for u alone. Here we follow a different approach which allows us to include also the case of nonlinear boundary conditions, which cannot be handled by the preceding device.

We consider the case of discontinuous F in the framework of maximal monotone graphs (cf. [6]). As known existence theorems for elliptic reaction-diffusion systems are given for locally Lipschitz nonlinearities, which is not the case here for $0 \leqslant p < 1$, it is necessary to prove existence in this more general situation. This can be done by following the same lines as [12] with a fixed-point argument using coupled sub- and supersolutions and the results in [8]. (Cf. [11] for the details.)

We also study the existence and non-existence of a dead core Ω_0 where $u = 0$ and then the existence of the free boundary $\partial\Omega_0$ (cf. [10] for the case of a single equation). Roughly speaking, such a dead core for (1.1) and (1.2) arises when diffusion is unable to supply enough reactant from outside Ω to reach the central region of Ω (cf. [14]). This can occur if the reaction rate $F(u)e^{\gamma(v-1)/v}$ remains high for small u. Thus, the existence of Ω_0 depends on three main factors: the reaction order p, the Thiele number μ^2, and the size of Ω. Moreover, we obtain some information about the location and size of the dead core Ω_0 and its dependence on μ. All these results are collected in Theorem 1 of the following paragraph.

Finally, we point out that C. Bandle, R.P. Sperb and I. Stakgold have obtained independently similar results for (1.1) and (1.2) by using a different method.

2. MAIN RESULTS

Here we state our main theorem concerning free boundaries, together with some indications about the method of proof and the meaning of the results.

We consider the system

$$\left.\begin{aligned} -\Delta u + \alpha(u)\, f(v) &= 0 \quad \text{in } \Omega \\ -\Delta v - \beta(u)\, g(v) &= 0 \quad \text{in } \Omega \end{aligned}\right\} \tag{2.1}$$

$$u = \phi_1 \qquad v = \phi_2 \qquad \text{on } \partial\Omega \tag{2.2}$$

where Ω is a bounded open subset of $\mathbb{R}^N$ with smooth boundary $\partial\Omega$. We assume that α and β are nondecreasing real continuous functions such that

$$\alpha(0) = \beta(0) = 0. \tag{2.3}$$

f and g are C^1 functions satisfying

$$f(s) \geq m_1 \geq 0 \quad \text{and} \quad 0 \leq g(s) \leq m_2 \quad \text{for } s \in \mathbb{R}. \tag{2.4}$$

$$\phi_1, \phi_2 \in C^1(\partial\Omega), \ \phi_1, \phi_2 \geq 0 \text{ on } \partial\Omega. \tag{2.5}$$

The main theorem in this paper is the following:

THEOREM 1. Assume that (2.3)-(2.5) are satisfied. Then there exists at least one solution (u,v) of (2.1) and (2.2) with $u,v \in W^{2,r}(\Omega)$ for any $1 \leq r < +\infty$, and $v > 0$ on Ω. Moreover, we have

(i) If $\alpha(s) = \mu^2|s|^{p-1}s$ and (u,v) is any solution of (2.1) and (2.2), then a dead core for u may exist only if $0 \leq p < 1$.

(ii) Let $\alpha(s) = \mu^2|s|^{p-1}$ with $0 < p < 1$ and let (u,v) be a solution of (2.1) and (2.2). For $\lambda > 0$ let $\Omega_\lambda = \{x \in \Omega | f(v(x)) \geq \lambda\}$. Then

$$\Omega_0 \supset \{x \in \Omega_\lambda | d(x, \partial\Omega_\lambda - (\partial\Omega - \text{supp } \phi_1)) \geq \left(\frac{M}{K_{\lambda,\mu}}\right)^{\frac{1-p}{2}}\} \tag{2.6}$$

where $M = ||\phi_1||_{L^\infty(\partial\Omega)}$ and

$$K_{\lambda,\mu} = \left(\frac{2N(1-p)+4p}{\lambda\mu^2(1-p)^2}\right)^{\frac{1}{p-1}}$$

Remark 2.1. The case $p = 0$ can be handled in a completely similar way by considering the graph $\alpha(s) = \mu^2 sg(s)$ and we also obtain the estimate (2.6) for some $M > 0$.

Remark 2.2. An estimate similar to (2.6) still holds in the case of non-homogeneous nonlinear boundary conditions if Ω has some geometrical properties and, in particular, if Ω is convex. Cf. [11] for the details.

The above theorem is particularly interesting if $m_1 > 0$ in (2.4) (this condition is satisfied in the case of the combustion system (1.1) and (1.2)). In this situation $\Omega_\lambda = \Omega$ for $0 < \lambda < m_1$ and estimate (2.6) reads

$$\Omega_0 \supset \{x \in \Omega | d(x,\partial\Omega) \geq \left(\frac{M}{K_{\lambda,\mu}}\right)^{\frac{1-p}{2}}\}$$

if $\phi_1 > 0$ on $\partial\Omega$. We see that $K_{\lambda,\mu} \to +\infty$ if $\mu \to +\infty$; therefore the existence of a dead core can only be guaranteed by estimate (2.6) if, for example

$$\delta(\Omega) \geq \left(\frac{M}{K_{\lambda,\mu}}\right)^{\frac{1-p}{2}}$$

where $\delta(\Omega)$ is the radius of the largest ball contained in Ω. Then, for a fixed Ω, it is clear that for μ large enough, Ω_0 has a positive measure.

The proof of Theorem 1 can be carried out by using results for a single nonlinear equation, but not in the usual way for the combustion problem. In fact, if (u,v) is a solution of (2.1) and (2.2) with $\alpha(s) = \mu^2|s|^{p-1}s$, then u satisfies

$$-\Delta u + \tilde{f}(x)\,\alpha(u) = F(x) \quad \text{in } \Omega \tag{2.7}$$

$$u = \phi_1 \quad \text{on } \partial\Omega \tag{2.8}$$

where $\tilde{f}(x) = f(v(x))$ a.e. on Ω and $F \equiv 0$.

Many authors have studied the subset Ω_0 for the problem (2.1) and (2.2) (cf., e.g., [4], [5], [7], [9], [10], [13], [15]) but, to the best of our knowledge, all the existing results concern the case $\tilde{f} \equiv$ constant. Our results here follow the ideas in [10].

The main ingredient for the proof of Theorem 1 is the following auxiliary result.

<u>LEMMA 1</u>. Let $F \in L^\infty(\Omega)$, $\phi \in C^1(\partial\Omega)$ and suppose that $u \in H^2(\Omega)$ satisfies

$$-\Delta u(x) + \mu^2\tilde{f}(x)|u(x)|^{p-1}u(x) = F(x) \quad \text{in } \Omega \tag{2.9}$$

$$u = \phi \quad \text{on } \partial\Omega \tag{2.10}$$

where $\tilde{f} \in L^\infty(\Omega)$, $\tilde{f} \geq 0$ on Ω and $0 < p < 1$. If we define $\Omega_\lambda = \{x \in \Omega | \tilde{f}(x) \geq \lambda\}$, $\lambda > 0$ we have the estimate

$$\Omega_0 = \{x \in \Omega | u(x) = 0\} \supset \{x \in \Omega_\lambda | d(x, \partial(\Omega_\lambda - \text{supp } F) - (\partial\Omega - \text{supp } \phi)) \geqslant \left(\frac{\tilde{M}}{K_{\lambda,\mu}}\right)^{\frac{1-p}{2}}\}. \tag{2.11}$$

Here $\tilde{M} = \max\left\{\left(\frac{\|F\|_{L^\infty(\Omega)}}{\lambda\mu^2}\right)^{\frac{1}{p}}, \|\phi\|_{L^\infty(\partial\Omega)}\right\}$ and $K_{\lambda,\mu}$ is given by

$$K_{\lambda,\mu} = \left(\frac{2N(1-p)+4p}{\lambda\mu^2(1-p)^2}\right)^{1/(p-1)}.$$

Sketch of the proof. Simple comparison arguments allow us to consider only the case $F \geqslant 0$, $\phi \geqslant 0$. If $u_\lambda \in H^2(\Omega)$ satisfies

$$\left.\begin{array}{lll} -\Delta u_\lambda + \lambda\mu^2|u_\lambda|^p \geqslant F & & \text{in } \Omega_\lambda \\ u_\lambda \geqslant \phi & & \text{on } \partial\Omega_\lambda \cap \partial\Omega \\ u_\lambda \geqslant \|u\|_{L^\infty(\Omega)} & & \text{on } \partial\Omega_\lambda - \partial\Omega \end{array}\right\} \tag{2.12}$$

it is not difficult to show by using comparison results in [10] that $0 \leqslant u(x) \leqslant u_\lambda(x)$ a.e. on Ω_λ (the same argument works on any subset of Ω_λ). Therefore estimate (2.11) will follow by constructing such functions u_λ. We look for u_λ of the form $u_\lambda(x) = h(|x-x_0|)$ for some $x_0 \in \Omega_\lambda$.

For $0 < \eta < 1$ fixed, let h_η be a solution of the Cauchy problem

$$\left.\begin{array}{l} h_\eta''(r) = \eta\lambda\mu^2|h_\eta(r)|^{p-1} h_\eta(r) \\ h_\eta(0) = h_\eta'(0) = 0. \end{array}\right\} \tag{2.13}$$

It is easy to check that

$$h_\eta(r) = L_\eta r^{\frac{2}{1-p}},$$

where L_η is a constant explicitly given, is a solution of (2.13). If

$$0 < \eta < \frac{p+1}{1+p+(N-1)(1-p)},$$

then for any $x \in \Omega_\lambda$,

$$-\Delta h_\eta(|x-x_0|) + \lambda\mu^2 h_\eta(|x-x_0|)^p \geqslant 0$$

and from this it follows that the function

$$u_\lambda(x) = K_{\lambda,\mu}|x - x_0|^{\frac{2}{1-p}}$$

satisfies

$$- \Delta u_\lambda + \lambda\mu^2 u_\lambda^p \geq 0 = F(x) \quad \text{in } \tilde{\Omega}$$

$$u_\lambda \geq 0 = \phi \quad \text{on } \partial\tilde{\Omega} \cap (\partial\Omega - \text{supp } \phi)$$

where $\tilde{\Omega} = \Omega_\lambda$ - supp F. Hence it is sufficient to have

$$u_\lambda \geq \max \{\phi, \|u\|_{L^\infty(\Omega)}\} \text{ on } \partial\tilde{\Omega} - (\partial\tilde{\Omega} \cap (\partial\Omega - \text{supp } \phi)) \tag{2.14}$$

to obtain $0 \leq u(x) \leq u_\lambda(x)$ on $\tilde{\Omega}$. But by the maximum principle $u \leq \tilde{M}$ on Ω and then (2.14) is satisfied if we choose x_0 such that

$$|x - x_0| \geq \left(\frac{\tilde{M}}{K_{\lambda,\mu}}\right)^{\frac{1-p}{2}} \tag{2.15}$$

for any $x \in \partial\tilde{\Omega} - (\partial\Omega \cap (\partial\tilde{\Omega} - \text{supp } \phi))$. The conclusion now follows from (2.14), (2.15) and $u_\lambda(x_0) = 0$.

REFERENCES

[1] H. Amann, Existence and stability of solutions for semilinear parabolic systems and applications to some diffusion-reaction equations, Proc. Roy. Soc., Edinburgh, 81A, 37-47 (1978).

[2] R. Aris, The Mathematical Theory of Diffusion and Reaction in Permeable Catalysts, Oxford, Clarendon Press (1975).

[3] J. Bebernes, K.N. Chueh and W. Fulks, Some applications of invariance to parabolic systems, Indiana Univ. Math. J. 28, 269-277 (1979).

[4] A. Besoussan, H. Brézis and A. Friedman, Estimates on the free boundary for quasi-variational inequalities, Comm. in Partial Differential Equations 2, 297-321 (1977).

[5] Ph. Bênilan, H. Brézis and M.G. Crandall, A semilinear elliptic equation in $L^1(\mathbb{R}^N)$, Ann. Scuola Norm. Sup. Pisa., Series IV-II, 523-555 (1975).

[6] H. Brézis, Opérateurs maximaux monotones et semigroupes de contractions dans les espaces de Hilbert, North Holland, Amsterdam (1973).

[7] H. Brézis, Solutions of variational inequalities with compact support, Uspekhi Mat. Nauk. 129, 103-108 (1974).

[8] H. Brézis and W. Strauss, Semilinear second-order elliptic equations in L^1, J. Math. Soc. Japan 25, 565-590 (1973).

[9] J.I. Díaz, Soluciones con soporte compacto para algunos problemas semilineales, Collect. Math. 30, 141-179 (1979).

[10] J.I. Díaz, Técnica de supersoluciones locales para problemas estacionarios no lineales: aplicaciones al estudio de flujos subsónicos, Memory no.XVI of the Real Academia de Ciencias, Madrid (1980).

[11] J.I. Díaz and J. Hernández, On the existence of a free boundary for a class of reaction-diffusion systems, Technical Summary Report 2330, M.R.C., Madison, Wisconsin (1982)(to appear, SIAM J. Math. Analysis).

[12] J. Hernández, Some existence and stability results for solutions of reaction-diffusion systems with nonlinear boundary conditions, in Nonlinear Differential equations: Invariance, Stability and Bifurcation. P. de Mottoni and L. Salvadori (eds.), Academic Press, 161-173 (1981).

[13] T. Nagai, Estimates for the coincidence sets of solutions of elliptic variational inequalities, Hiroshima Math. J. 9, 335-346 (1979).

[14] I. Stakgold, Estimates for some free boundary problems, in Ordinary and Partial Differential Equations Proceedings, Dundee, Scotland 1980, W.N. Everitt and B.D. Sleeman (eds.) Lecture Notes No. 846, Springer-Verlag, Berlin (1981).

[15] N. Yamada, Estimates on the support of solutions of elliptic variational inequalities in bounded domains, Hiroshima Math. J. 9, 7-16 (1979).

J.I. Díaz*
Universidad Complutense de Madrid,
Madrid-3,
Spain

J.H. Hernández
Departamento de Matemáticas,
Universidad Autónoma,
Madrid-34,
Spain

* Partially sponsored by the US Army under Contract No. DAAG29-80-0-0041.

M J ESTEBAN

Compactness results and existence of many solutions of nonlinear elliptic problems in strip-like domains

0. INTRODUCTION

We shall be interested here in finding nontrivial solutions u of the following secord-order nonlinear elliptic problem:

$$\left.\begin{aligned} -\Delta u &= f(u) \text{ in } \Omega \\ u &= 0 \text{ on } \partial\Omega, \end{aligned}\right\} \qquad (1)$$

where f is a given nonlinearity, and Ω is a strip in $\mathbb{R}^N$ i.e., $\Omega = 0 \times \mathbb{R}^p$, where 0 is a bounded domain of $\mathbb{R}^m$.

This kind of problem, which arises in many physical and mechanical problems, has been studied by many authors when Ω is bounded. The unboundedness of Ω makes it very difficult to find a global solution of (1), because of the lack of compactness.

Problem (1) was solved for $\Omega = \mathbb{R}^N$ (see Berestycki and Lions [3] and Strauss [12]), and for a few more cases, but the principal fact seemed to prove a certain sort of compactness which permitted the application of global methods.

In the case of $\Omega = \mathbb{R}^N$, for example, the solutions were sought in the subspace of $H^1(\mathbb{R}^N)$, which consists of the radial symmetric functions.

When Ω is a strip, it seemed to be important to obtain a certain compactness using the symmetries of the strip. This is the key result which enables us to obtain global existence results.

The structure of this paper is as follows. In the first section, we state our principal results and we give the fundamental compactness lemma and some other auxiliary ones. The second section applies the preceding lemmas to the proof of the existence of a positive solution or many solutions of (1). In the third section we give some alternative results for a natural eigenvalue problem related to (1). This third part of our work gives us a better idea of the situation regarding existence and quantity of solutions of (1).

For more details about these results see Esteban and Lions [7] and Esteban [6].

1. MAIN RESULTS AND AUXILIARY LEMMAS

Let Ω be the strip $\mathcal{O} \times \mathbb{R}^p$, where $\mathcal{O}$ is a bounded domain of $\mathbb{R}^m$, $m \geq 1$, and let f be a function which satisfies the following assumptions:

$$f(0) = 0,\ f(t) = g(t) + \nu t, \text{ with } \nu < \lambda_1(\mathcal{O}), \text{ where } \lambda_1(\mathcal{O}) \text{ is the first eigenvalue of } -\Delta \text{ acting on } H_o^1(\mathcal{O}), \tag{2}$$

$$-\infty < \underline{\lim}_{t\to 0^+} \frac{g(t)}{t} \leq \overline{\lim}_{t\to 0^+} \frac{g(t)}{t} = -k < 0; \tag{3}$$

$$\overline{\lim}_{t\to+\infty} \frac{g(t)}{t^\ell} \leq 0, \text{ where } \ell = \frac{N+2}{N-2} \text{ if } N > 2;\ \ell < +\infty \text{ if } N = 2; \tag{4}$$

$$G(t)\, t^{-\theta} \text{ is nondecreasing for } t \geq 0,\ 0 < \overline{\lim}_{t\to+\infty} G(t)\, t^{-\theta} < +\infty \text{ for some } \theta > 2, \text{ where } G(t) = \int_0^t g(s)\, ds. \tag{5}$$

Then we can prove the two following theorems:

THEOREM 1. Suppose that $p \geq 2$ and let f,g satisfy (2)-(5), then problem (1) has a solution $u \in W^{2,q}_{loc}(\Omega) \cap H_o^1(\Omega)$, $\forall q < +\infty$, satisfying:

(i) $u > 0$ in Ω;

(ii) u is axially symmetric, i.e., u has the form $u(x_1,x^1) = u(x_1,|x^1|)$, $\forall x_1 \in \mathcal{O}$, $\forall x^1 \in \mathbb{R}^p$; moreover, u is decreasing in $|x^1|$;

(iii) if f is locally Hölder continuous, $u \in C^2(\bar{\Omega})$.

THEOREM 2. If $p \geq 2$, if f, g satisfy all the assumptions made above, and if g is odd, then problem (1) possesses an infinity of distinct solutions which are axially symmetric and which are in $C^2(\bar{\Omega})$ whenever f is locally Hölder continuous.

Next we give two lemmas that we use in the proof of the above theorems. The first lemma, which generalizes results of Strauss [12], and Berestycki and Lions [3], is fundamental in all that follows.

DEFINITION. That $u \in H_0^1(\mathcal{O} \times \mathbb{R}^p)$ is axially symmetric if $u(x_1,x^1) = u(x_1,|x^1|)$ for all $x_1 \in \mathcal{O}$, $x^1 \in \mathbb{R}^p$.

Let us note by $H^1_{0,s}(\Omega)$ the set of all axially symmetric functions in $H_0^1(\Omega)$.

Then, we can prove the following.

LEMMA 3. If $p \geq 2$, the Sobolev imbedding of $H^1_{0,s}(\Omega)$ in $L^q(\Omega)$ is compact for every $q \in (2, \frac{2N}{N-2})$, where $N = m + p$.

Furthermore, if $\{u_n\}$ is a bounded sequence in $H^1_{0,s}(\Omega)$, and if F is a continuous function which satisfies:

$$F(t) = o(t^2) \text{ as } t \to 0$$

$$F(t) = o(t^{\frac{2N}{N-2}}) \text{ as } t \to +\infty,$$

then the sequence $\{F(u_n)\}$ is relatively compact in $L^1(\Omega)$.

The proof of this lemma is given in detail in [7]. It uses in a fundamental way the radial symmetric with respect to the x^1-variable.

LEMMA 4. If $\Omega = 0 \times \mathbb{R}^p$ and $0 \subset \mathbb{R}^m$ is bounded, then for all $u \in H^1_0(\Omega)$, we have:

$$\int_\Omega |u|^2 \, dx \leq (\lambda_1(0))^{-1} \int_\Omega |\nabla u|^2 \, dx,$$

where $\lambda_1(0)$ is again the first eigenvalue of $(-\Delta)$ acting on $H^1_0(0)$.

We prove this lemma defining

$$\lambda = \inf_{\substack{u \in H^1_0(\Omega) \\ u \not\equiv 0}} \frac{\int_\Omega |\nabla u|^2 \, dx}{\int_\Omega |u|^2 \, dx}$$

and noting that λ must be actually equal to $\lambda_1(0)$.

2. PROOF OF THEOREMS 1 AND 2

Proof of Theorem 1. First we modify g, defining

$$\tilde{g}(t) = \begin{cases} g(t) & \text{for } t \geq 0 \\ 0 & \text{for } t \leq 0. \end{cases}$$

It is clear that the positive solutions of (1) will be the same if we change g to $\tilde{g}$.

Then g satisfies the stronger conditions

$$\lim_{t\to 0^+} \frac{|g(t)|}{t} < +\infty,$$

$$\lim_{|t|\to+\infty} \frac{|g(t)|}{|t|^{\frac{2N}{N-2}}} = 0.$$

Now we define a functional S by

$$S(u) = \frac{1}{2}\int_\Omega |\nabla u|^2\, dx - \frac{\nu}{2}\int_\Omega |u|^2\, dx - \int_\Omega G(u)\, dx,$$

and we verify that S is a good C^1-functional on $E = H^1_{0,s}(\Omega)$.

The existence of a positive solution of (1) follows from the application of a critical point theorem of P. H. Rabinowitz (see Theorem 3.9 in [1]) to the functional S, for it is clear that critical points of S are solutions of (1).

Then, the proof of Theorem 1 consists in verifying that all conditions needed in the critical point theorem we want to apply are accomplished.

The most interesting point of the proof is the verification of the Palais-Smale condition in which we strongly use Lemma 4.

The decreasingness in $|x^1|$ of the solution is proved by the application of some results of Gidas and Nirenberg [8].

Proof of Theorem 2. This is very similar to the above. The only modifications are that now we do not replace g by $\tilde{g}$, and that the critical point result used is another one (Theorem 3.37 in [1]) which takes into account the evenness of S.

Remark. If f is locally Hölder continuous we use a bootstrap argument to prove that $u \in C^2(\bar{\Omega})$.

Remark. The case p = 1, $\Omega = 0 \times \mathbb{R}$, is a little different, because now we cannot apply any compactness result, and so we cannot use a global method to prove the existence of solutions of (1) in $\Omega = 0 \times \mathbb{R}$.

In this case we must use a 'local' method which consists in solving (1) in $\Omega_R = 0 \times (-R,R)$. Then we find for every $R \geqslant R_0$ a positive solution u_R of (1) in Ω_R.

Then we obtain a priori uniform estimates on the u_R and we prove the symmetry of u_R with respect to the $\mathbb{R}$-variable, and the decreasingness of $u_R(x_1,y)$ for $y \geqslant 0$.

Next we apply a version of Lemma 3 which provides us with the compactness we need. This lemma states the following:

LEMMA 5. Let $p = 1$, $q \in (2, \frac{2N}{N-2})$, $N = m+1$. Then, if $\Omega = 0 \times \mathbb{R}$, $0 \subset \mathbb{R}^m$ bounded, and if we denote by K the cone of $H_0^1(\Omega)$ defined by

$$K = \{u \in H_0^1(\Omega) | u \geqslant 0 \text{ in } \Omega;\ u(x,y) \text{ non-increasing in } y \text{ for } x \in 0,\ y \geqslant 0;\quad u(x,y) \text{ non-decreasing in } y \text{ for } x \in 0,\ y \leqslant 0\}.$$

Then, the Sobolev imbedding from $H_0^1(\Omega)$ into $L^q(\Omega)$ maps bounded closed sets in K into compact sets.

The application of this lemma enables us to pass to the limit as $R \to +\infty$, and to find a positive solution of (1) in $\Omega = 0 \times \mathbb{R}$.

3. AN EIGENVALUE PROBLEM RELATED TO (1)

Let us now focus our attention on the following problem:

$$\left.\begin{aligned} \Delta u + \lambda f(u) = 0 \text{ in } \Omega \\ u = 0 \text{ on } \partial\Omega \end{aligned}\right\} \tag{6}$$

where $\Omega = 0 \times \mathbb{R}^p$, $0 \subset \mathbb{R}^m$ is bounded, $p,m \geqslant 1$, and where f satisfies (3), (4) and

$$f(0) = 0,\ f(t) = 0\ \forall t \leqslant 0 \tag{7}$$

$$\exists \zeta > 0 \text{ such that } F(\zeta) > 0, \tag{8}$$

where $F(t) = \int_0^t f(s)\, ds$.

To solve (6) we consider the following minimization problem:

$$\text{Minimize } J(v) = \frac{1}{2}\int_\Omega |\nabla v|^2\, dx \text{ over } K_\eta, \tag{9}$$

where $K_\eta = \{v \in H_0^1(\Omega) | \int_\Omega F(v)\, dx \geqslant \eta\}$.

Under the above assumptions we can state the following.

THEOREM 6. For every $\eta > 0$ problem (9) has a solution $u \in H_0^1(\Omega) \cap W_{loc}^{2,q}(\Omega)$ $(\forall q < +\infty)$, which is positive and axially symmetric, i.e. $u \in H_{0,s}^1(\Omega)$, $u(x_1,x^1)$ is decreasing with respect to $|x^1|$ and u satisfies $\int_\Omega F(u)\, dx = \eta$.

Moreover, there exists a Lagrange multiplier $\lambda > 0$ for which (u,λ) is a solution of (6).

Remarks

(1) This theorem gives an extension of Pohozaev's results [11] for the case of strip-like domains.

(2) The fact that we can solve (6) without making assumption (5) on f suggests to us that (5) is no more than a technical hypothesis, which can probably be disregarded.

Proof. First we prove that for every $\eta > 0$, the minimizing set K_η is not empty. Then we take a minimizing sequence, and we consider the Steiner symmetrizations of its elements. This sequence will be another minimizing sequence, but now it will be in $H_{0,s}^1(\Omega)$. Then we can apply Lemma 3 to find a solution of problem (9).

To conclude we see that for this $u \in H_{0,s}^1(\Omega) \cap W_{loc}^{2,q}(\Omega)$ $(\forall q < +\infty)$ there exists $\lambda > 0$ such that (u,λ) is a solution of (6).

Finally, the positiveness of u follows from the strong maximum principle.

CONCLUSION

The results we give in this paper answer the question of the existence of solutions of (1). As we have already pointed out, it seems to us that the hypotheses made to solve (1) are not the best. Assumptions (2)-(4) seem to be optimal but it should be possible to find a different proof which would not use assumption (5), or that would consider a weakened version of it.

REFERENCES

[1] A. Ambrosetti and P.H. Rabinowitz, Variational methods for non-linear eigenvalue problems, in Eigenvalue of NL Problems, G. Prodi (ed.), C.I.M.E., Ediz. Cremonese, Roma (1974).

[2] A. Ambrosetti and P.H. Rabinowitz, Dual variational methods in critical points - theory and applications, J. Funct. Anal. 14, 349-381 (1973).

[3] H. Berestycki and P.L. Lions, Existence of solutions of NL equations. I. The ground state II. Existence of infinitely many bound states. Arch. Rat. Mech. Anal. (to appear).

[4] J. Bona, K. Bose and R.E.L. Turner, Finite amplitude waves in stratified fluid, M.R.C. technical report, Univ. of Wisconsin, Madison (1981).

[5] H.J. Brascamp, E.H. Lieb and J.M. Luttinger, A general rearrangement inequality for multiple integrals, J. Funct. Anal. 17, 227-237 (1974).

[6] M.J. Esteban, Nonlinear elliptic problems in strip-like domains. Symmetry of positive vortex rings, Nonlinear Analysis, vol.7 (1983).

[7] M.J. Esteban and P.L. Lions, A compactness lemma, Nonlinear Analysis, vol.7 (1983).

[8] B. Gidas, W. Ni and L. Nirenberg, Symmetry and related properties via the maximum principle, Comm. Math. Physics 68, 209-283 (1979).

[9] E.H. Lieb, Existence and uniqueness of the minimizing solution of Choquard's nonlinear equation, Studies Appl. Math. 57, 93-105 (1977).

[10] P.L. Lions, Minimization problems in $L^1(\mathbb{R}^3)$, J. Funct. Anal. (to appear).

[11] S. Pohozaev, Eigenfunctions of the equation $\Delta u + f(u) = 0$, Sov. Math. Doklady 5, 1408-1411 (1965).

[12] W.A. Strauss, Existence of solitary waves in higher dimensions, Comm. Math. Physics 55, 149-162 (1977).

M.J. Esteban
C.N.R.S. Laboratoire d'Analyse Numérique
Université Pierre et Marie Curie,
4 Place Jussieu - 75230
Paris Cedex 05
France

J FLECKINGER

Perturbation of the eigenvalues of operators of Schrödinger type

We obtain here some asymptotic estimates for the eigenvalues of an operator of Schrödinger type: $A_q = A + q$, defined on an unbounded domain Ω in $\mathbb{R}^N$, where the potential q is irregular.

1. INTRODUCTION

It is well known that the spectrum of $H_q = -\Delta + q$ defined on $\mathbb{R}^n$ is discrete when the potential q is a positive smooth function, tending to $+\infty$ at infinity. $N(s,H_q,\mathbb{R}^n)$, the number of eigenvalues of H_q less than s, is such that [12, 17]

$$N(s,H_q,\mathbb{R}^n) \sim c \int_{\{x\in\mathbb{R}^n/q(x)<s\}} (s-q(x))^{n/2}dx \qquad s \to +\infty.$$

This formula has been generalized by many authors to more general cases [1, 4, 5, 14, ...].

Let us consider, for example, an operator of Schrödinger type $A_q = A + q$ defined on an unbounded domain $\Omega \subset \mathbb{R}^n$; A is formally self-adjoint, elliptic, with smooth coefficients.

With suitable hypotheses on A, q and Ω one has the following formula for $N(s, A_q, \Omega)$, the number of eigenvalues less than s of A_q defined on Ω (with homogeneous Dirichlet boundary conditions):

$$N(s,A_q,\Omega) \sim \int_{\Omega_s} \mu(x)(s-q(x))^{n/2m}\, dx \qquad s \to +\infty \tag{0}$$

where

$$\Omega_s = \{x \in \Omega/q(x) < s\};$$

$$\mu(x) = (2\pi)^{-n} \text{ meas } \{\xi \in \mathbb{R}^n/A'(x,\xi) < 1\}$$

$A'(x,\xi)$ is the symbol of the leading part of A.

When $\Omega = \mathbb{R}^n$ sharp estimates are obtained for some of these operators in [8, 9, 13, 16].

The case of Neumann boundary conditions has been studied in [1] and an estimate like (0) for the singular numbers of Schrödinger operators with complex potentials has been obtained in [2, 3, 7].

Some results for nonsmooth potentials have been studied in [10, 14].

Notation. It will be convenient to define some notation: when

$$\alpha = (\alpha_1, \dots, \alpha_n) \in \mathbb{N}^n$$

is a multi index, $|\alpha| = \alpha_1 + \dots + \alpha_n$, and D^α denotes a derivative of order $|\alpha|$:

$$D^\alpha = (-1)^{|\alpha|} \partial_{x_1}^{\alpha_1}, \dots, \partial_{x_n}^{\alpha_n}.$$

For positive functions f and g, $f \approx g$ means that f/g and g/f are uniformly bounded.

When $x \in \mathbb{R}^n$, $\langle x \rangle = (1 + |x|^2)^{1/2}$.

2. RECALLS FOR THE SMOOTH CASE

Let m be a positive integer and Ω be an unbounded domain in $\mathbb{R}^n$.

(1) Let ρ and q be two continuous functions defined on Ω, real-valued, bounded below by 1,q, tending to $+\infty$ at infinity; we suppose that

$$\rho^{-m}(x) \sqrt{q(x)} \to +\infty \quad \text{when} \quad |x| \to +\infty.$$

Let us denote by $V_q^o(\Omega)$ the completion of $C_o^\infty(\Omega)$ with respect to the norm $\| \ \|_{q,\Omega}$ where:

$$\|u\|_{q,\Omega} = \{\int_\Omega [\sum_{|\alpha| \leq m} \rho^{|\alpha|}(x) |D^\alpha u(x)|^2 + q(x)|u(x)|^2] \, dx\}^{1/2}$$

$\alpha = (\alpha_1, \dots, \alpha_n) \in \mathbb{N}^n$ and D^α is a derivative of order $|\alpha| = \alpha_1 + \dots + \alpha_n$. $V_q^o(\Omega)$ is a Hilbert space and it is simple to verify that:

PROPOSITION 1. The imbedding $V_q^o(\Omega)$ into $L^2(\Omega)$ is compact.

(2) Let a_q be an integrodifferential form, continuous and coercive on $V_q^o(\Omega)$:

$$a_q(u,v)=(a+q)(u,v) = \int_\Omega \Big(\sum_{\substack{|\alpha|\leq m \\ |\beta|\leq m}} a_{\alpha\beta}(x)D^\alpha u(x)\,\overline{D^\beta v(x)}+q(x)u(x)\overline{v(x)}\Big)\,dx$$

for $(u,v) \in V_q^o(\Omega) \times V_q^o(\Omega)$.

(3) We suppose that : $a_{\alpha\beta} = \overline{a_{\beta\alpha}} \in C^o(\bar{\Omega})$ and that

$$\exists \gamma_1 > 0 \ |\rho^{-(\alpha+\beta)/2}(x)\ a_{\alpha\beta}(x)| \leq \gamma_1 \quad \forall x \in \bar{\Omega}, (\alpha,\beta) \in \mathbb{N}^n \times \mathbb{N}^n,$$

$$|\alpha|\leq m,\ |\beta|\leq m.$$

Let us denote by A_q^o the positive self-adjoint operator, unbounded in $L^2(\Omega)$, associated by the Lax-Milgram theorem with the variational problem $(V_q^o(\Omega), L^2(\Omega), a_q)$.

We deduce from the Proposition 1 that A_q^o has a discrete spectrum constituted by isolated eigenvalues:

$$0 < s_1 \leq s_2 \leq \dots \leq s_j \leq \dots \qquad s_j \xrightarrow[j \to +\infty]{} +\infty;$$

(each eigenvalue is repeated according to its multiplicity).

We denote by $N(s,A_q^o,\Omega)$ the number of eigenvalues of A_q^o less than s (each eigenvalue is counted according to its multiplicity).

In [6] we proved (0) for $N(s,A_q^o,\Omega)$ under the following assumptions:

(4) There exist two positive numbers ε_0 and s_0 such that ρ, q and $a_{\alpha\beta}$ can be extended to $\tilde{\Omega} = \{x \in \mathbb{R}^n/\text{dist}\,(x,\Omega) < \varepsilon_0\}$ and

$$\forall \varepsilon \in]0,\varepsilon_0[,\ \forall s \geq s_0 \quad \forall \eta \leq \eta_s = \sup_{\{x\in\tilde{\Omega}/q(x)>s\}} (\rho(x)\ q^{-1/2m}(x))^{1/2}$$

$$\forall (x,y) \in \tilde{\Omega} \times \tilde{\Omega} \quad q(x) < s,\ q(y) < s,\ |x-y| < \sqrt{n}\,\eta \Longrightarrow$$

$$|\rho(x) - \rho(y)| \leq \varepsilon\rho(x),$$

$$|q(x) - q(y)| \leq \varepsilon q(x),$$

$$|a_{\alpha\beta}(x) - a_{\alpha\beta}(y)| \leq \varepsilon |a_{\alpha\beta}(x)|.$$

When ω is a subset of $\tilde{\Omega}$, we denote by $V_q^1(\omega)$ the set of the restrictions to ω of functions in $V_q^o(\Omega)$ and by A_q^1 the realization of the variational problem $(V_q^1(\omega), L^2(\omega), a_q)$.

We suppose that a is uniformly coercive, i.e. :

$$\exists \gamma_2 > 0 \quad \forall \omega \subset \tilde{\Omega} \quad \forall u \in V_q^i(\omega) \quad a(u,u) \geqslant \gamma_2 \int_\omega \sum_{|\alpha| \leqslant m} \rho^\alpha(x) |D^\alpha u(x)|^2 \, dx. \tag{5}$$

For any positive number s, $\Omega_s = \{x \in \Omega / q(x) < s\}$ is a Lebesgue measurable set and

$$\exists s_1 \geqslant s_0 \quad \exists \gamma_3 > 0 \text{ such that } [\Omega_s] \leqslant \gamma_3 \, [\Omega_{s/2}] \text{ for all } s \geqslant s_1, \tag{6}$$

where

$$[\Omega_s] = \int_{\Omega_s} \rho^{-n/2}(x) \, dx.$$

We consider a partition of $\mathbb{R}^n$ into nonoverlapping cubes $(Q_\zeta)_{\zeta \in \mathbb{Z}^n}$ with side η and centres x_ζ and we suppose that

$$\frac{\sum_{\zeta \in I \setminus I^o} (\eta^2/\rho_\zeta)^{n/2}}{\sum_{\zeta \in I^o} (\eta^2/\rho_\zeta)^{n/2}} \underset{\eta \to 0}{\to} 0 \quad \forall s \geqslant s_1 \text{ where } \rho_\zeta = \rho(x_\zeta); \tag{7}$$

$I^o = \{\zeta \in \mathbb{Z}^n / \bar{Q}_\zeta \subset \Omega_s\}$; $I = \{\zeta \in \mathbb{Z}^n / \bar{Q}_\zeta \cap \Omega_s \neq \emptyset\}$.

Using a Tauberian theorem, it is possible to deduce from (0) an asymptotic estimate for s_j as j tends to $+\infty$.

Suppose, for example, that A is a differential operator of order 2m with bounded coefficients and that $q(x) = \langle x \rangle^k$; if $\Omega = \mathbb{R}^n$, then:

$$N(s, A_q, R^n) \sim c' \, s^{(n/2m)+(n/k)} \text{ as } s \to +\infty \text{ and}$$

$$s_j \sim c'' \, j^{2km/(n(k+2m))} \quad j \to +\infty.$$

In that case, when $\Omega = \mathbb{R}^n$, under suitable assumptions, it is possible to obtain sharp estimates. For example, the following result is proved in [16]:

Suppose that

$$\text{A is an uniformly elliptic operator of order 2 and that the coefficients } a_{\alpha\beta} \text{ are such that } |D^\gamma a_{\alpha\beta}| < c_\gamma \langle x \rangle^{-\gamma} \tag{8}$$

and that

> q is such that $q(x) \sim \langle x\rangle^k$; $|D^\gamma q| \leq c\langle x\rangle^{k-|\gamma|}$ and
> $x.\nabla q \geq c\langle x\rangle^k$ for $|x|$ big enough; (9)

PROPOSITION 2. If $n \geq 2$ and if (8) and (9) are satisfied, we have:

$$N(s,A_q,\mathbb{R}^n) = (2\pi)^{-n} \iint_{A(x,\xi)+q(x)<s} d\xi\, dx\, (1+O(s^{-(1/2)-(1/k)})) \quad (10)$$

as $s \to +\infty$, where $A(x,\xi)$ is the symbol of A.

When $A(x,\xi)$ is 'quasihomogeneous' of (x,ξ) and when q is a homogeneous polynomial of even order, (10) holds (see [8]).

3. PERTURBATION OF THE EIGENVALUES

We consider now a real potential h, and the associated operator A_h^o defined on Ω as previously. We suppose that

> there exist two positive numbers c_o and r in $]0,1[$ and a potential q such that q and A_q^o satisfy (1), (2) and (3) and are such that $|q-h|q^{-r} < c_o$. (11)

We deduce from (11) that

LEMMA 1. $q \approx h$ and $V_h^o(\Omega) = V_q^o(\Omega)$ with equivalent norms.

Proof: It follows from (11) that $|1 - h/q| \leq c_o\, q^{r-1} < c_o$. Hence

$$0 < 1 - c_o < h/q < 1+c_o \quad \text{and} \quad (1+c_o)^{-1} < q/h < (1-c_o)^{-1}$$

Therefore, $\int q|u|^2 dx < c' \int h|u|^2\, dx < c'' \int q|u|^2\, dx$. It is now easy to prove

LEMMA 2. $|a_q(u,u) - a_h(u,u)| < c_o\, (a_q(u,u))^r\, \|u\|^{2(1-r)}$.

Proof:

$$|a_q(u,u) - a_h(u,u)| = |\int (q-h)|u|^2 dx|^2 < c_o \int q^r\, |u|^2\, dx$$
$$< c_o\, (\int q|u|^2\, dx)^r (\int |u|^2 dx)^{1-r} \leq c_o(a_q(u,u))^r\, \|u\|^{2(1-r)}.$$

From Lemma 1, it is obvious that the spectrum of A_h^o is discrete; let us denote by $0 < \lambda_1 \leq \lambda_2 \leq \ldots \leq \lambda_j \leq \ldots$ the eigenvalues of A_h^o.

THEOREM 1. Assume that (11) holds: then λ_j/s_j tends to 1 as $j \to +\infty$.

Proof: We use the perturbation method of [11] as in [6]. We know from the max-min principle that

$$s_{j+1} = \inf_{u \perp E_j} \frac{a_q(u,u)}{\|u\|^2}$$

where E_j is the j dimensional space spanned by the eigenfunctions associated with the j first eigenvalues $s_1,\ldots,s_j$. Similarly

$$\lambda_{j+1} = \sup_{F \in G_j} \inf_{u \perp F} \frac{a_h(u,u)}{\|u\|^2}$$

where G_j is the set of all j dimensional subspaces. Hence:

$$\lambda_{j+k+1} \geq \inf_{u \perp E_{j+k}} \frac{a_h(u,u)}{\|u\|^2} \geq \inf_{u \perp E_j} \frac{a_q(u,u)}{\|u\|^2} \inf_{u \perp E_k} \frac{a_h(u,u)}{a_q(u,u)}$$

$$= s_{j+1} \inf_{u \perp E_k} \left(1 + \frac{a_h(u,u) - a_q(u,u)}{a_q(u,u)}\right).$$

Choosing the integer k big enough so that $c_o\, s_k < 1$, we deduce from Lemma 2 that

$$\frac{|a_q(u,u) - a_h(u,u)|}{a_q(u,u)} < c_o \left(\frac{a_q(u,u)}{\|u\|^2}\right)^{r-1}.$$

Hence:

$$\lambda_{j+k+1} \geq s_{j+1}(1 + O(s_k^{-(1-r)})). \tag{12}$$

To prove the converse inequality, we first notice that it follows from Lemma 1 that $(q-h)h^{-r}$ is bounded. After analogous calculations we obtain

$$s_{j+2k+1} \geq \lambda_{j+k+1}\,(1 + O(\lambda_k^{-(1-r)})). \tag{13}$$

We obtain Theorem 1 from (12) and (13) on letting j tend to $+\infty$ with k/j tending to 0.

We will now extend to nonsmooth potentials the sharp estimates of Proposition 2.

THEOREM 2. Suppose that (11) is satisfied and that

$$s_j = c\, j^d (1 + O(j^{-t})) \text{ as } j \to +\infty \tag{14}$$

where $0 < t < (1-r)d\,(1+(1-r)d)^{-1}$; then

$$\lambda_j = c\, j^d(1 + O(j^{-t})) \text{ as } j \to \infty .$$

Example. Consider $H = -\Delta + \langle x\rangle^{2k} + f(x)$ and $H_o = -\Delta + \langle x\rangle^{2k}$, with $|f(x)| < c\, \langle x\rangle^{2k/p}$, $p > 1$, $n > 2$. We know [8, 13, 16] that $N(s,H_o,R^n) = c\, s^{n(k+1)/2k}(1 + O(s^{-(k+1)/2k}))$ as $s \to +\infty$, so that

$$s_j = c\, j^{2k/((n(k+1))}(1 + O(j^{-1/n})) \text{ as } j \to +\infty. \tag{15}$$

Then (15) holds for λ_j when $p \geqslant \frac{2(n-1)k}{(n-2)k-n}$ and $k > \frac{n}{n-2}$.

Proof of Theorem 2. This is the same as in [7] or [11]. We first established

LEMMA 3. $s_{j+1}/s_j = 1 + O(1/j) + O\,(j^{-t})$, as $j \to \infty$.

We deduce from (12) and (13) that

$$s_{j+k+1}(1 + O(s_k^{r-1})) \leqslant \lambda_{j+2k+1} \leqslant s_{j+4k+1}(1 + O(s_k^{r-1})).$$

Using Lemma 3 and choosing $k/j = j^{-z}$, we therefore have $\lambda_{j+2k+1}/s_j = 1+O(j^{-\nu})$ where $\nu = \min\left(t, \frac{(1-r)d}{1+(1-r)d}\right)$.

Remark. (1) Theorem 3 holds when $s_j = c\, j^d \log j(1+a(j^{-t}))$, with the same bound for t; of course, in that case, λ_j has the same asymptotic estimate as s_j.

It is possible, with the same method to study complex potentials [7].

It is a simple consequence of the proof of Theorem 1 that:

COROLLARY 1. If (11) is satisfied and if $c_0 < 1$, then $\lambda_1 > s_1(1-c_0 s^{r-1}) > 0$.

REFERENCES

[1] D.E. Edmunds and W.D. Evans, On the distribution of eigenvalues of Schrödinger operators (to appear).

[2] D.E. Edmunds, W.D. Evans and J. Fleckinger, Spectres d'opérateurs de Schrödinger à potentiels complexes et répartition de leurs valeurs singuliéres, CRAS, Paris 295, 333 (1982).

[3] D.E. Edmunds, W.D. Evans and J. Fleckinger, On the spectrum and the distribution of singular values of Schrödinger operators with a complex potential (to appear).

[4] J. Fleckinger, Répartition des valeurs propres d'opérateurs de type Schrödinger, CRAS, Paris 292, 359 (1981).

[5] J. Fleckinger, Estimate of the number of eigenvalues for an operator of Schrödinger type, Proc. Roy. Soc. Edinburgh, 89A, 355 (1981).

[6] J. Fleckinger, Distribution of the eigenvalues of operators of Schrödinger type, Math. Studies 55 (North Holland), 173 (1981).

[7] J. Fleckinger, On the singular values of non self adjoint operators of Schrödinger type, Proc 1982 Dundee Conference on ODE and PDE (to appear).

[8] B. Helffer and D. Robert, Comportement asymptotique précisé du spectre d'opérateurs globalement elliptiques dans $\mathbb{R}^n$. CRAS, Paris 292, 363 (1981).

[9] L. Hormander, On the asymptotic distribution of eigenvalues of pseudo differential operators in $\mathbb{R}^N$, Arkiv für Mat. 17, 297 (1979).

[10] Pham The Lai, Comportement asymptotique des valeurs propres d'une classe d'opérateurs de type Schrödinger, J. Math. Kyoto Univ. 18, 353 (1978).

[11] A.G. Ramm, Spectral properties of some non self adjoint operators and some applications, Math Studies (North Holland) 55, 349 (1981).

[12] M. Reed and B. Simon, Mathematical Physics, Academic Press, New York, (1978).

[13] D. Robert, Propriétés spectrales d'opérateurs différentiels, Thèse, Nantes (1977).

[14] G.V. Rosenbljum, Asymptotics of the eigenvalues of the Schrödinger operator, Math USSR Sb, 22, 349 (1974).

[15] G.V. Rosenbljum, Asymtotics of the eigenvalues of the Schrödinger operator, Problemy Matematiceskogo Analiza, 5, 152 (1975).

[16] H. Tamura, Asymptotic formulas with sharp remainder estimates for eigenvalues of elliptic operators of second order. Duke Math. J. 49 (1), 87 (1982).

[17] E. Titchmarsh, Eigenfunctions expansions, Part 2, Oxford Univ. Press (1958).

J. Fleckinger
Université P. Sabatier
31077 Toulouse
FRANCE

M FREMOND, H GHIDOUCHE & N POINT

Coupled Stefan problem in wet porous media

1. THE PHYSICAL PROBLEM

A wet porous medium, for instance a soil, contains water which freezes when the temperature is below 0°C. In the unfrozen part, movements of water are induced by a depression which appears on the free surface (the 0°C isotherm). This depression, the cryogenic suction, causes vertical displacements of soil up to many metres.

We describe below a physical and mathematical model which takes into account the existence of zones with non-zero measure where the temperature is 0°C. They are called clouds or mushy regions [1], [5]: there can exist intermediate states between pure water and pure ice. So the free surface is not always a surface in the common sense.

2. THE EQUATIONS

They are obtained by using the conservation and constitutive laws - see [2], [4]. The porous medium considered during a time interval [0,T], occupies a regular domain $\Omega \subset \mathbb{R}^n$. The unknowns are the temperature $u(x,t)$, the water content $\mu(x,t)$ and the head of water $H(x,t)$ $(x \in \Omega,\ t \in (0,T))$.

We define the unfrozen part $Q_1(u(x,t) > 0)$, the frozen part $Q_2(u(x,t) < 0)$ and the cloud $Q_3(u(x,t) = 0)$ of $Q = \Omega \times]0,T[$.

To simplify we assume that all the physical constants are equal to 1. The energy is $E(u,\mu) = u + (\mu-1)$ with $\mu = 1$ if $u > 0$, $\mu = 0$ if $u < 0$ and $0 \leqslant \mu \leqslant 1$ if $u = 0$. The equations are:

$$\left.\begin{aligned} &\frac{\partial E}{\partial t} - \Delta u = r, && (1)\\ &- \Delta H = 0, && (2)\\ &\mu = 1, && (3)\end{aligned}\right\} \quad \text{in } Q_1,$$

$$\left.\begin{aligned} \frac{\partial E}{\partial t} - \Delta u &= r, \\ H &= 0, \\ \mu &= 0, \end{aligned}\right\} \text{ in } Q_2, \qquad \begin{aligned} &(4)\\ &(5)\\ &(6) \end{aligned}$$

$$\left.\begin{aligned} u &= 0, \\ \frac{\partial E}{\partial t} - \Delta H &= r, \\ 0 \leqslant \mu &\leqslant 1 \end{aligned}\right\} \text{ in } Q_3, \qquad \begin{aligned} &(7)\\ &(8)\\ &(9) \end{aligned}$$

where the rate of heat production r is a given function. On the free surface, we have

$$[\text{grad } u + \text{grad } H] \cdot \vec{N} + [E]\vec{W}\cdot\vec{N} = 0, \; u = 0, \; H = 0 \qquad (10)$$

where [A] is the discontinuity of the quantity A, $\vec{N}$ a unit normal vector to the free surface and $\vec{W}$ the velocity of the free surface.

The equations are completed by boundary and initial conditions. We assume that the boundary Γ of Ω is divided into three parts Γ^+ (unfrozen), Γ^- (frozen), Γ^a (adiabatic and impermeable) ($\Gamma^{\pm} = \Gamma^+ \cup \Gamma^-$).

We consider two types of boundary conditions, the Fourier boundary conditions and the Dirichlet ones.

The Fourier conditions are:

$$\frac{\partial u}{\partial n} + (u-\bar{u}) = 0, \; \frac{\partial H}{\partial n} + (H-\bar{H}) = 0 \text{ on } \Gamma^{\pm},$$

$$\frac{\partial u}{\partial n} = \frac{\partial H}{\partial n} = 0 \text{ on } \Gamma^a,$$

where $\bar{u}$ and $\bar{H}$ are given functions.

The Dirichlet conditions are

$$u = \bar{u}, \; H = \bar{H} \text{ on } \Gamma^{\pm}, \; \frac{\partial u}{\partial n} = \frac{\partial H}{\partial n} = 0 \text{ on } \Gamma^a$$

we suppose $\bar{u} > 0$, $\bar{H} > 0$ on Γ^+, $\bar{u} < 0$, $\bar{H} = 0$ on Γ^-. The initial condition is

$$E(u(0),\mu(0)) = E_0.$$

3. VARIATIONAL FORMULATION

3.1. Notation

Let us give some notation. For smooth functions ϕ and ψ, we define

$$(\phi,\psi) = \int_\Omega \phi\psi \, d\Omega, \quad |\phi| = (\int_\Omega \phi^2 \, d\Omega)^{\frac{1}{2}},$$

$$a(\phi,\psi) = \int_\Omega \text{grad } \phi \text{ grad } \psi \, d\Omega + \int_{\Gamma^\pm} \phi\psi \, d\Gamma,$$

$$b(\phi,\psi) = \int_\Omega \text{grad } \phi \text{ grad } \psi \, d\Omega,$$

$$R(\phi) = (r,\phi), \; L_1(\phi) = \int_{\Gamma^\pm} \bar{u}\phi \, d\Gamma, \; L_2(\phi) = \int_{\Gamma^\pm} \bar{H}\phi \, d\Gamma$$

$$\mathcal{L} = R + L_1 + L_2,$$

$$\Sigma^+ = \Gamma^+ \times]0,T[, \; \Sigma^- = \Gamma^- \times]0,T[, \Sigma^\pm = \Sigma^+ \cup \Sigma^-.$$

Let us define the following functional spaces (supp ϕ is the support of a function ϕ):

$V(Q^+)$ (respectively $V_0(Q^+)$) the closure in $L^2(0,T;H^1(\Omega))$ of the space

$$\{\phi \mid \phi \in C^\infty(\bar{Q}); \text{ supp } \phi \subset Q^+; \phi|_{\Sigma^-} = 0\}$$

(respectively of the space

$$\{\phi \mid \phi \in C^\infty(\bar{Q}); \text{ supp } \phi \subset Q^+; \phi|_{\Sigma^\pm} = 0\}),$$

$$W = \{\psi \mid \psi \in L^2(0,T;H^1(\Omega)); \frac{d\psi}{dt} \in L^2(Q); \psi(T) = 0\},$$

$$W_0 = \{\psi \mid \psi \in W ; \psi|_{\Sigma^\pm} = 0\}.$$

3.2. The mathematical variational formulation

We intend to give a variational formulation of the problem (equations (1) to (10)) for the Fourier boundary conditions (it is easy to give it for the Dirichlet boundary conditions). It turns out that the sets Q_1, Q_2 and Q_3 are not well adapted to a variational formulation. We need sets Q^+ and Q^- slightly different from the physical ones Q_1, Q_2 and Q_3.

The set Q^+ is the set of the unfrozen points (x,t) where no phase change occurs ($E \geq 0$, $\mu = 1$). It contains, at least, Q_1 and possibly a part of the cloud. The set Q^- is the set of the frozen points and of the points where a phase change can occur ($E \leq 0$, $0 \leq \mu \leq 1$). More precisely the unknowns are the functions u, μ, H and two open sets Q^+, Q^- which satisfy the following conditions:

$$\text{supp } u^+ \subset C\, Q^- \subset \text{supp } u^+ \cup Z \cup N^+, \; u \leq 0 \text{ on } Q^- \tag{11}$$

$$\text{supp } u^- \subset C\, Q^+ \subset \text{supp } u^- \cup Z \cup N^-, \; u \geq 0 \text{ on } Q^+, \tag{12}$$

where N^+ and N^- are two sets of measure zero and

$$Z = \{(x,t);\; u(x,t) = 0\},$$

$$\mu = 1 \text{ on } Q^+,\; \mu = 0 \text{ on } \overset{\circ}{\widehat{\text{supp}\, u^-}} \text{ (the interior of the support of } u^-). \tag{13}$$

Note that the sets Q^+ and Q^- are linked to the water content μ and not to the temperature u, as Q_1, Q_2 and Q_3 are. The variational formulation is obtained in the following manner.

Let us consider a smooth test function ϕ whose support is contained in Q^+ and multiply equation (2) by ϕ. On integrating by parts we obtain:

$$\int_0^T a(H,\phi)\, dQ + \int_{\Sigma^+} H\phi\, d\Sigma = \int_0^T L_2(\phi)\, dt. \tag{14}$$

Let us add the equations (1) and (2) to obtain

$$\frac{\partial E}{\partial t} - \Delta u - \Delta H = r. \tag{15}$$

Let us multiply this equation (15) by a smooth function ψ ($\psi(T) = 0$). Integration by parts and the equation (10) lead to

$$\int_0^T \{ -\left(E, \frac{d\psi}{dt}\right) + a(u,\psi) + a(H,\psi)\}\, dQ = \int_0^T \mathcal{L}(\psi) dt + (E_0, \psi(0)). \tag{16}$$

The unknowns are the functions u, μ and H and the sets Q^+ and Q^-. The equations are (11)-(14) and (16).

4. THE MAIN RESULTS

We have the two following theorems, the first one for the Fourier boundary conditions, the second for the Dirichlet boundary conditions.

THEOREM 1. We assume that the open set Ω is regular, $E_0 \in L^2(\Omega)$, $r \in L^2(Q)$, $\bar{u}, \bar{H} \in L^2(\Sigma^{\pm})$, $\bar{H}_{|_{\Sigma^-}} = 0$. Then

(i) There exist two open sets Q^+, Q^- of $\bar{Q}$ and three functions u, μ, H which verify

$$u, H \in L^2(0,T;H^1(\Omega)), \; u \in L^\infty(0,T;L^2(\Omega)), \; \mu \in L^\infty(Q),$$

$$\text{supp } u^+ \subset \complement\, Q^- \subset \text{supp } u^+ \cup Z \cup N^+, \; u \leq 0 \text{ on } Q^-$$

$$\text{supp } u^- \subset \complement\, Q^+ \subset \text{supp } u^- \cup Z \cup N^-, \; u \geq 0 \text{ on } Q^+$$

$$0 \leq \mu \leq 1, \; \mu_{|_{Q^+}} = 1, \; \mu = 0 \text{ on } \widehat{\overset{\circ}{\text{supp}}\, u^-}$$

$$H = 0 \text{ on } \overset{\circ}{\text{supp}}\, u^-, \; H_{|_{\Sigma^-}} = 0.$$

If $\bar{H} \geq 0$ then $H \geq 0$.

(ii) $\forall \phi \in V, \displaystyle\int_0^T a(H,\phi)\, dt = \int_0^T L_2(\phi)\, dt,$

$$\forall \psi \in W, \int_0^T \left(-\left(E(u,\mu), \frac{d\psi}{dt}\right) + a(u,\psi) + a(H,\psi)\right) dt = \int_0^T \mathcal{L}(\psi) dt + (E_0, \psi(0))$$

where

$$E(u,\mu) = u + (\mu - 1).$$

THEOREM 2. We suppose that the hypotheses of Theorem 1 are satisfied and that there exist two functions E, q and a number $\gamma > 0$ such that

$$E \in L^2(0,T;H^1(\Omega)), \; \frac{dE}{dt} \in L^2(Q), \; q \in L^\infty(Q), \; \text{grad } q \in L^\infty(Q),$$

$$\bar{u}_{|_{\Sigma^+}} = E_{|_{\Sigma^+}} \geq \gamma, \; \bar{u}_{|_{\Sigma^-}} = (E_{|_{\Sigma^-}} - 1) \leq -\gamma,$$

$$q_{|_{\Sigma^+}} = \frac{\bar{H}}{\bar{u}}, \; \|q\|_{L^\infty(Q)} \leq c, \; \|\text{grad } q\|_{L^\infty(Q)} \leq c,$$

where c is a strictly positive constant which depends on the data (but not on q!).

The conclusions (i) of Theorem 1 are satisfied and we have

$$\forall\phi \in V_0, \int_0^T b(H,\phi)\, dt = 0,\ H_{|_{\Sigma^\pm}} = \bar{H}$$

$$\forall\psi \in W_0, \int_0^T \{-\ (E(u,\mu), \frac{d\psi}{dt}) + b(u,\psi) + b(H,\psi)\}\, dt$$

$$= \int_0^T R(\psi)\, dt + (E_0,\psi(0)),$$

$$u_{|_{\Sigma^\pm}} = \bar{u}.$$

5. OUTLINE PROOF OF THEOREM 1

We use a regularization of E and a penalization of the condition H = 0. We introduce two regularizations $P_n(x)$ and $M_n(x)$ of the symmetric function of the Heaviside function such that $|M_n'(x)| \leqslant nP_n(x)$ (Figure 1).

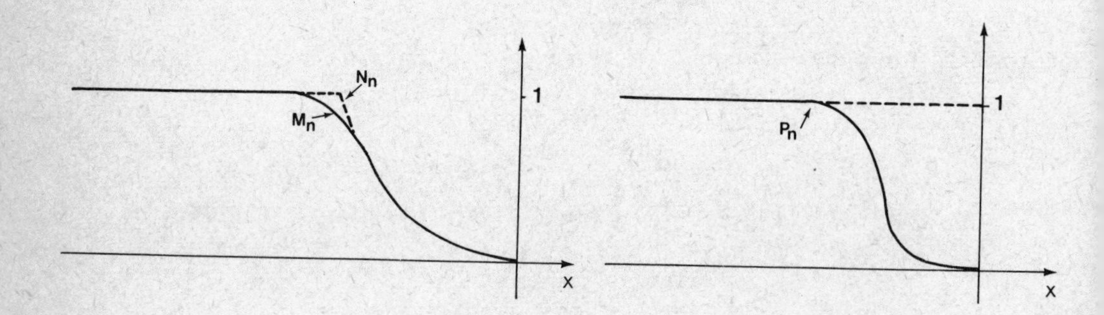

Figure 1. The functions P_n and M_n ($N_n(x) = \text{Inf}\ \{1,n \int_x^0 P_n(y)\, dy\}$).

We define the regularized energy

$$E_n(u) = u - M_n(u).$$

We introduce the following problem:

Find u_n and H_n such that

$$\forall\psi \in W, \int_0^T \{ -\left(E_n(u_n), \frac{d\psi}{dt}\right) + a(u_n,\psi) + a(H_n,\psi)\} \, dt$$

$$= \int_0^T \mathcal{L}(\psi) \, dt + (E_0,\psi(0)) \tag{17}$$

$$\forall\phi \in V, \int_0^T a(H_n,\phi) + n^2(P_n^2(u_n)\text{grad } H_n, \text{ grad } \phi) + n^2(P_n^2(u_n)H_n,\phi)$$

$$+ n^2 \int_{\Sigma^-} H_n\phi \, d\Sigma^- = \int_0^T L_2(\phi)dt. \tag{18}$$

Note that the condition $H = 0$ in Q_2 is penalized by the terms $n^2P_n^2(u_n)\text{grad } H_n$ and $n^2P_n^2(u_n)H_n$. By using the Galerkin method we show that the problem (17)-(18) has a solution which verifies

$$\|E_n(u_n)\|_{L^\infty(0,T;L^2(\Omega))} \leq c, \quad \|u_n\|_{L^2(0,T;H^1(\Omega))} \leq c, \tag{19}$$

$$\|H_n\|_{L^2(0,T;H^1(\Omega))} \leq c, \ |P_n \text{ grad } H_n|_{L^2(Q)} \leq \frac{c}{n},$$

$$|P_nH_n|_{L^2(P)^n} \leq \frac{c}{n} \ , \ |H_n|_{L^2(\Sigma^-)} \leq \frac{c}{n} \tag{20}$$

where c is a constant which does not depend on n.

These a priori estimates imply that we can extract subsequences (u_n) and (H_n) which converge to u and H in $L^2(0,T;H^1(\Omega))$ equipped with the weak topology.

We define the approximated set Q_n^+ and Q_n^- by

$$Q_n^+ = \complement F_n^-, \ F_n^- = \text{supp } v_n^-, \ Q_n^- = \complement F_n^+, \ F_n^+ = \text{supp } w_n^+$$

where

$$v_n^- = \text{supp } \{-v_n,0\}, \ w_n^+ \text{ sup } \{w_n,0\}$$

and

$$v_n(x,t) = \sup \{u_{n+p}(x,t), u(x,t), p \geq 0\}$$

$$w_n(x,t) = \inf \{u_{n+p}(x,t), u(x,t), p \geq 0\}.$$

The a priori estimates (19)-(20) do not imply that u_n converges almost everywhere. By using the equation (15) integrated with respect to t, we prove the convergence almost everywhere of a subsequence of u_n.

Let us consider a subsequence of the sequence u_n such that the F_n^- and F_n^+ are not empty. The subsequences $(F_n^\pm)$ are compact for the Hausdorff topology [6]. Then there exist F^+ and F^- such that subsequences F_n^- and F_n^+ tend towards F^+ and F^-.

By using the following lemma and the a priori estimates (19)-(20) and by letting $n \to +\infty$ we prove Theorem 1.

LEMMA (from [6]) Let $\mathcal{O}$ be an open set of $\bar{Q}$ and $\mathcal{O}_n$ a subsequence of open sets of $\bar{Q}$ such that $C\mathcal{O}_n$ tends towards $C\mathcal{O}$ for the Hausdorff topology. Then

$$\forall\phi \in V(\mathcal{O}) \text{ (resp. } V_0(\mathcal{O})) \quad \exists\phi_n \in V(\mathcal{O}_n) \text{ (resp. } \phi_n \in V_0(\mathcal{O}_n))$$

such that $\phi_n \to \phi$ in $L^2(0,T;H^1(\Omega))$ equipped with the strong topology.

6. OUTLINE PROOF OF THEOREM 2

By using Galerkin's method, we prove the existence of a solution u_n, H_n of the problem:

$$\forall\phi \in W, \int_0^T \Big(-\Big(E_n(u_n), \frac{d\phi}{dt}\Big) + b(u_n,\phi) + b(H_n,\phi)\Big)dt$$

$$= \int_0^T (R(\phi)\, dt + (E_0,\phi(0)), \qquad (21)$$

$$u_n\big|_{\Sigma^\pm} = \bar{u}$$

$$\forall\phi \in L^2(0,T;H^1(\Omega)),$$

$$\int_0^T (b(H_n,\phi) + n^2(P_n^2(u_n) \text{ grad } H_n, \text{ grad } \phi) + n^2(P_n^2(u_n)H_n,\phi))\, dt$$

$$+ n^2 \int_{\Sigma^+} (H_n-\bar{H})\phi\, d\Sigma + n^2 \int_{\Sigma^-} H_n\phi\, d\Sigma = 0. \qquad (22)$$

The functions u_n and H_n satisfy the a priori estimates (19)-(20) and also

$$|H_n-\bar{H}|_{L^2(\Sigma^\pm)} < \frac{c}{n},$$

where c is a constant which does not depend on n.

Let n tend towards infinity in (21) and (22); we prove Theorem 2 in the same way as we proved Theorem 1.

7. CONCLUSION

The notion of free surface (the isotherm 0°C) which seems to offer a simple way of describing the physical problem is in fact not used. It is more convenient to use the zone Q^+ which is the set where the water diffusion equation is valid ($\mu = 1$) than to use the zone Q_1 where the temperature is positive.

REFERENCES

[1] J. Aguirre-Puente and M. Fremond, Frost and water propagation in porous media. Second conference on soil-water in cold regions. Edmonton (1976).

[2] M. Fremond, H. Ghidouche and N. Point, Congelation d'un milieu poreux humide alimenté en eau. Modèle et théorèmes d'existence. Note aux C.R.A.S., Paris (15 March 1982).

[3] M. Fremond, H. Ghidouche and N. Point (to appear).

[4] P. Germain, Cours de Mécanique des Milieux Continus, Masson, Paris (1973).

[5] J. Ockendon, W. Hodgkins (eds.) Moving Boundary Problems in Heat Flow and Diffusion, Clarendon Press (Oxford 1974).

[6] O. Pironneau and C. Saguez, Rapport de Recherche No. 218, I.N.R.I.A. (1977).

M. Fremond
Service de Mathématiques,
Laboratoire Central des Ponts et Chaussées,
58 Bld Lefebvre,
75732 Paris Cedex 15,
France

H. Ghidouche
Université de Paris-Nord,
Centre Scientifique Polytechnique,
Av. Jean-Baptiste Clément,
93430 Villetaneuse,
France

N. Point
Convervatoire Nationale des Arts et Métiers,
292 Rue St-Martin,
75003 Paris
France

B HANOUZET & J L JOLY

Bilinear maps compatible with a system

0. INTRODUCTION

In studying the behaviour of solutions to second-order elliptic equations with strongly oscillating coefficients, Murat [5] and Tartar [6] have observed and used the following result:

$$(u,v) \to \sum_{i=1}^{3} u_i v_i$$

is weakly continuous on bounded subsets of

$$\{u \in (L^2_{loc}(\mathbb{R}^3))^3 \mid \text{div } u \in L^2_{loc}(\mathbb{R}^3)\}$$

$$\times \{v \in (L^2_{loc}(\mathbb{R}^3))^3 \mid \text{curl } v \in (L^2_{loc}(\mathbb{R}^3))^3\}$$

without any of the $(u,v) \to u_i v_i$, $1 \leqslant i \leqslant 3$, being individually weakly continuous.

In [1] Ball proved a similar result:

$$(u^1, u^2,\dots,u^n) \to u^{i_1} \wedge u^{i_2} \wedge \dots \wedge u^{i_k}$$

is weakly continuous on bounded subsets of

$$(u \in (L^2_{loc}(\mathbb{R}^n))^n, \text{ curl } u \in (L^2_{loc}(\mathbb{R}^n))^N\}^n.$$

Some nonlinear (semilinear) hyperbolic systems with interesting properties can often be written as

$$A(D)u = D_t u + \sum_{i=1}^{n} B_i D_{x_i} u = f(u,u),$$

where f is bilinear ($u \to f(u,u)$ is quadratic). In various examples (Dirac system, Yang-Mills equations,...) we have an algebraic relation between f and $A(\xi)$ (see Section 1: we say that f is compatible with A). This relation can be used for the definition of $f(u,u)$ with continuity properties even when u is not a regular function; it may be useful in studying the global Cauchy problem, the asymptotic behaviour of solutions, and in other situations.

The results of Sections 1 and 2 are announced in [2]; for more details, see [3].

1. MULTILINEAR MAPS WHICH ARE COMPATIBLE WITH A DIFFERENTIAL SYSTEM

We shall study in this section relations between first-order systems with constant coefficients

$$A(D) = \sum_{i=1}^{n} A_i D_i, \; A_i \in M_{M,N}(\mathbb{C})$$

and k-linear maps f

$$f : (\mathbb{C})^k \to \mathbb{C}.$$

We note that (X,Y being distribution spaces)

$$W(A,X,Y) = \{u \in X \mid A(D)\, u \in Y\}$$

and we want to give a good definition (by this we mean one with continuity properties) of the k-linear map (induced by f and still denoted by f)

$$f : (W(A,X,Y))^k \to \mathcal{D}'$$

$$(u_1,\ldots,u_k) \to f(u_1,\ldots,u_k).$$

(a) Products with continuity properties

First let us mention the following results which are more or less classical.

If Γ_1 and Γ_2 are two closed conic subsets of $T^*(\mathbb{R}^n)\setminus 0$ with $0 \notin \Gamma_1 + \Gamma_2$, we know [4] that we have a continuous product

$$\mathcal{D}'_{\Gamma_1} \times \mathcal{D}'_{\Gamma_2} \to \mathcal{D}'.$$

As for Sobolev spaces we have the following result: let $s_1 + s_2 \geqslant 0$ and let us denote by $s_1 * s_2$ the interval of the real line

$$s_1 * s_2 = \{s_1 \wedge s_2 \wedge (s_1 + s_2 - \frac{n}{2} - \varepsilon) \;\; \varepsilon > 0\}.$$

Then the product can be continuously extended:

$$H^{s_1}_{loc} \times H^{s_2}_{loc} \to H^{s_1 * s_2}_{loc}.$$

Clearly this product is not weakly continuous if $s_1 + s_2 = 0$. But it still may be continuously defined for negative index s_1, s_2 if we get some additional estimates. For instance, using microlocal estimates it is possible to prove that if $0 \notin \Gamma_1 + \Gamma_2$, $s_1 + t_2 \geqslant 0$, $s_2 + t_1 \geqslant 0$, $t_1 + t_2 \geqslant 0$, then the product:

$$(H^{s_1}_{loc} \cap H^{t_1}_{\Gamma_1}) \times (H^{s_2}_{loc} \cap H^{t_2}_{\Gamma_2}) \to H^{s_1 * s_2}$$

is continuous. In this case $s_1 + s_2 < 0$ is possible.

These results can be applied for some spaces $W(A,X,Y)$. For example, if u_1, $u_2 \in \mathcal{D}'(\mathbb{R}^2)$ are such that $D_i u_{i_2} \in C^\infty(\mathbb{R}^2)$, we may define $u_1 u_2 \in \mathcal{D}'(\mathbb{R}^2)$. In Sobolev spaces, if u_1, $u_2 \in H^s_{loc}(\mathbb{R}^2)$ and $D_i u_i \in H^t_{loc}(\mathbb{R}^2)$ with $s + t + 1 \geqslant 0$ then $u_1 u_2 \in H^{s*s}_{loc}(\mathbb{R}^2)$. All these products are continuous.

However, these results, using standard microlocal arguments, do not cover many interesting examples. For instance, if $u, v \in (\mathcal{D}')^2$ and $\text{div}\, u = 0$, $\text{curl}\, v = 0$, it is not readily possible to give a continuous definition for the particular bilinear form $u_1 v_1 + u_2 v_2$.

(b) Multilinear form compatible with a system

We begin by recalling a result of Murat and Tartar:

PROPOSITION. Let $f:(\mathbb{C}^N)^k \to \mathbb{C}$ be a k-linear form. If f is a continuous k-linear map

$$(W(A,(\mathcal{D}')^N, (C^\infty)^M))^k \to \mathcal{D}',$$

then the following condition is verified

$$\left.\begin{aligned} &\xi_1 + \dots + \xi_k = 0, \quad \xi_i \in \mathbb{R}^n, \\ &|\xi_1| + \dots + |\xi_k| \neq 0, \\ &\Rightarrow f(\ker A(\xi_1),\dots,\ker A(\xi_k)) = 0. \end{aligned}\right\} \qquad (1)$$

Assuming (1) is true, we say that f is A-compatible.

Remarks. 1. The proof of (1) is easy. Consider $p_j \in \ker A(\xi_j)$ and let $u_j^\nu = e^{i\nu\langle x,\xi_j\rangle} p_j$ be a sequence of monochromatic plane waves polarized by p_j, so that each of the u_j^ν verifies

$$A(D)u_j^\nu = 0.$$

Clearly $u_j^\nu \to 0$ if $\xi_j \neq 0$ and $u_j^\nu = p_j$ if $\xi_j = 0$. By assumption the sequence $f(u_1^\nu,\ldots,u_k^\nu) = f(p_1,\ldots,p_k)$ must converge to 0, from which it follows that $f(p_1,\ldots,p_k) = 0$.

2. If A(D) is an elliptic system, then all k-linear maps are A-compatible.

3. Suppose there exists some subset $V \cap \mathbb{R}^n \setminus 0$ with dim V = m and such that $\mathbb{C}^N = \mathrm{lin}\,(\bigcup_{\xi\in V} \ker A(\xi))$. Then if f is A-compatible and k-linear with k > m, f has to be identically zero.

4. Let now A(D) be a strongly hyperbolic system

$$A(D) = ID_t + \sum_{i=1}^n B_i D_{x_j}$$

where $B(\xi) = \sum_{i=1}^n \xi_i B_i = \sum_{i=1}^N \tau_i(\xi)\, \pi_i(\xi)$, ($\tau_1(\xi) \geqslant \ldots \geqslant \tau_N(\xi)$ are real numbers, $\pi_i(\xi)\, \pi_j(\xi) = \delta_{ij}\, \pi_i(\xi)$). We can choose for V in Remark 3 the subset

$$\{(\xi^0,\tau_1(\xi^0)),\ldots,(\xi^0,\tau_N(\xi^0))\}, \quad \xi_0 \neq 0.$$

The dimension of V is two, from which it follows that $k \leqslant 2$ if $f \not\equiv 0$.

(c) Some examples

(1) The only quadratic form compatible with the first-order system related to the wave equation is the Lagrangian one

$$u_t^2 - |\nabla_x u|^2.$$

(2) The linear elastic waves in an isotropic medium form a first-order system if we introduce the speed field v and the constraint field σ. The quadratic form compatible with this system is the elastic Lagrangian

$$|v|^2 - (E^{-1}\sigma,\sigma).$$

(3) For the Dirac system we obtain the sesquilinear form

$$u_1\bar{v}_1 + u_2\bar{v}_2 - u_3\bar{v}_3 - u_4\bar{v}_4.$$

(4) If A(D) is obtained by the exterior differentiation d of differential forms, then the k-linear maps which are compatible with d are indeed the coefficients of all exterior products. For instance, we have the 'div, curl system and Yang-Mills system.

These examples and others can be found in [6] and [4].

2. A-COMPATIBLE BILINEAR MAPS ON $W(A,(H^s_{loc})^N, (H^t_{loc})^M)$

Let f be a bilinear form $\mathbb{C}^N \times \mathbb{C}^N \to \mathbb{C}$. In [5], [6] the following theorem is proved:

$$f : (W(A,(L^2_{loc})^N, (L^2_{loc})^M))^2 \to \mathcal{D}'$$

is continuous iff f is A-compatible. Here we also obtain this result as a consequence of properties on Sobolev spaces with negative index. Let s_1, s_2, t_1, t_2 be real numbers satisfying $s_1 + t_2 + 1 \geq 0$, $s_2 + t_1 + 1 \geq 0$, $s_i - 1 \leq t_i \leq s_i$, $i = 1,2$. We have

<u>THEOREM</u> If f is regularly A-compatible, then f is a (strongly) continuous bilinear map

$$W(A,(H^{s_1}_{loc})^N, (H^{t_1}_{loc})^M) \times W(A,(H^{s_2}_{loc})^N,(H^{t_2}_{loc})^M) \to H^{s_1 * s_2}_{loc}.$$

Let us denote by $F \in M_{N,N}(\mathbb{C})$ the matrix such that

$$f(p,q) = (p,Fq).$$

The A-compatibility reads

$$\forall\xi \neq 0, \quad \exists G(\xi), D(\xi) \in M_{M,N}(\mathbb{C}) \quad \text{s.t.}$$

$$F = {}^tA(\xi)\, D(\xi) + {}^tG(\xi)\, A(\xi).$$

We now say that f is regularly A-compatible if we can choose $D(\xi)$ and $G(\xi)$ homogeneous of degree -1 and bounded on S^{n-1}. Now, the theorem is proved by using the regularizing effect of the operators $D(\xi)$ and $G(\xi)$. In all usual examples A-compatible bilinear forms are actually regularly A-compatible. For the 'div, curl' system we have for instance:

$$F = \left(\begin{array}{c:c} 0 & I \\ \hdashline I & 0 \end{array}\right) = {}^tA(\xi)\, C(\xi) + {}^tC(\xi)\, A(\xi)$$

with $\quad C(\xi) = \dfrac{1}{|\xi|^2} \left(\begin{array}{c:c} 0 & \text{div } \xi \\ \hdashline \text{curl } \xi & 0 \end{array}\right).$

Remarks. 1. If $s_1 + t_2 > 0$ or $s_2 + t_1 > 0$, using Rellich's theorem we obtain that f is sequentially weakly continuous.

2. Negative values for s are possible in the result. Take for instance

$$s_1 = t_1 = s_2 = t_2 = s > -\frac{1}{2}.$$

3. $H^{s_1 * s_2}$ does not depend on the values of t_i. This is the same situation as the one for the product in microlocal Sobolev spaces.

4. This result is not the best possible for elliptic or hyperbolic systems, but it is optimal for general systems, for instance the 'div, curl' system.

3. SOME REMARKS ABOUT HYPERBOLIC SYSTEMS

(a) Let us consider a strictly hyperbolic system given by the symbol

$$A(\tau, \xi) = \tau . I + \sum_{i=1}^{n} \xi_i B_i$$

the symbol $B(\xi) = \sum_{i=1}^{n} \xi_i B_i$ may be written

$$B(\xi) = \sum_{k=1}^{N} \tau_k(\xi)\, \pi_k(\xi),$$

where the real eigenvalues τ_k are ordered in such a way that

$$\tau_1(\xi) > \tau_2(\xi) > \dots > \tau_N(\xi) \text{ for } \xi \neq 0$$

and where $\pi_k(\xi)$ denotes the kth eigen projector

$$\pi_k(\xi).\pi_\ell(\xi) = \delta_{k\ell}\, \pi_k(\xi).$$

Note that

$$\tau_k(\xi) = -\tau_{k'}(-\xi), \; \pi_k(\xi) = \pi_{k'}(-\xi)$$

if $k' = N + 1 - k$.

Now, $f : (\mathbb{C})^2 \to \mathbb{C}$ is A-compatible if and only if

$$\tau_k(\xi) + \tau_\ell(-\xi) = 0,$$

which implies

$$f(\pi_k(\xi), \; \pi_\ell(-\xi) = 0.$$

For any $\phi \in E'(\mathbb{R}^n)$, let u denote the solution of the Cauchy problem

$$\begin{cases} D_t u + B(D_x)u = 0, \\ u(0) = \phi. \end{cases}$$

We know that such a solution is given (mod C^∞) by

$$u(x,t) = \sum_{k=1}^{N} \int e^{i(\langle x,\xi\rangle - t\tau_k(\xi))} \pi_k(\xi)\, \hat{\phi}(\xi)\, d\xi.$$

If u and v are any two solutions of the Cauchy problem corresponding to initial data ϕ and ψ, we therefore get (at least formally), for $f(u,v)$ the expression

$$f(u,v)(x,t) = \sum_{k,\ell} \iint e^{i[\langle x,\xi+\eta\rangle - t(\tau_k(\xi)+\tau_\ell(\eta))]} f(\pi_k(\xi)\hat{\phi}(\xi), \; \pi_\ell(\eta)\hat{\psi}(\eta))\, d\xi\, d\eta.$$

The phase function

$$\Phi_{k_\ell}(x,t,\xi,\eta) = \langle x,\xi+\eta\rangle - t(\tau_k(\xi) + \tau_\ell(\eta))$$

is singular at those points (ξ,η) which satisfy

$$\frac{\partial \Phi_{k\ell}}{\partial x} = \xi + \eta = 0$$

$$\frac{\partial \Phi_{k\ell}}{\partial t} = \tau_k(\xi) + \tau_\ell(\eta) = 0.$$

If we assume that f is A-compatible, we notice that at those critical points (ξ,η) the amplitude

$$f(\pi_k(\xi), \pi_\ell(\eta))$$

vanishes.

Let us assume now there is only one space variable. In this special case it is readily seen, using homogeneity arguments, that indeed $\Phi_{k\ell}$ has no critical point in the conic support of the amplitude $f(\pi_k(\xi),\pi_\ell(\eta))$, so that standard results apply. For instance, it is possible to show the following continuity property of any A-compatible f:

$$f : X_s^2 \to H^s(\mathbb{R}_x \times]0,T[), \quad s \geqslant 0$$

where X_s denote the space of the solutions of the Cauchy problem

$$\begin{cases} D_t u + B(D_x)u = g \in H^s(\mathbb{R}_x]0,T[) \\ u(0) = \phi \in H^s(\mathbb{R}_x). \end{cases}$$

(b) We consider now the general n-dimensional space case, and assume that u and v are regular wave packets, that is, solutions of the previous Cauchy problem with initial data ϕ and ψ which satisfy $\hat{\phi},\hat{\psi} \in \mathcal{D}'(\mathbb{R}^n-\{0\})$.

In such a situation, the representation formula is exact:

$$u(x,t) = \sum_{k=1}^{N} \int e^{i(\langle x,\xi\rangle - t\,\tau_k(\xi))} \pi_k(\xi)\, \hat{\phi}(\xi)\, d\xi.$$

Let us denote by u_k each of the N integrals in the right-hand member of the preceding equality.

Outside of the rays corresponding to $\xi \in \operatorname{supp} \pi_k(\xi)\hat{\phi}(\xi)$:

$$\{(x,t) \in \mathbb{R}^{n+1} \mid x = t\,\tau_k'(\xi),\ \xi \in \operatorname{supp} \pi_k(\xi)\, \hat{\phi}(\xi)\},$$

we know that u_k decreases more quickly than any power of x and t:

$$u_k = O((|x| + |t|)^{-\infty}).$$

In the neighbourhood of the rays, however, u_k decreases as $|t|^{-(n-1)/2}$, provided τ_k satisfies some additional regularity properties we shall not mention here.

More precisely, for (x,t) in a sufficiently small neighbourhood of the kth sheet of the wave cone, we get the asymptotic behaviour of u_k,

$$u_k(x,t) = \left(\frac{c}{t}\right)^{(n-1)/2} \int_0^\infty e^{i\rho t[\langle x/t,\theta^k(x/t)\rangle - \tau_k \theta^k(x/t)]}$$

$$\times\, e^{i\,\mathrm{sign}\,\tau''(\theta^k(x/t)).\pi/4}\, \pi_k(\theta^k(x/t))\hat{\phi}(\rho\theta^k(x/t))$$

$$\times\, \det \tau''(\rho\theta^k(x/t))^{-\frac{1}{2}}\, d\rho$$

where $y \to \theta^k(y) \in S^{n-1}$ is a uniquely defined differentiable mapping.

Therefore, there exist $\psi_k(x,t) \in C^\infty(\mathbb{R}^{n+1})$, with mutually disjoint supports, equal to one in a neighbourhood of the kth sheet of the wave cone (for large t), and such that

$$u_k(x,t) = \left(\frac{1}{t}\right)^{(n-1)/2} \pi_k(\theta^k(x/t))\ \psi_k(x,t) + O\left(\left(\frac{1}{t}\right)^{(n+1)/2}\right).$$

It is clear that if f is any bilinear map we get for $f(u,v)$ (u and v being regular wave packets) and large t an $O\left(\left(\frac{1}{t}\right)^{n-1}\right)$ estimate.

This estimate can be improved, however, for A-compatible mappings.

THEOREM. For a bilinear f to be A-compatible it is necessary and sufficient that for any regular wave packets u and v, f satisfies

$$f(u,v) = O\left(\left(\frac{1}{t}\right)^n\right).$$

Sketch of the proof. Indeed

$$f(u,v) = \sum_{k,\ell} f(u_k,v_\ell).$$

Neglecting interactions between disjoint sheets of the wave cone, we show the principal terms in the right-hand sum to be

$$\sum_k f(u_k,v_k) + \sum_k f(u_k, v_{k'}) + O\left(\left(\frac{1}{t}\right)^n\right) \quad (k' = N + 1 - k).$$

After noticing that

$$\theta^k(x/t) = -\theta^{k'}(x/t)$$

$$\pi_k(\theta^k(x/t)) = \pi_{k'}(-\theta^{k'}(x/t)).$$

We conclude the proof using the A-compatibility, which says that

$$f(\pi_k(\theta^k(x/t)), \pi_k(\theta^k(x/t))) = 0$$

and

$$f(\pi_k(\theta^k(x/t)), \pi_{k'}(\theta^{k'}(x/t))) = 0.$$

REFERENCES

[1] J.M. Ball, Convexity conditions and existence theorem in nonlinear elasticity, Arch. Rational Mech. Anal. 63, 337-403 (1977).

[2] B. Hanouzet and J.L. Joly, Applications bilinéaires sur certains sous-espaces de type Sobolev; C.R.A.S. Paris (to appear).

[3] B. Hanouzet and J.L. Joly, Formes multilinéaires sur des sous-espaces de distributions. Publications Bordeaux I, No. 8203.

[4] L. Hörmander, Fourier integral operators I, Acta Math. 127, 79-183 (1971).

[5] F. Murat, Compacité par compensation. Ann. Scuola Norm. Sup. Pisa Sci. Fis. Mat. (IV), 5 (1978).

[6] L. Tartar, Compensated compactness and applications to partial differential equations. Research Notes in Mathematics, 39, R.J. Knops (ed.), Pitman 136-212 (1979).

B. Hanouzet and J.L. Joly
U.E.R. de Mathématiques et Informatique,
Laboratoire associé au C.N.R.S. No. 040226,
Université de Bordeaux I,
351, Cours de la Libération,
33405 Talence Cedex,
France

M A HERRERO

On the growth of the interfaces of a nonlinear degenerate parabolic equation

1. INTRODUCTION

This paper deals with the following Cauchy problem

$$u_t - (u^m)_{xx} + cu^n = 0 \quad \text{in} \quad Q = \mathbb{R} \times (0,\infty) \tag{1.1}$$

$$u(x,0) = u_0(x) \qquad \text{in } \mathbb{R} \tag{1.2}$$

under the hypotheses:

(H1) $n \geq m > 1, \quad c > 0$

(H2) $u_0(x)$ is a bounded continuous nonnegative function whose support is an interval $I = [a_1,a_2]$.

Equation (1.1) is usually referred to as the one-dimensional nonlinear heat equation with absorption. When $c = 0$ we obtain the porous media equation

$$u_t = (u^m)_{xx} \quad \text{in} \quad Q.$$

The problem (1.1), (1.2) need not have classical solutions. We define a weak solution of (1.1), (1.2) as being a bounded nonnegative and continuous function $u(x,t)$ such that (1.2) holds and the following integral identity is satisfied:

$$\int_{t_0}^{t_1}\int_{x_0}^{x_1} [u^m f_{xx} + uf_t - cu^n f]\, dx\, dt - \int_{x_0}^{x_1} uf\, dx\Big|_{t_0}^{t_1} - \int_{t_0}^{t_1} u^m f_x\, dt\Big|_{x_0}^{x_1} = 0$$

whenever $0 \leq t_0 < t_1$ and $x_0 < x_1$, and for all $f \in C^1(\bar{Q})$ such that $f_{xx} \in C(\bar{Q})$ and $f = 0$ for $x = x_0, x_1$. Existence and uniqueness of weak solutions for (1.1), (1.2) is obtained in [4]. It is also known ([4], [8]) that if (H1), (H2) hold, the set

$$P[u] = \{(x,t) \in Q \mid u(x,t) > 0\}$$

is bounded by two curves usually called interfaces or free boundaries:

$$\zeta_1(t) = \inf \{x \in \mathbb{R} \mid u(x,t) > 0\}, \ \zeta_2(t) = \sup \{x \in \mathbb{R} \mid u(t,x) > 0\}$$

such that $-\infty < \zeta_1(t) < \zeta_2(t) < \infty$, $\zeta_1(0) = a_1$, $\zeta_2(0) = a_2$ and $(-1)^i \zeta_i(t)$ is continuous nondecreasing for $i = 1,2$. Thus

$$P[u] = \{(x,t) \in Q \mid \zeta_1(t) < x < \zeta_2(t)\}.$$

We are interested here in the growth of the $\zeta_i(t)$ $(i = 1,2)$ as $t \to \infty$. For simplicity we drop the index i in what follows and state our results for the right interfaces denoted by $\zeta(t)$, the extension to the left ones being straightforward. It is known that $|\zeta(t)| \to \infty$ as $t \to \infty$ ([7], [8]). We prove

THEOREM. There exists a positive constant M depending only on m such that for $t > 2$:

(a) If $n = m$:

$$\zeta(t) \leqslant \zeta(2) + \frac{2M}{\ell n\ 2}\left(\frac{1}{c(n-1)}\right)^{\frac{1}{2}} \ell n\ t \tag{1.3}$$

(b) If $m < n < m + 2$:

$$\zeta(t) \leqslant \zeta(2) + \frac{2M}{(2^\beta - 1)}\left(\frac{1}{c(n-1)}\right)^\gamma t^\beta \tag{1.4}$$

where $\beta = \dfrac{n-m}{2(n-1)}$, $\gamma = \dfrac{m-1}{2(n-1)}$.

It is known ([7], [8]) that when $n \geqslant m + 2$ for large t $\zeta(t) \leqslant at^{1/m+1}$ for some $a > 0$ depending on m, n and u_0. In fact in this case we can obtain for some $B > 0$ depending only on m:

$$\zeta(t) \leqslant \zeta(2) + \frac{B\, \|u_0\|_1^{m-1/(m+1)}}{(2^\delta - 1)} t^\delta, \quad \delta = \frac{1}{m+1}, \quad t > 2.$$

We prove the theorem in Section 2.

After the completion of this work we were informed that bounds for $\zeta(t)$ with the same order of growth in t as (1.3), (1.4) have been obtained independently by a different method in [3]. For lower bounds for $\zeta(t)$ the reader is referred to [7], [8] and [3].

2. THE RESULTS

We first introduce some material and recall several results which are useful in the following. For each fixed $t_0 > 0$ we consider the following Cauchy problem:

$$w_t = (w^m)_{xx} \text{ in } Q_{t_0} = \mathbb{R} \times (t_0,\infty) \tag{2.1}$$

$$w(x,t_0) = u(x,t_0) \tag{2.2}$$

where $u(x,t)$ is the generalized solution of (1.1), (1.2). Now we put

$$v = \frac{m}{m-1} u^{m-1}, \quad \bar{v} = \frac{m}{m-1} w^{m-1}.$$

As $u(x,t)$ is compactly supported for each $t \geqslant 0$, in problem (2.1), (2.2) two interfaces defined for $t \geqslant t_0$ appear ([9]). In the sequel we restrict our attention to the right-hand one, which we denote by $\zeta_0(t)$. When (2.1) is the equation for the (scaled) density of an ideal gas flowing on a homogeneous porous medium, $\bar{v}$ is essentially the pressure and the interface is expected to move with the local velocity of the gas, given by $-\bar{v}_x$ (see [1]). In fact it is known ([1], [8]) that $\zeta_0(t)$ is Lipschitz for $t > t_0$, and

$$\zeta_0'(t+) = -\bar{v}_x(\zeta_0(t),t) \tag{2.3}$$

where $h'(t+)$ is the right-hand derivative of h at t,

$$\bar{v}_x(\zeta_0(t),t) = \lim \bar{v}_x(x,t) \text{ when } x \to \zeta_0(t),\ x \in P[w]$$

and (2.3) holds for each $t > t_0$ (resp. $t \geqslant t_0$ if $\bar{v}_x(x,t_0)$ is bounded, in which case $\zeta_0(t)$ is also Lipschitz at t_0).

In fact for our purposes a weaker version of (2.3) will suffice, namely

$$\zeta'(t) = -\bar{v}_x(\zeta(t),t) \quad \text{a.e.} \quad t > t_0. \tag{2.4}$$

The proof to be presented here proceeds by noting that for fixed t_0 $\zeta(t)$ increases slower than $\zeta_0(t)$ when $t \geqslant t_0$. The growth of $\zeta_0(t)$ can be estimated in terms of $\bar{v}_x$, by writing $\zeta_0(t)$ as the integral of its derivative and using, for instance, (2.4). We then use a sharp bound for $\bar{v}_x$ which is part of the results of [10]. This bound depends on t_0 and $\|u(x,t_0)\|_\infty$. Then we take $t_0 = t/2$ for arbitrarily large t and use estimates on the decay in time of $\|u(x,t)\|_\infty$ related to the absorption term. This gives the result.

We first give the bound for $\bar{v}_x$.

LEMMA 1 ([10]): There exists a positive constant M depending only on m such that, for $t > t_0 \geq 0$,

$$|\bar{v}_x(x,t)| \leq M \|u(x,t_0)\|_\infty^{(m-1)/2} (t-t_0)^{-\frac{1}{2}}. \tag{2.5}$$

Proof: For completeness we sketch the proof. Without loss of generality we can assume $t_0 = 0$. It is known ([5]) that for each $\tau > 0$

$$|\bar{v}_x(x,t)| \leq C \quad \text{if} \quad t \geq \tau, \tag{2.6}$$

where $C = C(m,\tau, \|u_0\|_\infty)$. On the other hand ([6]), if $w(x,t)$ solves (2.1), the same is true for

$$w_{K,L}(x,t) \equiv Tw = Kw(Lx, K^{m-1}L^2t),$$

with $K,L > 0$. Now for each fixed $t > 0$ take $K = \|u_0\|_\infty$, $K^{m-1}L^2t = 1$ and define $\hat{w}$ by $w = T\hat{w}$. Thus using (2.6) and writing $\hat{v} = \frac{m}{m-1}\hat{w}^{m-1}$, we get

$$|\bar{v}_x(x,t)| = \frac{m}{m-1} K^{m-1}L|\hat{v}_x(Lx,1)| \leq C_1K^{m-1}L$$

with $C_1 = C_1(m,1,1) > 0$. This gives (2.5).

We next state a calculus lemma.

LEMMA 2. Let $\zeta(t)$ be a nondecreasing real function and assume that, for each $t > 0$ and some $k > 0$,

$$\zeta(t) \leq \zeta\left(\frac{t}{2}\right) + k;$$

then, for $t > 2$,

$$\zeta(t) \leq \zeta(2) + \left(\frac{k}{\ell n 2}\right) \ell n\, t. \tag{2.7}$$

If, for each $t > 0$ and some positive α and β,

$$\zeta(t) \leqslant \zeta\left(\frac{t}{2}\right) + \alpha\left(\frac{t}{2}\right)^{\beta},$$

then for $t > 2$

$$\zeta(t) \leqslant \zeta(2) + \frac{\alpha t^{\beta}}{2^{\beta}-1} . \tag{2.8}$$

Proof: Let us prove (2.7). Given an arbitrary $t \geqslant 2$ there exists a natural $n \geqslant 1$ such that $1 \leqslant \frac{t}{2^n} \leqslant 2$. Thus n satisfies

$$n \leqslant \frac{\ln t}{\ln 2} . \tag{2.9}$$

Now we write

$$\zeta(t) \leqslant \zeta\left(\frac{t}{2}\right) + k,$$

$$\zeta\left(\frac{t}{2}\right) \leqslant \zeta\left(\frac{t}{2^2}\right) + k,$$

....

$$\zeta\left(\frac{t}{2^{n-1}}\right) \leqslant \zeta\left(\frac{t}{2^n}\right) + k.$$

Summing all these and using (2.9), we get

$$\zeta(t) \leqslant \zeta\left(\frac{t}{2^n}\right) + nk \leqslant \zeta(2) + \left(\frac{k}{\ln 2}\right) \ln t.$$

The proof of (2.8) is similar and we omit it.

Proof of the Theorem. Using the monotonicity of the interfaces and the fact that $w(x,t) \geqslant u(x,t)$ for $t \geqslant t_0$ ([8]), we obtain for each $t_0 > 0$ and $t \geqslant t_0$

$$0 \leqslant \zeta(t) - \zeta(t_0) \leqslant \zeta_0(t) - \zeta_0(t_0) = \int_{t_0}^{t} \zeta_0'(s)\, ds. \tag{2.10}$$

Now we fix $t > 0$, take $t_0 = \frac{t}{2}$ and use equation (2.4) and estimate (2.5) in (2.10) to get

$$\zeta(t) - \zeta\left(\frac{t}{2}\right) \leqslant M \int_{t/2}^{t} \left\|u\left(x,\frac{t}{2}\right)\right\|_{\infty}^{(m-1)/2} \left(s - \frac{t}{2}\right)^{-\frac{1}{2}} ds. \tag{2.11}$$

Next we use the fact that, for $t > 0$ and $n \geqslant 1$,

$$|u(x,t)| \leq \left(\frac{1}{c(n-1)}\right)^{1/n-1} t^{-1/n-1} \qquad (2.12)$$

(see [11]). Suppose first that n = m. Then substituting (2.12) into (2.11) gives

$$\zeta(t) - \zeta\left(\frac{t}{2}\right) \leq M\left(\frac{1}{c(n-1)}\right)^{\frac{1}{2}} \left(\frac{t}{2}\right)^{-\frac{1}{2}} \int_{t/2}^{t} \left(s - \frac{t}{2}\right)^{-\frac{1}{2}} ds$$

$$= 2M\left(\frac{1}{c(n-1)}\right)^{\frac{1}{2}}.$$

Then (1.3) follows from this and (2.7) with $k = 2M\left(\frac{1}{c(n-1)}\right)^{\frac{1}{2}}$. If $m < n < m + 2$, the same substitution gives

$$\zeta(t) - \zeta\left(\frac{t}{2}\right) \leq 2M\left(\frac{1}{c(n-1)}\right)^{\frac{m-1}{2(n-1)}} \left(\frac{t}{2}\right)^{\frac{n-m}{2(n-1)}},$$

so that (1.4) follows from (2.8) with

$$\alpha = 2M\left(\frac{1}{c(n-1)}\right)^{\frac{m-1}{2(n-1)}}, \ \beta = \frac{m-1}{2(n-1)}.$$

We finally observe that, for $t > 0$ and arbitrary nonnegative n and c, bounds for w(x,t) (starting from $t_0 = 0$) are also bounds for u(x,t), so that

$$|u(x,t)| \leq A \, \|u_0\|_1^{2/(m+1)} \, t^{-\frac{1}{m+1}} \qquad (2.13)$$

for some $A > 0$ (see [2], [11]). Note that exponents in (2.12), (2.13) agree if $n = m + 2$, whereas for large times estimate (2.12) (resp. (2.13)) is sharper if $n < m + 2$ (resp. $n > m + 2$). If we substitute (2.13) into (2.11) we arrive at

$$\zeta(t) - \zeta\left(\frac{t}{2}\right) \leq 2M \cdot A^{\frac{m-1}{2}} \, \|u_0\|_1^{m-i/(m+1)} \left(\frac{t}{2}\right)^{\frac{1}{m+1}}$$

and thus (1.5) follows from (2.8) with $B = 2MA^{(m-1)/2}$.

REFERENCES

[1] D.G. Aronson, Regularity properties of flows through porous media: The interface, Arch. Rat. Mech. Anal. 37 1-10 (1970).

[2] P. Bênilan, Opérateurs accretifs et semi-groupes dans les espaces L^p $(1 \leqslant p \leqslant \infty)$, in Functional Analysis and Numerical Analysis, France-Japan Seminar, H. Fujita (ed.) Jap. Soc. for the Promotion of Sciences Tokyo (1978).

[3] M. Bertsch, R. Kersner and L.A. Peletier, Sur le comportement de la frontière libre dans une équation en théorie de la filtration. C.R. Acad. Sc. Paris (to appear).

[4] A.S. Kalashnikov, The propagation of disturbances in problems of non-linear heat conduction with absorption, Z. Vycisl. Mat Fiz. 14, 891-905 (1974).

[5] A.S. Kalashnikov, On the differential properties of generalized solutions of equations of the nonsteady state filtration type, Vestnik Moskov. Univ. Ser. I Mat. Mekh 29, 62-68 (1974).

[6] S. Kamenomotskaya (Kamin), The asymptotic behaviour of the solution of the filtration equation, Israel J. of Math. 14, 76-87 (1973).

[7] R. Kersner, On the behaviour when $t \to \infty$ of the solutions of a quasi-linear degenerate parabolic equation, Acta Mat. Acad. Sci. Hung 34, 157-163 (1979) (in Russian).

[8] B.F. Knerr, The behaviour of the support of solutions of the equation of nonlinear heat equation with absorption in one dimension, Trans. Amer. Math. Soc. 249, 409-424 (1979).

[9] O.A. Oleinik, A.S. Kalashnikov and C. Yu Lin, The Cauchy problem and boundary problems for equations of the type of nonstationary filtration, Izv. Akad. Nauk. URSS. Ser. Mat 22, 667-704 (1958) (in Russian).

[10] J.L. Vazquez, One dimensional flow of gases in porous media: behaviour of the velocity (to appear).

[11] L. Veron, Effets régularisants de semi-groupes non linéaires dans des espaces de Banach, Ann. Fac. Sci. Toulouse 1, 171-200 (1979).

M.A. Herrero
Departamento de Ecuaciones Funcionales,
Facultad de Matemáticas
Universidad Complutense,
Madrid - 3,
Spain

P L LIONS

On the concentration–compactness principle

1. INTRODUCTION

We want to present here a general method, called the concentration-compactness method, for solving various problems of the calculus of variations. This method is introduced in Lions [4], [5]. To explain the goal of this method, let us first mention that most of the standard methods for treating minimization problems in functional spaces consisting of functions defined on bounded domains of $\mathbb{R}^N$ use heavy compactness arguments and results like for example the famous Rellich Theorem. On the other hand many minimization problems arising in mathematical physics are set on unbounded domains (like $\mathbb{R}^N$, half-spaces, strips ...) and therefore standard methods do not apply because of the lack of global compactness. The concentration-compactness principle enables us to solve these minimization problems (with constraints) in unbounded domains (assuming only some form of a priori estimates which would be sufficient to solve the problem 'if it were set in a bounded region').

To be more specific we will treat completely one simple example, hoping that the reader will convince himself of the generality of the method. We will thus consider the following minimization problem:

$$I^V = \inf \{E^V(u) \,\Big|\, u \in H^1(\mathbb{R}^3),\ |u|_{L^2} = 1 \tag{1}$$

where $E^V(u)$ is the following functional defined on $H^1(\mathbb{R}^3)$:

$$E^V(u) = \int_{\mathbb{R}^3} \frac{1}{2}|\nabla u|^2 + \frac{1}{2}V(x)u^2\,dx - \frac{1}{4}\int_{\mathbb{R}^3}\int_{\mathbb{R}^3} u^2(x)u^2(y)|x-y|^{-1}\,dx\,dy, \tag{2}$$

V being a given potential in $\mathbb{R}^3$. For simplicity (and to make sure E^V is well defined on $H^1(\mathbb{R}^3)$) we will always assume in the following:

$$V \in L^p(\mathbb{R}^3) + L^q(\mathbb{R}^3) \text{ with } \frac{3}{2} \leqslant p,q < +\infty. \tag{3}$$

This problem, known as the Pekar-Choquard problem, arises in various problems in theoretical physics - see Lieb [3], Donsker and Varadhan [2].

Lieb showed us how the case of a Coulomb potential:

$$V(x) = - \sum_{i=1}^{m} z_i |x-x_i|^{-1} \; ; \; z_i > 0, \; x_i \in \mathbb{R}^3, \tag{4}$$

is of interest for Hartree-Fock theory. The case $V \equiv 0$ has been treated in Lieb [3], Lions [6] by a method which does not extend to the case $V \not\equiv 0$.

Before stating our main result on this problem, let us recall that $(u_n)_{n \geq 1}$ is said to be a minimizing sequence for problem (1) if (u_n) satisfies: $u_n \in H^1(\mathbb{R}^3)$, $|u_n|_{L^2} = 1$, $E(u_n) \underset{n}{\longrightarrow} I^V$.

THEOREM 1. (i) If $V \equiv 0$, then every minimizing sequence (u_n) is relatively compact in $H^1(\mathbb{R}^3)$ up to a translation i.e. $\exists y_n \in \mathbb{R}^3$, $(u_n(\cdot + y_n))_n$ is relatively compact in $H^1(\mathbb{R}^3)$. In particular there exists a minimum in problem (1).

(ii) If $V \not\equiv 0$, then every minimizing sequence is relatively compact in $H^1(\mathbb{R}^3)$ if and only if the following inequality holds:

$$I^V < I^0. \tag{5}$$

(iii) In particular if (5) holds, there exists a minimum in problem (1). Furthermore (5) holds if $V \leq 0$, $V \not\equiv 0$ and (5) does not hold if $V \geq 0$.

Remarks. (1) The above result shows that if V is given by (4) then (5) holds and thus not only a minimum of (1) exists but every minimizing sequence is relatively compact in $H^1(\mathbb{R}^3)$.

(2) Let us also point out that when $V \geq 0$, $V \not\equiv 0$, by the above result, there exist minimizing sequences which are not relatively compact in H^1. Actually it is easy to show that in this case there is no minimum of (1).

(3) The proof of this result is given in Section 2 below and will actually show that, for all V, every minimizing sequence is relatively compact in H^1 up to a translation. However, except when $V \equiv 0$, and thus when E is translation invariant, this of course leads to a minimum only when we know that the translation remains bounded.

As we said before this result is a quite simple application of the general concentration-compactness argument, and after proving Theorem 1 in Section 2, we will come back in Section 3 to the generality of the argument involved.

PROOF OF THEOREM 1

Before going into the proof let us give some notation and make a few preliminary observations: first if $\lambda > 0$, we define

$$I^V_\lambda = \inf \{E^V(u) \big| u \in H^1(\mathbb{R}^3),\ |u|^2_{L^2} = \lambda\}; \tag{1'}$$

of course we have: $I^V = I^V_1$. Next, in what follows, $(u_n)_{n\geq 1}$ will denote a minimizing sequence corresponding to (1'). We claim that I^V_λ is finite and that (u_n) remains bounded in $H^1(\mathbb{R}^3)$: indeed let $C > 0$ be such that $I^V_\lambda \leq E^V(u_n) \leq C \quad \forall n \geq 1$. By well-known convolution inequalities we deduce that

$$\int_{\mathbb{R}^3} \frac{1}{2} |\nabla u_n|^2 dx \leq \int_{\mathbb{R}^3} (-V) \frac{1}{2} u_n^2 \, dx + C_0 \|u_n\|^4_{L^{12/5}}$$

and using (3) and the Hölder inequalities, this yields

$$\int_{\mathbb{R}^3} |\nabla u_n|^2 \, dx \leq \varepsilon \int_{\mathbb{R}^3} |\nabla u_n|^2 \, dx + C_\varepsilon \int_{\mathbb{R}^3} u_n^2 \, dx + C_0 |u_n|^3_{L^2} |u_n|_{L^6}$$

and our result follows by using Sobolev inequalities and the fact that $|u_n|^2_{L^2} = \lambda$.

We now turn to the proof of Theorem 1, which is divided in three steps: Steps 1 and 2 below correspond to general arguments and are the basis of the concentration-compactness method.

Step 1: Subadditivity inequalities: we show that the following inequalities hold:

$$I^V_\lambda \leq I^V_{\lambda-\alpha} + I^0_\alpha, \quad \forall \alpha \in]0,\lambda] \tag{6}$$

(where, by convention, $I^V_0 = 0$).

Let us first show the case $\alpha = \lambda$ in (6): let $\varepsilon > 0$, and let $u_\varepsilon \in H^1(\mathbb{R}^3)$, $|u_\varepsilon|^2_{L^2} = \lambda$, $E^0(u_\varepsilon) < I^0_\lambda + \varepsilon$. By density, we may assume without loss of generality that $u_\varepsilon \in \mathcal{D}(\mathbb{R}^3)$. Next let e be a unit vector in $\mathbb{R}^3$ and let $u^n_\varepsilon(x) = u_\varepsilon(x+ne)$. Since E^0 is translation invariant, we still have $|u^n_\varepsilon|^2_{L^2} = \lambda$, $E^0(u^n_\varepsilon) < I^0_\lambda + \varepsilon$. On the other hand $(u^n_\varepsilon)^2$ converges weakly, as n goes to $+\infty$, to 0 in any $L^\beta (1 < \beta < \infty)$, and this gives

$$\int_{\mathbb{R}^3} \frac{1}{2} V(u_\varepsilon^n)^2 \, dx \xrightarrow[n]{} 0.$$

Thus for n large enough, we have: $E^V(u_\varepsilon^n) < I_\lambda^0 + \varepsilon$; and our result follows since $I_\lambda^V \leqslant E^V(u_\varepsilon^n)$.

The general case $(0 < \alpha < \lambda)$ in (6) is proved by similar arguments: let $\varepsilon > 0$, $u_\varepsilon, v_\varepsilon \in \mathcal{D}(\mathbb{R}^3)$ be such that

$$|u_\varepsilon|^2_{L^2} = \alpha, \ |v_\varepsilon|^2_{L^2} = \lambda - \alpha, \ E^0(u_\varepsilon) < I_\alpha^0 + \varepsilon, \ E^V(v_\varepsilon) < I_{\lambda-\alpha}^V + \varepsilon.$$

We introduce again $u_\varepsilon^n(x) = u_\varepsilon(x+ne)$. Then for n large enough we have

$$|u_\varepsilon^n + v_\varepsilon|^2_{L^2} = \lambda, \ E^V(u_\varepsilon^n + v_\varepsilon) < I_\alpha^0 + I_{\lambda-\alpha}^V + 2\varepsilon;$$

indeed note that $E^V(u_\varepsilon^n + v_\varepsilon) - \{E^0(u_\varepsilon^n) + E^V(v_\varepsilon)\} \xrightarrow[n]{} 0$ since

$$\int_{\mathbb{R}^3}\int_{\mathbb{R}^3 \times \mathbb{R}^3} (u_\varepsilon^n(x))^2 \, (v_\varepsilon(y))^2 \, |x-y|^{-1} \, dx \, dy \xrightarrow[n]{} 0, \ \int_{\mathbb{R}^3} V(u_\varepsilon^n)^2 \, dx \xrightarrow[n]{} 0,$$

and we conclude easily that $I_\lambda^V \leqslant I_\alpha^0 + I_{\lambda-\alpha}^V$.

Let us also mention the continuity of I_λ^V with respect to λ; indeed, if $I_\lambda^V \leqslant E^V(u) \leqslant I_\lambda^V + \varepsilon$, $|u|^2_{L^2} = \lambda$, then u is bounded in H^1 (and the bound depends only on bounds on ε and λ), and for $R > \theta > 0$,

$$I_{\theta\lambda}^V \leqslant E^V(\theta^{\frac{1}{2}}u) \leqslant E^V(u) + C_R|\theta - 1| \leqslant I_\lambda^V + \varepsilon + C_R|\theta - 1|.$$

This proves the continuity of I_λ^V with respect to λ.

Step 2: The concentration lemma. The proof made in the preceding step shows in fact that if for some $\alpha \in]0,\lambda]$ (6) becomes an inequality, then there exists a minimizing sequence which is not relatively compact in H^1 (and if $\alpha \in]0,\lambda[$ this sequence is not even relatively compact up to a translation).

We now prove the converse: namely, if the inequalities in (6) become strict inequalities, i.e.

$$I_\lambda^V < I_{\lambda-\alpha}^V + I_\alpha^0, \quad \forall \alpha \in]0,\lambda], \tag{S.1}$$

then any minimizing sequence is relatively compact in H^1. We will also prove that, if

$$I_\lambda^V < I_{\lambda-\alpha}^V + I_\alpha^0, \quad \forall\alpha \in]0,\lambda[, \tag{S.2}$$

then any minimizing sequence is relatively compact in H^1 up to a translation.

To prove this claim, the main ingredient is the following lemma:

LEMMA 2. (1) Let $(p_n)_{n\geq 1}$ be a sequence in $L^1_+(\mathbb{R}^N)$ satisfying $\int_{\mathbb{R}^N} p_n(x)dx = \lambda$. Then there exists a subsequence (that we still denote by p_n for simplicity) such that one of the following three properties hold for all $n \geq 1$:

(i) (vanishing) p_n goes weakly to 0 in the following sense:

$$\forall R < \infty, \quad \sup_{y\in\mathbb{R}^N} \int_{y+B_R} p_n(\xi)\, d\xi \xrightarrow[n]{} 0; \tag{7}$$

(ii) (compactness) p_n is relatively weakly compact up to a translation, i.e.

$$\exists y_n \in \mathbb{R}^N, \quad \forall\varepsilon > 0, \quad \exists R < \infty, \quad \int_{y_n+B_R} p_n(\xi)\, d\xi \geq \lambda - \varepsilon; \tag{8}$$

(iii) (dichotomy) there exists $\alpha \in]0,\lambda[$ such that: $\forall\varepsilon > 0$, $\exists n_0 \geq 1$, $\forall n \geq n_0$, $\exists p_n^1 \in L^1_+(\mathbb{R}^N)$, $\exists p_n^2 \in L^1_+(\mathbb{R}^N)$ satisfying:

$$\begin{cases} \|p_n - (p_n^1 + p_n^2)\|_{L^1} \leq \varepsilon, \ \left|\int_{\mathbb{R}^N} p_n^1\, dx - \alpha\right| \leq \varepsilon, \ \left|\int_{\mathbb{R}^N} p_n^2\, dx - (\lambda-\alpha)\right| \leq \varepsilon, \\ \text{dist}\,(\text{Supp}\, p_n^1, \text{Supp}\, p_n^2) \xrightarrow[n]{} +\infty. \end{cases} \tag{9}$$

(2) If in (1), $p_n = (u_n)^2$ where u_n is bounded in $H^1(\mathbb{R}^N)$ then in (iii) we may take $p_n^i = (u_n^i)^2$ $(i = 1,2)$ where u_n^1, u_n^2 satisfy:

$$\|u_n-(u_n^1+u_n^2)\|_{L^p} \leq \varepsilon \text{ for } 2 \leq p < \frac{2N}{N-2}, \ |\nabla u_n|^2_{L^2} \geq |\nabla u_n^1|^2_{L^2} + |\nabla u_n^2|^2_{L^2} + \varepsilon. \tag{10}$$

Remark. There are many variants of this lemma depending on the functional space in which the minimization problem is set.

We first admit this lemma and conclude the proof of the above claim.

We consider any minimizing sequence $(u_n)_{n \geq 1}$; we already know that u_n is bounded in $H^1(\mathbb{R}^N)$. We next apply Lemma 2 with $p_n = (u_n)^2$. Therefore on a subsequence (still denoted by p_n) either (i) occurs, or (ii) or (iii) (if $p_n^i = (u_n^i)^2$ and (9), (10) hold). We will eliminate (i) and (iii) using (S.2). If (i) occurs, then we prove that

$$\int_{\mathbb{R}} |V(x)| u_n^2 \, dx + \iint_{\mathbb{R}^3 \times \mathbb{R}^3} u_n^2(x) u_n^2(y) |x-y|^{-1} \, dx \, dy \xrightarrow[n]{} 0.$$

Indeed:

$$\int_{\mathbb{R}^3} |V| u_n^2 \, dx \leq \int_{\mathbb{R}^3} (|V| \wedge M) u_n^2 \, dx + \int_{\mathbb{R}^3} (|V|-M)^+ \, u_n^2 \, dx, \quad \forall M > 0$$

$$\leq \int_{B_R} (|V| \wedge M) u_n^2 \, dx + \int_{\mathbb{R}^3} (|V|-M)^+ + 1_{(\mathbb{R}^3 - B_R)} (|V| \wedge M) \} u_n^2 \, dx,$$

$$\forall M, R > 0$$

$$\leq M \int_{B_R} u_n^2 \, dx + C \, \| (|V|-M)^+ + 1_{(\mathbb{R}^3 - B_R)} (|V| \wedge M) \|_{L^p},$$

$$\forall M, R > 0,$$

where $V \in L^p$ (if $V = V_1 + V_2$ with $V_1 \in L^p$, $V_2 \in L^q$, we argue similarly for V_1 and V_2). We conclude our proof by letting $n \to +\infty$, then $R \to +\infty$ and $M \to \infty$. Next, since

$$\iint_{\mathbb{R}^3 \times \mathbb{R}^3} u_n^2(x) u_n^2(y) \; |x-y|^{-1} \; 1_{(|x-y| \geq R)} \, dx \, dy \leq \frac{1}{R} \lambda^2,$$

we just have to prove:

$$\iint_{|x-y| \leq R} u_n^2(x) u_n^2(y) \; |x-y|^{-1} \, dx \, dy \xrightarrow[n]{} 0.$$

In addition if $\delta \in]0, R[$, we have

$$\iint_{|x-y| \leq \delta} u_n^2(x) u_n^2(y) \; |x-y|^{-1} \, dx \, dy \leq \varepsilon(\delta) \, \|u_n\|_{H^1}^4 \leq C\varepsilon(\delta)$$

where $\varepsilon(\delta) \to 0$ as $\delta \to 0_+$.

Finally we have

$$\iint_{\delta\leq|x-y|\leq R} u_n^2(x)u_n^2(y)|x-y|^{-1}\, dx\, dy \leq \frac{1}{\delta}\int u_n^2(x)\,\{\int_{x+B_R} u_n^2(y)\, dy\}\, dx$$

$$\leq \frac{\lambda}{\delta} \sup_{x\in\mathbb{R}^3} \int_{x+B_R} u_n^2(y)\, dy \xrightarrow[n]{} 0.$$

Therefore, in conclusion, if (i) occurs we have: $\underline{\lim}_n\, E^V(u_n) \geq 0$; and this yields: $I_\lambda^V \geq 0$. But on the other hand we know by Step 1 that $I_\lambda^V \leq I_\lambda^0$ and if $u \in \mathcal{D}(\mathbb{R}^3)$, $|u|^2_{L^2} = \lambda$, then choosing σ small

$$I_\lambda^0 \leq E^0(\sigma^{-3/2}\, u(\tfrac{\cdot}{\sigma}))$$

$$= \frac{2}{\sigma^2} \int_{\mathbb{R}^3} \frac{1}{2}\,|\nabla u|^2\, dx - \frac{1}{\sigma} \iint_{\mathbb{R}^3\times\mathbb{R}^3} u^2(x)u^2(y)|x-y|^{-1}\, dx\, dy < 0.$$

And thus (i) cannot occur.

If (iii) occurs, we deduce from Lemma 2 that

$$E^V(u_n) \geq E^V(u_n^1) + E^V(u_n^2) + \delta(\varepsilon) + \varepsilon_n \tag{11}$$

where $\varepsilon_n \xrightarrow[n]{} 0$, $\delta(\varepsilon) \to 0$ as $\varepsilon \to 0_+$.

Indeed, from (10), we deduce easily that

$$E^V(u_n) \geq E^V(u_n^1 + u_n^2) + \delta(\varepsilon).$$

To complete the proof of our claim, we observe that

$$E^V(u_n^1 + u_n^2) \geq E^V(u_n^1) + E^V(u_n^2) + \varepsilon_n,$$

since

$$\iint_{\mathbb{R}^3\times\mathbb{R}^3} (u_n^1(x))^2(u_n^2(y))^2\, |x-y|^{-1}\, dx\, dy \leq \frac{1}{d_n}\,\lambda^2,$$

where d_n = dist (Supp p_n^1, Supp p_n^2).

In addition, since $d_n \xrightarrow[n]{} +\infty$, we may assume for example, that

$$\inf \{|x| \mid x \in \text{Supp } p^1_{n_k}\} \xrightarrow[k]{} +\infty \text{ for some subsequence } n_k.$$

We then deduce, as in Step 1, that

$$|E^V(u^1_{n_k}) - E^0(u^1_{n_k})| \xrightarrow[k]{} 0.$$

And thus (11) finally yields $I^V_\lambda \geqslant I^0_\alpha + I^V_{\lambda-\alpha} + \delta(\varepsilon)$, and since $\varepsilon > 0$ is arbitrary, this contradicts (S.2). The contradiction shows that (iii) cannot occur.

Therefore we proved that (ii) occurs. But since u_n is bounded in $H^1(\mathbb{R}^3)$, we may take a subsequence (still denoted by u_n) such that $u_n(\cdot+y_n)$ converges a.e. and weakly in $L^p(\mathbb{R}^3)$ (for $2 \leqslant p \leqslant 6$) and in $H^1(\mathbb{R}^3)$ to some $\tilde{u} \in H^1(\mathbb{R}^3)$. But in view of (8), we see that

$$\int_{B_R} \tilde{u}^2(\xi)\, d\xi = \lim_n \int_{B_R} u_n^2(y_n + \xi)\, d\xi \geqslant \lambda - \varepsilon,$$

and this shows that $|\tilde{u}|^2_{L^2} = \lambda$ and $u_n(\cdot+y_n) \xrightarrow[n]{} \tilde{u}$ in $L^2(\mathbb{R}^3)$. By Hölder inequalities we deduce that $u_n(\cdot+y_n) \xrightarrow[n]{} \tilde{u}$ in $L^p(\mathbb{R}^3)$ for $2 \leqslant p < 6$. It is then easy to complete the proof for the case when $V \equiv 0$ and (S.2) holds (or when (S.2) holds and $I^V_\lambda = I^0_\lambda$).

Finally, using the case $\alpha = \lambda$ in (S.1), we will show that y_n is bounded. Indeed, if it were not the case, there would exist a subsequence n_k such that $|y_{n_k}| \xrightarrow[k]{} +\infty$. Recall that $|V| = V_1 + V_2$ with $V_1 \in L^p(\mathbb{R}^3)$, $V_2 \in L^q(\mathbb{R}^3)$ and $\frac{3}{2} \leqslant p < q < \infty$; if $p = \frac{3}{2}$ we may write V_1 in the form $W^\varepsilon_1 + W^\varepsilon_2$, where $W^\varepsilon_2 \in L^q(\mathbb{R}^3)$ and $W^\varepsilon_1 \in L^{3/2}(\mathbb{R}^3)$, $\|W^\varepsilon_1\|_{L^{3/2}(\mathbb{R}^3)} \leqslant \varepsilon$. Next, since $V_2(\cdot+y_{n_k}) \xrightarrow[k]{} 0$ weakly in $L^q(\mathbb{R}^3)$, $V_1(\cdot+y_{n_k}) \xrightarrow[k]{} 0$ weakly in $L^p(\mathbb{R}^3)$ if $p > \frac{3}{2}$ or $W^\varepsilon_2(\cdot+y_{n_k}) \xrightarrow[k]{} 0$ weakly in $L^q(\mathbb{R}^3)$, and since $u^2_{n_k}(\cdot+y_{n_k})$ converges in $L^{p'} \cap L^{q'}$ (if $p > \frac{3}{2}$) we deduce that if $p > \frac{3}{2}$, then

$$\int_{\mathbb{R}^3} |V(x)| u_{n_k}^2 (x)\, dx = \int_{\mathbb{R}^3} |V(x+y_{n_k})| u_{n_k}^2 (x+y_{n_k})\, dx \underset{k}{\longrightarrow} 0,$$

while if $p = \frac{3}{2}$

$$\limsup_k \int_{\mathbb{R}^3} |V(x)| u_{n_k}^2 (x)\, dx \leqslant \|W_1^\varepsilon\|_{L^{3/2}} \limsup_k \|u_{n_k}\|_{L^6}^2$$

$$\leqslant C\varepsilon.$$

Therefore in all cases we see that

$$I_\lambda^V = \lim_k E^V(u_{n_k}) = \lim_k E^0(u_{n_k}) \geqslant I_\lambda^0,$$

and this contradicts (S.1). Therefore y_n is bounded and u_n converges to some u in $L^p(\mathbb{R}^3)$ for $2 \leqslant p < 6$ and weakly in $H^1(\mathbb{R}^3)$. It is then easy to complete the proof by standard arguments.

We now turn to the

PROOF OF LEMMA 2

The proof is based upon the notion of concentration function of the measure $p_n(x)\, dx$, namely:

$$Q_n(t) = \operatorname*{Sup}_{y \in \mathbb{R}^N} \int_{y+B_t} p_n(\xi)\, d\xi.$$

Obviously $(Q_n(\cdot))_{n \geqslant 1}$ is a sequence of nondecreasing continuous functions on $\mathbb{R}_+$ such that $Q_n(0) = 0$ and $Q_n(+\infty) = \lim_{t \uparrow +\infty} Q_n(t) = \lambda$. By a well-known lemma, there exist a nondecreasing nonnegative function Q and a subsequence (that we still denote by n) such that: $Q_n(t) \underset{n}{\longrightarrow} Q(t) \quad \forall t \in \mathbb{R}_+$. Of course $\alpha = \lim_{t \to +\infty} Q(t) = Q(+\infty)$ lies in $[0,\lambda]$. Obviously if $\alpha = 0$, $Q \equiv 0$ and case (i) occurs. Next, if $\alpha = \lambda$, it is well-known (see for example Parthasarathy [7]) that (ii) occurs: indeed let $0 < \varepsilon < \frac{\lambda}{2}$; then there exists $(y_n^\varepsilon)_{n \geqslant 1}$ in $\mathbb{R}^N$ and R_ε such that:

$$\int_{y_n^\varepsilon + B_{R_\varepsilon}} p_n(\xi) d\xi = \lambda - \varepsilon.$$

We set $y_n = y_n^{\varepsilon_0}$, where $\varepsilon_0 = \frac{\lambda}{4}$, for example, then we claim that if $0 < \varepsilon < \frac{\lambda}{2}$, then we have for $R'_\varepsilon = 2R_\varepsilon + R_{\varepsilon_0}$

$$\int_{y_n + B_{R'_\varepsilon}} p_n(\xi)\, d\xi \geq \lambda - \varepsilon;$$

indeed observe that $|y_n^\varepsilon - y_n| \leq R_\varepsilon + R_{\varepsilon_0}$. And this proves (ii). Finally, if $0 < \alpha < \lambda$, let $\varepsilon > 0$: there exists R_0 such that for $n \geq n_0$ $(=n_0(\varepsilon))$ $Q_n(R_0) \in]\alpha-\varepsilon, \alpha+\varepsilon[$. Next let y_n be such that

$$Q_n(R_0) = \int_{y_n + B_{R_0}} p_n(\xi)\, d\xi.$$

We then set $p_n^1 = p_n\, 1_{(y_n + B_{R_0})}$. In addition it is quite clear that there exists $R_n \xrightarrow[n]{} +\infty$ such that $Q_n(R_n) \leq \alpha + \varepsilon$. Then for $n \geq n_1$, we see that $R_n > R_0$, and we may set $p_n^2 = p_n\, 1_{\mathbb{R}^N - (y_n + B_{R_n})}$. It is then clear that (9) holds, since we have

$$\int_{R_0 \leq |x-y| \leq R_n} p_n(\xi)\, d\xi \leq Q_n(R_n) - \int_{y_n + B_{R_0}} p_n(\xi)\, d\xi = Q_n(R_n) - Q_n(R_0)$$

$$\leq 2\varepsilon.$$

To conclude the proof of Lemma 2, we observe that if p_n is given by $p_n = u_n^2$, where u_n is bounded in H^1, then we may argue as follows. Let $\chi \in \mathcal{D}_+(\mathbb{R}^3)$, $\chi \equiv 1$ if $|x| \leq 1$, $\chi \leq 1$ on $\mathbb{R}^3$, $\chi \equiv 0$ if $|x| \geq 2$ and let $\phi \in C_b^\infty(\mathbb{R}^3)$, $\phi \equiv 1$ if $|x| \geq 1$, $0 \leq \phi \leq 1$ on $\mathbb{R}^3$ and $\phi \equiv 0$ if $|x| \leq \frac{1}{2}$. Then we set $u_n^1 = u_n\, \chi(\frac{x}{R'_0})$, $u_n^2 = u_n\, \phi(\frac{x}{R_n})$, where R_n is chosen as above, and $R'_0 \geq R_0$ is such that

$$\| \nabla u_n^1 |^2_{L^2(\mathbb{R}^3)} - \int_{\mathbb{R}^3} \chi\left(\frac{x}{R'_0}\right)^2 |\nabla u_n|^2\, dx | \leq \varepsilon.$$

Remarking that for n large enough we have $R_n \geq 4R'_0$,

$$Q_n(R_0) = \int_{y_n + B_{R_0}} p_n(\xi)\, d\xi \leq \int_{y_n + B_{R'_0}} p_n(\xi)\, d\xi \leq Q_n(R'_0) \leq \alpha + \varepsilon$$

and

$$\| \nabla u_n^2 |^2_{L^2(\mathbb{R}^3)} - \int_{\mathbb{R}^3} \phi\left(\frac{x}{R_n}\right)^2 \quad |\nabla u_n|^2 \, dx| \xrightarrow[n]{} 0;$$

it is easy to obtain (10). This concludes the proof of Lemma 2.

Step 3: Conclusion. To conclude the proof of Theorem 1, we just have to show that (S.2) always holds for all V and thus that (S.1) holds if and only if we have: $I_\lambda^V < I_\lambda^0$. But this is an immediate consequence of the following:

LEMMA 3.

If $\theta > 1$, $\lambda > 0$; we have: $I_{\theta\lambda}^V < \theta I_\lambda^V < 0$. If Lemma 3 is proved, we see that

If Lemma 3 is proved, we see that if $0 < \lambda-\alpha \leqslant \alpha < \lambda$ we have

$$I_\lambda^V < \frac{\lambda}{\alpha} I_\alpha^V = I_\alpha^V + \frac{\lambda-\alpha}{\alpha} I_\alpha^V \leqslant I_\alpha^V + I_{\lambda-\alpha}^V \leqslant I_\alpha^0 + I_{\lambda-\alpha}^V$$

and thus (S.2) holds. Next, to prove Lemma 3, we observe that

$$I_{\theta\lambda}^V = \inf \{E^V(\theta^{\frac{1}{2}}v) \mid v \in H^1, |v|^2_{L^2} = \lambda\}$$

$$= \inf_{|v|^2_{L^2}=\lambda} \{\theta[\int_{\mathbb{R}^3} \frac{1}{2} |\nabla v|^2 + V(x) \frac{v^2}{2} \, dx] - \frac{\theta^2}{4} \iint_{\mathbb{R}^3\times\mathbb{R}^3} \frac{v^2(x)v^2(y)}{|x-y|} dxdy\}$$

and Lemma 3 is proved as soon as we know that in I_λ^V the infimum may be restricted to functions v satisfying in addition:

$$\iint_{\mathbb{R}^3 \times \mathbb{R}^3} v^2(x) \, v^2(y) \, |x-y|^{-1} \, dx \, dy \geqslant \alpha > 0$$

for some $\alpha > 0$. And this follows easily from the inequalities that we have already proved: $I_\lambda^V < 0$.

3. EXTENSIONS

In this short section we want to explain the main features of the concentration compactness method as they appeared in the preceding proofs. To do so we will outline formally the basic ideas that underlie this method. We emphasize that what follows is only a heuristic principle and that on each specific problem some precise argument has to be made along the lines which follow. We indicate at the end a list of problems which can be treated by such methods.

First, let H be a functional space on $\mathbb{R}^N$ and let J, E be functionals defined on H by (for example):

$$E(u) = \int_{\mathbb{R}^N} e(x, Au(x))\,dx, \quad J(u) = \int_{\mathbb{R}^N} j(x, Bu(x))\,dx$$

where, for example, A, B are operators from H into E, F (functional spaces of functions defined on $\mathbb{R}^N$ with values in $\mathbb{R}^m$, $\mathbb{R}^n$) commuting with translations and where $e(x,p)$, $j(x,q)$ are real functions defined on $\mathbb{R}^N \times \mathbb{R}^m$, $\mathbb{R}^N \times \mathbb{R}^n$. We consider the following minimization problem:

$$\inf \{E(u) \mid u \in H,\ J(u) = 1\}. \tag{M}$$

To solve (M) we first imbed (M) in a one-parameter family of problems (M_λ):

$$I_\lambda = \inf \{E(u) \mid u \in H,\ J(u) = \lambda\}. \tag{M_λ}$$

We then assume for simplicity that $J(0) = 0$, $E(0) = 0$ and that natural assumptions on H, A, B, e, j (adapted to each problem) hold, ensuring in particular that $\{u \in H \mid J(u) = 1\} \neq \emptyset$, and that we have a priori estimates in H on minimizing sequences. Our main assumption is then

$$j(x,q) \to j^\infty(q),\ e(x,p) \to e^\infty(p) \text{ as } |x| \to \infty$$

(in a convenient sense adapted to each problem).

We next introduce the problems 'at infinity' (M_λ^∞):

$$I_\lambda^\infty = \inf \{E^\infty(u) \mid u \in H,\ J^\infty(u) = \lambda\}. \tag{M_λ^∞}$$

where

$$E^\infty(u) = \int_{\mathbb{R}^N} e^\infty(Au(x))\,dx, \quad J^\infty(u) = \int_{\mathbb{R}^N} J^\infty(Bu(x))\,dx.$$

With the above notations and assumptions, the heuristic concentration-compactness principle states that <u>every minimizing sequence of (M_λ) is relatively compact if and only if the following condition holds:</u>

$$I_\lambda < I_{\lambda-\alpha} + I^\infty_\alpha \ , \ \ \forall \alpha \in]0,\lambda] \tag{S.1}$$

In addition if e,j do not depend on x (then $I_\lambda = I^\infty_\lambda$), every minimizing sequence of (M_λ) is relatively compact up to a translation if and only if the following condition holds:

$$I^\infty_\lambda < I^\infty_\alpha + I^\infty_{\lambda-\alpha}, \ \forall \alpha \ \in]0,\lambda[. \tag{S.2}$$

In each problem, the proof of this principle is based upon Lemma 2 or variants of it (adapted to H as (2) in Lemma 2 is adapted to H^1); and the main lines of the arguments are those developed in Steps 1 and 2. Therefore the main (and often difficult) question which remains in each problem is to decide whether (S.1) (or (S.2)) holds; but let us emphasize that since they are necessary and sufficient conditions one has to check them in order to be able to say that the problem is well-posed (in the above sense).

We want to conclude by two remarks: the first concerns the possibility of treating unbounded domains distinct from $\mathbb{R}^N$. The method can be easily adapted (see Lions [5] for some relevant examples). The second one is the fact that it may happen that (M_λ) has a minimum but still (S.1) is violated: then it only means that the problem has a minimum and that some minimizing sequences are not relatively compact - see Lions [5] for an example of such a phenomenon.

Finally let us mention a list of problems which can be solved by the use of the above considerations: minimization problems in $L^1(\mathbb{R}^N)$ - including problems arising in astrophysics and quantum mechanics; nonlinear fields equations; problems in strips or in half-spaces arising in fluid dynamics, such as the question of the existence of solitary waves or the so-called global vortex rings problem. We refer to Lions [4], [5] for more precise statements. Let us also indicate that this method yields the orbital stability of some standing waves in nonlinear Schrödinger equations, as was observed in Cazenave and Lions [1].

REFERENCES

[1] T. Cazenave and P.L. Lions, Orbital stability of standing waves for some nonlinear Schrödinger equations, Comm. Math. Phys. 85, 549-561 (1982).

[2] Donsker and S.R.S. Varadhan, personal communication.

[3] E.H. Lieb, Existence and Uniqueness of the Minimizing Solution of Choquard's Nonlinear Equation, Studies in Appl. Math. 57, 93-105 (1977)

[4] P.L. Lions, Principe de concentration-compacité en calcul des variations Comptes-Rendus Acad. Sci. Paris, 294, 261-264 (1982).

[5] P.L. Lions, work in preparation.

[6] P.L. Lions, The Choquard Equation and Related Questions, Nonlinear Anal. T.M.A. 4, 1063-1073 (1980).

[7] K.R. Parthasarathy, Probability Measures on Metric Spaces, Academic Press, New York (1967).

P.L. Lions
Université Paris 9 Dauphine,
Place de Lattre de Tassigny,
75775 Paris Cedex 16,
France

M PIERRE

Solutions of the porous medium equation in R^N under optimal conditions on initial values

Let us consider the nonlinear Cauchy problem

$$\left.\begin{aligned} u_t &= \Delta(|u|^{m-1}u) \quad \text{on} \quad \mathbb{R}^N \times (0,T) \\ u(\cdot,0) &= u_0 \end{aligned}\right\} \text{(IVP)}$$

where $m > 1$ and $T > 0$. Assume the initial datum u_0 has a singularity at infinity; for instance, $u_0 = |x|^\alpha$ with $\alpha > 0$ or $u_0 = \Sigma\, \alpha_n \delta_{x_n}$ where $(\alpha_n) \subset \mathbb{R}$ and δ_{x_n} is the Dirac mass at $x_n \in \mathbb{R}^N$ with $\lim |x_n| = +\infty$. Then, under which conditions on α or (α_n), (x_n), does (IVP) have a solution on some interval $(0,T)$? More generally, what is the largest class of initial data u_0 for which (IVP) has a local solution in time?

In the case when $m = 1$, that is

$$\left.\begin{aligned} u_t &= \Delta u \quad \text{on} \quad \mathbb{R}^N \times (0,T) \\ u(.,0) &= u_0 \in L^1_{loc}(\mathbb{R}^N), \end{aligned}\right\} \tag{1}$$

it is well-known that (1) has a solution on some interval $(0,T)$ if

$$\int_{\mathbb{R}^N} |u_0(x)|\, e^{-c|x|^2}\, dx < \infty \text{ for some } c > 0. \tag{2}$$

Moreover, if $u_0 \geqslant 0$, (2) is also necessary for (1) to have a solution $u \geqslant 0$ on some time interval.

Our purpose is to state a nonlinear version of this result for (IVP). To do so, we require some definitions.

Let $M(\mathbb{R}^N)$ denote the set of Radon measures on $\mathbb{R}^N$. For each μ in $M(\mathbb{R}^N)$ and $r > 0$, let

$$||\mu||_r = \sup_{R \geqslant r} \frac{1}{R^N} \int_{|x| \leqslant R} \frac{d|\mu|(x)}{(1+|x|^2)^{1/(m-1)}}. \tag{3}$$

Set

$$X = \{\mu \in M(\mathbb{R}^N) \mid \|\mu\|_1 < \infty\}$$

and equip X with the norm $\|\cdot\|_1$. Note that, if $\|\mu\|_r$ is finite for some $r > 0$, then it is finite for all $r > 0$ and $\|\cdot\|_r$ is an equivalent norm on X.

If $\mu \in X$, we define

$$\ell(\mu) = \lim_{r\uparrow\infty} \|\mu\|_r.$$

Next for $\alpha > 0$, define

$$\rho_\alpha(x) = (1 + |x|^2)^{-\alpha}.$$

<u>THEOREM</u>. Let $m > 1$ and $\mu \in X$. Then there exist $T > 0$ and $u \in C((0,T);X)$ such that

(i) $[t \mapsto u(t)\rho_{1/(m-1)}] \in L^\infty_{loc}((0,T);L^\infty(R^N))$.

(ii) $u_t = \Delta(|u|^{m-1}u)$ in the sense of distributions on $\mathbb{R}^N \times (0,T)$

(iii) $\lim_{t\downarrow 0} \int_{\mathbb{R}^N} u(t)f = \int_{\mathbb{R}^N} f\, d\mu$, for all $f : \mathbb{R}^N \to \mathbb{R}$ continuous with compact support.

Moreover there exists a constant C depending only on N and m such that

$$T \geqslant C/\ell(\mu)^{m-1}. \tag{4}$$

This result has been proved by Bénilan et al. in [3] to which we also refer for other related results. Let us emphasize the fact that the condition $\mu \in X$ is optimal with respect to the existence problem. Indeed, it was proved by Aronson and Caffarelli in [1] that if $\mu \geqslant 0$, the condition $\mu \in X$ is *necessary* for (IVP) to have a solution.

As a consequence of the above results (see also [4]), if $u_0(x) = (a+b|x|^2)^\alpha$, then (IVP) has a local solution if and only if $\alpha \leqslant 1/(m-1)$. If

$$u_0(x) = (a + b|x|^2)^{1/(m-1)},$$

then explicit solutions are known (see [2], [5]) and given by

$$u(x,t) = \left(a \frac{\tau^k}{(\tau-t)^k} + \frac{k}{2mN} \frac{|x|^2}{(\tau-t)}\right)^{1/(m-1)}$$

$$k = \frac{N(m-1)}{N(m-1)+2}, \quad \tau = \frac{k}{2mN}\frac{1}{b}.$$

This solution blows up at $T = \tau = k/2mN(\ell(u_0))^{m-1}$. This verifies that (4) is sharp as regards the functional dependence on u_0.

If $u_0(x) = (a + b|x|^2)^\alpha$ with $\alpha < 1/(m-1)$, then $\ell(u_0) = 0$ and the solution exists on $(0,\infty)$. In fact, it is proved in [3] that when $u_0 \geqslant 0$, a solution to (IVP) exists on $(0,\infty)$ if and only if $\ell(u_0) = 0$. This also shows the sharpness of (4).

Note that

$$L^1(\rho_\alpha) = \{u_0 \in L^1_{loc}(\mathbb{R}^N) \;\Big|\; \int_{\mathbb{R}^N} |u_0|\rho_\alpha < \infty\}$$

is included in X if and only if $\alpha \leqslant 1/(m-1) + N/2$. As soon as $\alpha < 1/(m-1)+N/2$, then

$$u_0 \in L^1(\rho_\alpha) \Longrightarrow \ell(u_0) = 0.$$

This is also the case when $u_0\rho_{1/(m-1)} \in L^p(\rho_{N/2})$ with $1 < p < \infty$ as proved in [3]. These remarks show how the above theorem gives global existence results which go far beyond the known global solvability results concerning $u_0 \in L^1(\mathbb{R}^N)$ or $u_0 \in L^1 + L^\infty(\mathbb{R}^N)$.

Let us finally mention that a uniqueness result for (IVP) is also proved in [3] in the class of functions u_0 belonging to X.

REFERENCES

[1] D.G. Aronson and L. Caffarelli (to appear).

[2] G.I. Barenblatt, On certain non-stationary motions of liquids and gases in porous media, Prikl. Mat. Mekh. 16, 67-78 (1952).

[3] Ph. Bénilan, M.G. Crandall and M. Pierre, Solutions of the porous medium equation in $\mathbb{R}^N$ under optimal conditions on initial values, Tech. Report, M.R.C. Un. of Wisconsin, Madison (1982).

[4] A.S. Kalashnikov, The Cauchy problem in a class of growing functions for equations of unsteady filtration type, Vestnik Moskov. Univ. Ser. VI, Mat. Mech. 6 17-27 (1963) (in Russian).

[5] R.E. Pattle, Diffusion from an instantaneous point source with concentration dependent coefficient, Q.J. Mech. Appl. Math. 12, 407-409 (1959).

Michel Pierre
Université Scientifique et Médicale de Grenoble,
Institut Fourier, B.P. 116,
38402- Saint Martin-d'Hères Cedex,
France

M SCHATZMAN

Stationary solutions and asymptotic behaviour of an equation like the porous media equation

1. INTRODUCTION

Consider the equation

$$u_t(x,t) - \Delta(a(u))(x,t) = f(x)\, u(x,t), \tag{1.1}$$

where x belongs to $\mathbb{R}^N$, a is increasing, $a'(0) = 0$, and f takes positive values at some point of $\mathbb{R}^N$, but is negative at infinity.

The object of this paper is to study stationary and non-stationary solutions of (1.1), which are nonnegative and have a compact support in $\mathbb{R}^N$.

This kind of equation has been introduced by Gurney and Nisbet [17], Gurtin and MacCamy [18, 19] and Namba [21] to describe the evolution of a population of density u; the term $\Delta(a(u))$ describes a diffusion biased towards the less densely populated areas, and the right-hand side f describes the effect of the environment on u: it is hostile whenever $f < 0$, and favourable whenever $f > 0$.

Gurney and Nisbet announce in [17], where the proof is only sketched, that, if f is positive on a bounded region, and negative elsewhere, there exists a nonnegative, stationary solution with compact support. Namba exhibits such a solution in [21]; if $x \in \mathbb{R}$, $a(u) = u^2$, $f(x) = A - B\,x^2$, then

$$\left.\begin{aligned} &u(x) = B(x - L)^2\,(x + L)^2/28 \text{ for } |x| < L \\ &u(x) = 0 \quad \text{for } |x| \geqslant L, \end{aligned}\right\} \tag{1.2}$$

with $L = \surd(7A/B)$, satisfies the above requirements.

In Section 2, we prove that the stationary equation

$$-\Delta(a(u)) = f.u \tag{1.3}$$

possesses nontrivial nonnegative solutions with compact support, under a light smoothness assumption on f, and a condition on a which means that at 0, it is flat enough. A technique of sub- and supersolutions is employed.

We then show, by a simple combinatorial argument that (1.3) may possess a very large number of distinct solutions, when f has more than one region of positivity, or bump.

It must be noticed that the boundary of the support of u is a free boundary and of course, an unknown of the problem.

In Section 3, we prove that the Cauchy problem for (1.1) possesses a unique solution for all time, under suitable assumptions on the initial data. If the initial data are nonnegative, essentially bounded and compactly supported, they keep these properties uniformly for all time. Then, under some smoothness assumptions, there exists a Lyapunov function for (1.1). Unfortunately, this is not enough to conclude that the ω-limit set of the trajectories is a set of stationary solutions, because a number of results of functional analysis are missing. In the particular case $a(u) = u^m$, $m > 1$, and $N = 1$, it is proved in Section 4 that the operator $-\Delta(a(\cdot)) - f$, has a smoothing effect, and that the ω-limit set of trajectories is indeed a set of stationary solutions.

Some of the results described here have appeared in [23]; when a proof is not technical, or improves the results of [23], it is completely given here, with the exception of a few cases where the holes can be easily filled by the reader. All technical proofs, and in particular the proofs of the results of Section 4, are to be found in [23].

Notice that if $a(u) = u^m$, $m > 1$, and $f = 0$, (1.1) is just the equation of porous media, which has been studied by many authors. In particular, the asymptotic behaviour of the solutions of the porous media equation has been studied by Friedman and Kamin [16]; ideas very close to mine have been used by Aronson et al. [1], for studying a porous media equation with a nonlinear right-hand side. Moreover, I learned from I. Díaz and J. Hérnandez, during the meeting in Madrid, that they have started to study fairly systematically the properties of quasilinear parabolic equations by methods of super- and subsolutions. Lastly, S. Kamin mentioned to me a number of papers concerned with the same family of problems.

2. THE STATIONARY PROBLEM

This section is devoted to a proof of an existence theorem (Subsection 2.2), and to an estimate of the number of solutions (Subsection 2.3).

2.1 Hypotheses and preliminaries

Let there be given a function a from $\mathbb{R}^+$ to itself, such that

$$a(0) = 0, \tag{2.1}$$

$$\lim_{r\to\infty} (a(r)/r) = +\infty, \tag{2.2}$$

$$a \text{ is strictly increasing and continuous on } \mathbb{R}^+ - \{0\}. \tag{2.3}$$

Let $b = a^{-1}$ be the reciprocal of a, and let B be the primitive of b which vanishes at zero:

$$B(r) = \int_0^r b(s)\,ds,\ r > 0. \tag{2.4}$$

We assume that

$$(B(r))^{-1/2} \text{ is integrable in a neighbourhood of zero.} \tag{2.5}$$

Two examples of such a function a are

$$a(r) = r^m,\ m \in (1,\infty); \tag{2.6}$$

$$a(r) = \exp(-k\, r^{-p}),\quad r,k,p \in \mathbb{R}^+ - \{0\}. \tag{2.7}$$

For k a positive constant, consider the ordinary differential equation

$$v''(y) = k\, b(v(y)),\quad y > 0; \tag{2.8}$$

$$v(0) = v'(0) = 0. \tag{2.9}$$

If (2.8) has a nontrivial solution, which is twice differentiable and non-negative, we can multiply (2.8) by $v'(y)$ and integrate, so as to obtain

$$v'(y)^2 = 2\,kB(y),$$

so that

$$\int_0^{v(y)} B(w)^{-1/2}\, dw = (2k)^{1/2} y.$$

According to (2.5), we can define the function c as the reciprocal of the function $v \mapsto \int_0^v B(w)^{-1/2}\, dw$, and we have then

$$v(y) = c((2k)^{1/2} y).$$

One checks that the only nontrivial solutions of (2.8)-(2.9) are of the form, for arbitrary nonnegative y_0,

$$v(y) = 0,\ y \leqslant y_0; \quad v(y) = c((2k)^{1/2}(y-y_0)),\ \ y \geqslant y_0. \tag{2.10}$$

Then, one checks that

$$\left.\begin{aligned} &c'(r) = B(c(r))^{1/2} \\ &c''(r) = b(c(r))/2 \\ &c(0) \ = c'(0) = c''(0) = 0. \end{aligned}\right\} \tag{2.11}$$

Finally, we make the following assumption on f:

$$f \in C^{0,\alpha}(\mathbb{R}^N), \tag{2.12}$$

$$\overline{\lim_{|x| \to \infty}}\ f(x) < 0, \tag{2.13}$$

$$\{x | f(x) > 0\} \neq \emptyset. \tag{2.14}$$

2.2 The existence theorem

THEOREM 2.1. Under hypotheses (2.1)-(2.5) on a, and (2.12)-(2.14) on f, there exists a function u from $\mathbb{R}^N$ to $\mathbb{R}$, with compact support in $\mathbb{R}^N$, such that

$$\left.\begin{aligned} &u \in C^{1,\lambda}(\mathbb{R}^N), \ \ \forall\lambda \in (0,1), \\ &u \in W^{2,p}(\mathbb{R}^N), \ \ \forall p \in [1,\infty) \\ &u \geqslant 0,\ u \not\equiv 0; \end{aligned}\right\} \tag{2.15}$$

$$-\Delta u = f.\ b(u). \tag{2.16}$$

Moreover, if b is Hölder-continuous of order β, then the second derivatives of u are Hölder continuous of order min (α,β).

Proof. We shall exhibit a compactly supported supersolution, and a nontrivial subsolution of (2.16). We take the supersolution to be radially symmetric:

$$\bar{u}(x) = \bar{v}(|x|),$$

with $\bar{v}$ of the form

$$\left.\begin{aligned} &\bar{v}(r) = k_1 - k_2 r^2, && r \leqslant r_1, \\ &\bar{v}''(r) = 0, && r_1 \leqslant r \leqslant r_2, \\ &\bar{v}(r) = c(k_3(r_3-r)), && r_2 \leqslant r \leqslant r_3, \\ &\bar{v}(r) = 0, && r_3 \leqslant r. \end{aligned}\right\} \tag{2.17}$$

We shall adjust the k_i and the r_i $i = 1,2,3$, so that $\bar{u}$ is in $C^{1,1}(\mathbb{R}^N)$, $\bar{u}$ in $L^\infty(\mathbb{R}^N)$, and

$$-\Delta\bar{u} \geqslant f.b(\bar{u}). \tag{2.18}$$

Let us compute indeed $-\Delta\bar{u}$ from definition (2.17), or rather $-\bar{v}'' - (N-1)\bar{v}'/r$:

$$-\bar{v}'' - (N-1)\bar{v}/r = \begin{cases} 2N\, k_2 \text{ if } 0 \leqslant r \leqslant r_1, \\ -(N-1)\bar{v}'(r_1)/r \text{ if } r_1 \leqslant r \leqslant r_2, \\ -(k_3)^2\, c''(k_3(r_3-r)) + (N-1)k_3\, c'(k_3(r_3-r))/r \text{ if} \\ \qquad r_2 \leqslant r \leqslant r_3. \end{cases} \tag{2.19}$$

Thanks to hypotheses (2.12)-(2.14), there exist positive constants M, M' and R such that

$$f(x) \leqslant M \quad \text{if } |x| \leqslant R$$

$$f(x) \leqslant -M' \text{ if } |x| > R.$$

If we choose $r_1 = R$ and $k_2 = k_1/(4R^2)$, then, we may satisfy (2.18) on $\{0 \leqslant r \leqslant r_1\}$, for k_1 large enough: relation (2.18) is implied by the following relation:

$$2N\, k_1/(4R^2) \geq Mb(k_1 - k_1 r^2/(4R^2)), \text{ for } 0 \leq r \leq r_1.$$

According to (2.2), $\lim_{r\to\infty} b(r)/r = 0$, and the above relation holds indeed for large k_1.

On $\{r_1 \leq r \leq r_2\}$, $-\Delta\bar{u}$ is positive so that (2.18) holds. On the last corona, $\{r_2 \leq r \leq r_3\}$, we have to satisfy

$$-(k_3)^2\, c''(k_3(r_3-r)) + r^{-1}\,(N-1)\, k_3\, c'(k_3(r_3-r)) \geq -M'b[c(k_3(r_3-r))].$$

If we set $c(k_3(r_3-r)) = s$, we can rewrite the above inequality with the help of (2.11) as

$$(k_3)^2\, b(s)/2 - (N-1)k_3\, B(s)^{1/2}\, r^{-1} \leq M'b(s),$$

or, if we divide both sides by $b(s)$:

$$(k_3)^2/2 - (N-1)k_3\, B(s)^{1/2}\, b(s)^{-1}\, r^{-1} \leq M'.$$

As B and b are nonnegative, (2.18) holds on $\{r_2 \leq r \leq r_3\}$ whenever $k_3 \leq (2M')^{1/2}$. We fix a k_3 satisfying this requirement.

A simple convexity argument now shows that, if $h(r) = k_1 - k_1 r^2/(4R^2)$ for $r \leq R$ and $h(r) = k_1((5/4) - (r/2R))$ for $r \geq R$, then there exists a unique r_2 such that

$$h(r) \leq c(k_3(r_3 - r)) \quad \forall r \in [0,r_3],$$

and

$$h(r_2) = c(k_3(r_3 - r_2)) \quad \text{for some } r_2 \text{ in } [r_1,r_3].$$

Clearly, this yields a $\bar{v}$ with the desired regularity.

Now, to exhibit a nontrivial, nonnegative subsolution $\underline{u}$, we shall use assumptions (2.12) and (2.14); there exist $x_0 \in \mathbb{R}^N$, $k_4 \geq$ and ρ such that

$$f(x) \geq k_4 \text{ for } |x - x_0| \leq \rho.$$

Let then, for some $\beta > 0$,

$$\underline{u}(x) = c(\beta(\rho^2 - |x - x_0|^2)) \text{ for } |x - x_0| \leq \rho,$$

$$\underline{u}(x) = 0 \quad \text{for } |x - x_0| > \rho.$$

The function c is by construction twice continuously differentiable (see 2.11), and its second derivative vanishes at zero; therefore, $\underline{u}$ belongs to $C^2(R^N)$, and has compact support. Inside $\{|x - x_0| < \rho\}$ we want to find a suitable β to satisfy

$$-\Delta\underline{u} \leq f.b(u) \tag{2.20}$$

Let

$$\underline{v}(r) = c(\beta(\rho^2 - r^2)), \quad r = |x - x_0|.$$

Then,

$$-\Delta\underline{u}(x) = -\underline{v}''(r) - (N-1)\,\underline{v}'(r)/r$$

$$= -4r^2\,\beta^2\,c''(\beta(\rho^2 - r^2)) + 2\,N\beta\,c'(\beta(\rho^2 - r^2)),$$

and according to (2.11),

$$-\Delta\underline{u}(x) = -2r^2\,\beta^2\,b(\underline{v}(r)) + 2N\beta B(\underline{v}(r))^{1/2},$$

and it must satisfy (2.20) which is implied by

$$-2r^2\,\beta^2\,b(\underline{v}(r)) + 2N\,\beta\,(B(\underline{v}(r)))^{1/2} \leq k_4\,b(\underline{v}(r)).$$

If we divide both sides of the above inequality by $b(\underline{v}(r))$, we obtain:

$$-2\,r^2\,\beta^2 + 2N\,\beta\,B(\underline{v}(r))^{1/2}\,b(\underline{v}(r))^{-1} \leq k_4. \tag{2.21}$$

Now, we can observe that

$$\frac{dB(s)^{1/2}}{ds} = b(s)\,(2B(s)^{1/2})^{-1},$$

so that, for any s positive, there is a $t(s)$ in the interval $[0,s]$ such that

$$B(s)^{1/2} = b(t(s)) \int_0^s B(r)^{-1/2}\,dr/2.$$

Therefore,

$$\lim_{s\to 0} B(s)^{1/2}\, b(s)^{-1} = 0.$$

Thus, (2.21) can be satisfied if β is sufficiently small, and moreover we can have, if β is taken even smaller,

$$\underline{u} \leqslant \bar{u}.$$

Thus, we have enough ground to apply the 'théorème 2.1' of Puel [22], and the rest of the proof is standard. See [23] for details.

Remark 2.2. If we assume instead of (2.12) that ess sup $f < +\infty$, and that there is a positive k_4 such that $\{f > k_4\}$ contains an open ball, then, except the last one, the conclusions of Theorem 2.1 still hold.

2.3 The number of solutions in some particular cases

At first sight, there is no reason why the solutions of (2.15)-(2.16) should be unique. We shall show, by a combinatorial argument, that they can indeed exist in very large numbers, if the set $\{f > 0\}$ is multiply connected.

Let first u be a solution of (2.12)-(2.16) such that the set $\{u > 0\}$ has exactly one connected component, which we denote by C. If n is an arbitrary integer, we can construct a u_n and an f_n such that $\{u_n > 0\}$ has exactly n connected components, and u_n is the solution of (2.15)-(2.16) relative to f_n: let R be the radius of a ball containing C, and let a sequence x_i, $1 \leqslant i \leqslant n$ be such that $|x_i - x_j| \geqslant 5R$, for $i \neq j$. Then, we define f_n as follows:

$$f_n(x) = f(x - x_i) \text{ on } B_R(x_i), \quad 1 \leq i \leq n;$$

$$f_n(x) = -1 \text{ on } R^N \setminus \bigcup_{i=1}^{n} B_{2R}(x_i);$$

f_n is continuous and takes only negative values on

$$\bigcup_{i-1}^{n} (B_{2R}(x_i) \setminus B_R(x_i)).$$

The function u_n defined by

$$u_n(x) = u(x - x_i) \text{ on } B_R(x_i);$$

$$u_n(x) = 0 \text{ on } \bigcup_{i=1}^{n} B_R(x_i),$$

is obviously a solution of (2.15)-(2,16).

Let now u be a solution of (2.7) such that $\{u > 0\}$ has exactly n connected components; denote by C_i these components ($1 \leq i \leq n$), and by u_i the function equal to u on C_i, and zero elsewhere. Then (2.15)-(2.16) has at least C_n^p (binomial coefficient) solutions which do not vanish on p components; therefore (2.15)-(2.16) has at least 2^{n-1} distinct nontrivial solutions. But it can be observed that if u vanishes on a set where f is positive, then u is not asymptotically stable, as a solution of the associated evolution problem; see the end of Section 3.

3. THE EXISTENCE, UNIQUENESS, AND LYAPUNOV FUNCTION FOR THE EVOLUTION EQUATION

3.1 Existence, uniqueness and estimates

The existence theorem for the Cauchy problem for the evolution equation

$$u_t - \Delta a(u) - f.u = 0, \quad x \in \mathbb{R}^N, \; t > 0; \tag{3.1}$$

$$u(x,0) = u_0(x) \tag{3.2}$$

is a consequence of the theory of nonlinear semigroups in Banach spaces and m-accretive operators, which can be found in various references, such as [4], [10] and [15]. We have only to prove that if a function g is defined by

$$\omega = \max f \text{ and } g = \omega - f, \tag{3.3}$$

then the operator $u \mapsto -\Delta a(u) + g.u$ is m-accretive in $L^1(\mathbb{R}^N)$. We need an extension of a to all of $\mathbb{R}$; we shall assume, for simplicity, that a is continuous, odd and increasing. Let an operator A be defined by

$$\begin{cases} Au = \{z \in L^1(\mathbb{R}^N) \mid \exists w \text{ such that } w = a(u) \text{ a.e. in } \mathbb{R}^N \text{ and} \\ -\Delta w = z \text{ in } \mathcal{D}'(\mathbb{R}^N)\}; \quad D(A) = \{u \in L^1(\mathbb{R}^N) \mid Au \neq \emptyset\}. \end{cases} \tag{3.4}$$

Bénilan et al. have shown in [6] that A is m-accretive in $L^1(\mathbb{R}^N)$. Now define an operator B by

$$D(B) = \{u \in L^1(\mathbb{R}^N) \mid g.u \in L^1(\mathbb{R}^N)\}; \quad Bu = g.u. \tag{3.5}$$

We have the following result:

LEMMA 3.1. Let g belong to $L^\infty_{loc}(\mathbb{R}^N)$, and assume that g is nonnegative almost everywhere on $\mathbb{R}^N$. Then, A + B is m-accretive in $L^1(\mathbb{R}^N)$.

Proof. If g belongs to $L^\infty(\mathbb{R}^N)$, this is a consequence of Theorem 3.2 of Chapter III of Barbu's book [4], because, in this case, B is m-accretive and continuous from $L^1(\mathbb{R}^N)$ to itself.

In the general case, we truncate g, and we let, for n a positive integer,

$$g_n = \min (g,n); \quad B_n u = g_n.u. \tag{3.6}$$

Then $A + B_n$ is m-accretive, according to the result which we just quoted. Therefore, according to the definition of m-accretivity, for every positive λ, and for every v in $L^1(\mathbb{R}^N)$, there exists a solution u_n of

$$u_n + \lambda\, g_n\, u_n + \lambda A\, u_n = v. \tag{3.7}$$

Therefore, there exist w_n and z_n in $L^1(\mathbb{R}^N)$ such that

$$w_n + \lambda g_n w_n + \lambda A w_n = |v| \tag{3.8}$$

$$z_n + \lambda g_n z_n + \lambda A z_n = - |v|. \tag{3.9}$$

It is shown in [4] that A satisfies a maximum principle; thus, if $m \geq n$, we have

$$w_n \geq w_m \geq 0 \quad \text{a.e.} \tag{3.10}$$

$$z_n \leq z_m \leq 0 \quad \text{a.e.,} \tag{3.11}$$

and of course, for all n,

$$z_n \leq u_n \leq w_n \quad \text{a.e. on } \mathbb{R}^N \text{ and } \max(|z_n|_{L^1}, |w_n|_{L^1}) \leq |v|_{L^1}. \tag{3.12}$$

It follows from (3.10) that the sequence w_n converges strongly and almost everywhere to some function w of $L^1(\mathbb{R}^N)$. Moreover,

$$Aw_n \to Aw \text{ in } \mathcal{D}'(\mathbb{R}^N). \tag{3.13}$$

But, on the other hand, for any bounded ball B_R of centre 0 and radius R, we know that $g_n = g$ for n large enough, so that

$$g_n w_n \to gw \text{ in } L^1(B_R) \text{ and a.e. on } B_R.$$

If we integrate (3.8) on $\mathbb{R}^N$, we obtain

$$\int_{\mathbb{R}^N} w_n (1 + \lambda g) \, dx = \int_{\mathbb{R}^N} |v| \, dx \tag{3.14}$$

and, thus, by Fatou's lemma,

$$\lambda \int g \, w \, dx \leq \int |v| \, dx. \tag{3.15}$$

It follows that

$$w + \lambda g \, w + \lambda A \, w = |v| \text{ in } \mathcal{D}'(\mathbb{R}^N). \tag{3.16}$$

As w, v and g.w belong to $L^1(\mathbb{R}^N)$, w belongs to D(A); moreover, if we integrate (3.16) on $\mathbb{R}^N$, we get

$$\int w(1 + \lambda g)\, dx = \int |v|\, dx.$$

Consider now the expression $\int |g_n w_n - g w|\, dx$:

$$\int |g_n w_n - g w|\, dx = \int (g_n w_n - g w)\, dx + 2 \int (g_n w_n - g w)^- \, dx$$

$$= \int (g_n w_n - g w)\, dx + 2 \int_{\substack{g \geq n \\ gw \geq g_n w_n}} (g w - n w_n)\, dx.$$

But we know from (3.14), (3.16) and the convergence of w_n towards w in $L^1(\mathbb{R}^N)$ that

$$\int (g_n w_n - g w)\, dx \to 0.$$

On the other hand,

$$\int_{g > n} g w\, dx \to 0, \text{ as } n \to \infty ,$$

from Lebesgue's theorem. Therefore,

$$g_n w_n \to g w \text{ in } L^1(\mathbb{R}^N) \text{ and almost everywhere in } \mathbb{R}^N. \tag{3.17}$$

In the same fashion, one shows that

$$\left.\begin{array}{l} z_n \to z \quad \text{in } L^1(\mathbb{R}^N), \\ Az_n \to Az \text{ in } L^1(\mathbb{R}^N), \\ g_n z_n \to g z \text{ in } L^1(\mathbb{R}^N), \end{array}\right\} \tag{3.18}$$

and

$$z + \lambda A z + \lambda g z = - |v|.$$

If we consider now the sequence $(u_n)_n$, the properties of accretive operators give

$$\int |u_m - u_n|\, dx + \int (g_n u_n - g_m u_m) \operatorname{sgn} (u_n - u_m)\, dx \leq 0.$$

Assume m > n. Then,

$$\int (g_n u_n - g_m u_m) \operatorname{sgn}(u_n - u_m)\, dx = \int_{g<n} g|u_n - u_m|\, dx$$

$$+ \int_{E_n^+} (g_n u_n - g_m u_m)\, dx + \int_{E_n^-} (g_m u_m - g_n u_n)\, dx,$$

where $E_n^+ = \{g > n \text{ and } u_n > u_m\}$ and $E_n^- = \{g > n \text{ and } u_n < u_m\}$. Thus,

$$\int |u_n - u_m|\, dx \leqslant \int_{E_n^+} (g_m u_m - g_n u_n)^+\, dx + \int_{E_n^-} (g_n u_n - g_m u_m)^+\, dx.$$

But, we observe that, on E_n^+, $g_m u_m - g_n u_n$ can be positive if $u_m > 0$, so that

$$\int_{E_n^+} (g_m u_m - g_n u_n)\, dx \leqslant \int_{E_n^+} g_m u_m\, dx \leqslant \int_{g>n} g_m w_m\, dx,$$

which tends to zero as m, n tend to infinity, according to (3.18). An analogous observation holds on E_n^-. Therefore, $(u_n)_n$ is a Cauchy sequence in $L^1(\mathbb{R}^N)$, and it converges to some u in $L^1(\mathbb{R}^N)$; after possibly extracting a subsequence, it converges almost everywhere to u in $\mathbb{R}^N$, and thus

$$Au_n \to Au \text{ in } \mathcal{D}'(\mathbb{R}^N).$$

From the inequality

$$g_n |u_n| \leqslant \max (g_n w_n, g_n |z_n|)$$

it follows that $g_n u_n$ converges to $g u$ in $L^1(\mathbb{R}^N)$. Therefore, in the limit,

$$u + \lambda g u + \lambda A u = v, \tag{3.19}$$

and Au belongs to $L^1(\mathbb{R}^N)$, which proves Lemma 3.1, as (3.19) has obviously a unique solution in $L^1(\mathbb{R}^N)$.

We have now a number of results which are easy consequences of Lemma 3.1, and of general results of the theory of semigroups generated by m-accretive operators.

THEOREM 3.2. Assume that f belongs to $L^{\infty}_{loc}(\mathbb{R}^N)$, with ess sup f < + ∞; then, for any u_0 in $L^1(\mathbb{R}^N)$, there exists a function u in $C^0([0,\infty); L^1(\mathbb{R}^N))$ such that (3.1) is satisfied in the sense of distributions, with the initial condition (3.2).

Proof. Apply the Crandall and Liggett theorem, [10].

Now note $u(\cdot,t) = S(t)u$. The properties of the semigroup S are summarized in

THEOREM 3.3. (i) S is a strongly continuous semigroup.

(ii) let ω = ess sup f. For any $t > 0$, u and $\hat{u}$ in $L^1(\mathbb{R}^N)$, we have

$$|S(t)u - S(t)\hat{u}|_{L^1(\mathbb{R}^N)} \leqslant e^{\omega t} |u - \hat{u}|_{L^1(\mathbb{R}^N)}.$$

(iii) For any u, $\hat{u}$ in $L^1(\mathbb{R}^N)$ such that $u \geqslant \hat{u}$ a.e. in $\mathbb{R}^N$, and for any $t > 0$,

$$S(t)\, u \geqslant S(t)\hat{u} \text{ a.e. in } \mathbb{R}^N.$$

(iv) For any essentially bounded u_0, and any nonnegative t,

$$\min\,(0, \text{ess inf } u_0)\, e^{\omega t} \leqslant u(x,t) \leqslant \min\,(0, \text{ess sup } u_0)\, e^{\omega t}, \text{ a.e. in } \mathbb{R}^N.$$

Proof. This is a direct consequence of [10], and of the maximum principle for A shown in [6].

Then, we have a corollary:

COROLLARY 3.4 Assume that u_0 is essentially bounded; then u is essentially bounded on $\mathbb{R}^N \times [0,T]$, for all finite T; if, moreover, f is essentially bounded, then u is unique in the class

$$\{u \in L^{\infty}([0,T] \times \mathbb{R}^N) \cap L^1([0,T] \times \mathbb{R}^N) / \operatorname*{ess\,lim}_{t \downarrow 0} \int |u(x,t) - u_0(x)|\, dx = 0\}.$$

Proof. The first assertion is a consequence of (iv) of Theorem 3.3, and the second comes from a straightforward generalization of Theorem 3 of [8].

Lastly, if the initial condition is nonnegative, compactly supported, and essentially bounded, it retains these properties for all time:

COROLLARY 3.5. Assume that u_0 is nonnegative, compactly supported, and essentially bounded, and that a satisfies hypotheses (2.1)-(2.5). Then, there exists a compact set K such that the support of $u(\cdot,t)$ is included in K for all nonnegative time; moreover, u is nonnegative, and essentially bounded on $\mathbb{R}^N \times [0,\infty)$.

Proof. Given u_0 satisfying the assumptions of the corollary, we can build with the techniques of Theorem 2.1 a supersolution of (2.16), $\bar{v}$, such that

$$-\Delta\bar{v} \geqslant f\, b(\bar{v}),$$

$$\bar{v} \geqslant a(u_0) \text{ a.e. on } \mathbb{R}^N,$$

$$\bar{v} \in C^{1,1}(\mathbb{R}^N), \quad \bar{v} \geqslant 0, \ \bar{v} \text{ has compact support.}$$

Then, if $\bar{u} = b(\bar{v})$, we can see that the maximum principle implies

$$0 \leqslant u(\cdot,t) \leqslant \bar{u}, \quad \forall t > 0, \quad \text{a.e. on } \mathbb{R}^N,$$

which proves Corollary 3.5.

For detailed proofs of the above results, see [23].

3.2 The Lyapunov function

We can define formally a Lyapunov function for equation (3.1), given by

$$V(u) = \int_{\mathbb{R}^N} (|\nabla(a(u))|^2/2 - f\, h(u))\, dx, \tag{3.20}$$

where

$$h(u) = \int_0^u s\, a'(s)\, ds. \tag{3.21}$$

To write a complete proof of the above assertion, we shall assume

$$a \text{ is absolutely continuous} \tag{3.22}$$

and we shall approximate a by

$$a_\varepsilon(u) = a(u) + \varepsilon u.$$

Bénilan and Crandall have shown in [7] that if an m-accretive operator A_ε is defined by

$$A_\varepsilon u = -\Delta(a_\varepsilon(u)),$$

then there is a space X such that, for all positive λ, and all u in X,

$$(I + \lambda A_\varepsilon)^{-1}u \to (I + \lambda A)^{-1}u \text{ in } X. \tag{3.23}$$

The space X is given by

$$\begin{cases} X = L^1(\mathbb{R}^N) \text{ if either } N = 1 \text{ or } 2, \text{ or if } N \geq 3 \text{ and } r^{N-1}\, b(r^{-N+2}) \\ \qquad \text{is not integrable in a neighbourhood of infinity.} \end{cases}$$

or

$X = L^1(\mathbb{R}^N, \rho\, dx)$ if $N \geq 3$ and $r^{N-1}\, b(r^{-N+2})$ is integrable in some neighbourhood of infinity; the weight ρ is $\rho(x) = (1 + |x|^2)^{-\alpha}$, where α is some positive number.

As we are only concerned with compactly supported solutions, the two cases for X are not different. Moreover, we can assume, for the same reason, that f is essentially bounded. Then, we have

<u>LEMMA 3.6</u>. Assume that f is essentially bounded, and define g as in (3.3), and B as in (3.5). Then, for every u in X, as ε tends to zero,

$$(I + A_\varepsilon + B)^{-1}\, u \to (I + A + B)^{-1}\, u \text{ in } X. \tag{3.24}$$

<u>Proof</u>. See [23].

Let S be the semi-group generated by $A + B - \omega I$; we deduce from (3.24) and Proposition I.23 of [5] that for all u in X, and all finite T,

$$S_\varepsilon(\cdot)\, u \to S(\cdot)\, u \text{ in } C^0([0,T];X).$$

For the approximating equation, which is (2.16), with a replaced by a_ε, we let

$$V_\varepsilon(u) = \int_{\mathbb{R}^N} (|\nabla a_\varepsilon(u)|^2/2 - f\, h_\varepsilon(u))\, dx,$$

with

$$h_\varepsilon(u) = \int_0^u s\, a'_\varepsilon(u)\, ds = h(u) + \varepsilon u^2/2.$$

Then, thanks to the regularity results of [20], V_ε is a Lyapunov function for the approaching equation, and

$$V_\varepsilon(u_0) - V_\varepsilon(S(t)u_0) - \int_0^t \int_{\mathbb{R}^N} (\Delta a_\varepsilon(u_\varepsilon) + fu_\varepsilon)^2\, a'_\varepsilon(u_\varepsilon)\, dx\, dt = 0. \quad (3.25)$$

Then, we have a proposition which describes the domain of validity of the Lyapunov function (3.20) for equation (2.16).

THEOREM 3.7. Assume that u belongs to $L^\infty(\mathbb{R}^N)$, is nonnegative a.e. and has compact support; let a satisfy hypotheses (2.1)-(2.5), (3.22), and assume that

$$\nabla(a(u)) \in (L^2(\mathbb{R}^N))^N. \quad (3.26)$$

Then, if V is defined by (3.20), we have

$$V(S(t)u_0) \text{ is finite for all positive } t; \quad (3.27)$$

$$1_{\{u>0\}}\, \Delta(a(u))\, a'(u)^{\frac{1}{2}} \in L^2(\mathbb{R}^N \times [0,\infty)), \quad (3.28)$$

and the following inequality is fulfilled for all positive t:

$$V(u_0) - V(S(t)u_0) \geq \int_0^t \int_{\mathbb{R}^N} (\Delta a(u) + fu)^2\, 1_{\{u>0\}} a'(u)\, dx\, ds \quad (3.29)$$

Proof. See [23].

4. ASYMPTOTIC BEHAVIOUR IN A SPECIAL CASE

One main difficulty to overcome in the study of the asymptotic behaviour is that we lack functional analysis results which would yield enough compactness in our problem. We have indeed the three following methods available:

(1) if $S(\cdot)$ is equicontinuous in $u \in Y$, uniformly with respect to time, if the orbit $S(\cdot)u$ is compact in Y, and if V is lower semi-continuous on Y, then one may use the invariance principle [13, 14].

(2) if S is not equicontinuous in $u \in Y$, uniformly with respect to time, but if $V(u) - V(S(1)u)$ is l.s.c. on Y, we have a more refined form of the invariance principle [3], which allows one to complete the proof if the orbit is compact. Unfortunately, Section 2 does not allow us to employ these results! Therefore, a last possibility remains

(3) if S is not equicontinuous in $u \in Y$, uniformly with respect to time, and if the orbit is compact in Y, and V is continuous on the orbit, we can complete the proof.

Therefore, we have to prove a smoothing effect of S: it will hold under the following restrictions:

$$a(u) = u^m, \text{ with } m > 1; \tag{4.1}$$

$$f \in C^{2,\alpha}(\mathbb{R}^N), \text{ for some } \alpha \in (0,1). \tag{4.2}$$

<u>THEOREM 4.1</u>. Assume (2.1)-(2.5), (4.1) and (4.2); assume moreover that u_0 is nonnegative a.e., essentially bounded, and compactly supported. Then, for all positive r,

$$u_t \in L^\infty([r,\infty);L^1(\mathbb{R}^N))$$

and

$$a(u) \in L^\infty([r,\infty); W^{2,1}(\mathbb{R}^N)).$$

<u>Proof</u>. The proof is unfortunately very technical; it relies on ideas of [2], improved in [11, 12], and the details can be found in [23] when the proof differs from that given in [11].

With the help of Theorem 4.1, we can prove a statement on the ω-limit set of a trajectory of (3.1)-(3.2):

<u>THEOREM 4.2</u>. Assume the hypotheses of Theorem 4.1, together with (3.22), and

$$N = 1. \tag{4.3}$$

Then, the ω-limit set of $S(\cdot)u$ as t tends to ∞, defined by

$$\omega^+ S(\cdot)u = \bigcap_{s>0} \overline{\{S(t)u/t > s\}} ,$$

is a set of stationary solutions.

Proof. Essentially standard; see [23] for the details.

Before concluding, let us mention that under the conditions of Theorem 4.2, if u is a stationary solution which vanishes on a set A where f is positive, and if int(A) is not empty, then u is not asymptotically stable, because V has certainly not a local minimum at such a u; the proof of this statement is in [23].

The restrictions under which Theorem 4.2 was proved are very unsatisfactory; it would be interesting to show that the ω-limit set of u is always a set of stationary solutions, because the flow is gradient-like; a large amount of functional analysis seems necessary to complete this programme; the techniques of [9] might prove useful in this direction.

ACKNOWLEDGEMENTS

I wish to thank M. Gurtin for attracting my attention to Problems (1.1) and (1.3), A. Haraux for some helpful discussions, and P. L. Lions, for attracting my attention to [2].

REFERENCES

[1] D. Aronson, M.G. Crandall and L.A. Peletier, Stabilization of solutions of a degenerate nonlinear diffusion problem, M.R.C. technical report no 2220, Madison, Wisconsin (May 1981).

[2] D.G. Aronson and P. Bênilan, Régularité des solutions de l'équation des milieux poreux dans $\mathbb{R}^N$, C.R. Acad. Sci. Paris, Série A-B, 288, 103-105 (1979).

[3] J.M. Ball, On the asymptotic behavior of generalized processes, with applications to nonlinear evolution equations, J. Differential Equations 27 224-265 (1978).

[4] V. Barbu, Nonlinear semi-groups and differential equations in Banach spaces, Editura Academiei Bucuresti România, and Noordhoff Int. Publishing, Leyden (1976).

[5] Ph. Bênilan, Thèse d'Etat, Mathématiques, Université Paris 11, Orsay (1972).

[6] Ph. Bênilan, H. Brézis and M.G. Crandall, A semilinear elliptic equation in $L^1(\mathbb{R}^N)$, Ann. Scuola Norm. Sup. Pisa, Serie IV, 2, 523-555 (1975).

[7] Ph. Bênilan and M.G. Crandall, The continuous dependence on ϕ of solutions of $u_t - \Delta\phi(u) = 0$, Indiana University Math. J., 30, 161-177 (1981).

[8] H. Brézis and M.G. Crandall, Uniqueness of solutions of the initial value problem for $u_t - \Delta\phi(u) = 0$, J. Math. Pures Appl. 58, 153-163 (1979).

[9] L.A. Caffarelli and A. Friedman, Continuity of the density of a gas flow in a porous medium, Transactions A.M.S. 252, 99-113 (1979).

[10] M.G. Crandall and T.M. Liggett, Generation of semi-groups of nonlinear transformations on general Banach spaces, Amer. J. Math. 93 265-298 (1971).

[11] M.G. Crandall and M. Pierre, Regularity effects for $u_t = \Delta\phi(u)$, M.R.C. technical summary report no. 2166, Madison, Wisconsin (January 1981).

[12] M.G. Crandall and M. Pierre, Regularity effects for $u_t = A\phi(u)$ in L^1, M.R.C. technical summary report no. 2187, Madison, Wisconsin (1981).

[13] C.M. Dafermos, Uniform processes and semicontinuous Liapunov functionals, J. Differential Equations, 11, 401-415 (1972).

[14] C.M. Dafermos, Applications of the invariance principle for compact principles II, Asymptotic behavior of solutions of a hyperbolic conservation law, J. Differential Equations, 11, 416-424 (1972).

[15] L.C. Evans, Applications of nonlinear semigroup theory to certain partial differential equations, in Nonlinear Evolution Equations, M.G. Crandall (ed.), Academic Press, N.Y., 163-188 (1978).

[16] A. Friedman and S. Kamin, Transactions A.M.S. 262, 551-563 (1980).

[17] W.S.C. Gurney and R.M. Nisbet, The regulation of inhomogeneous populations, J. Theor. Biol. 52, 441-457 (1975).

[18] M.E. Gurtin and R.C. MacCamy, On the diffusion of biological populations, Math. Biosciences, 33, 34-49 (1977).

[19] M.E. Gurtin and R.C. MacCamy, Product solutions and asymptotic behavior for age-dependent, dispersing populations, Preprint (1981).

[20] O.A. Ladyženskaja, V.A. Solonnikov and N.N. Ural'tseva, Linear and quasilinear equations of parabolic type, Translations of Math. Monographs, Vol. 23, A.M.S., Providence (1968).

[21] T. Namba, Density dependent dispersal and spatial distribution of a population, J. Theor. Biol. 86, 351-363 (1980).

[22] J.P. Puel, Existence, comportement à l'infini et stabilité dans certains problèmes quasilinéaires elliptiques et paraboliques d'ordre 2, Ann. Scuola Sup. Pisa., Serie IV, 3, 89-119 (1976).

[23] M. Schatzman, Stationary solutions and asymptotic behavior of a quasilinear degenerate parabolic equation, Rapport interne no. 76, Centre de Math. Appli. Ecole Polytechnique, Palaiseau, France (March 1982).

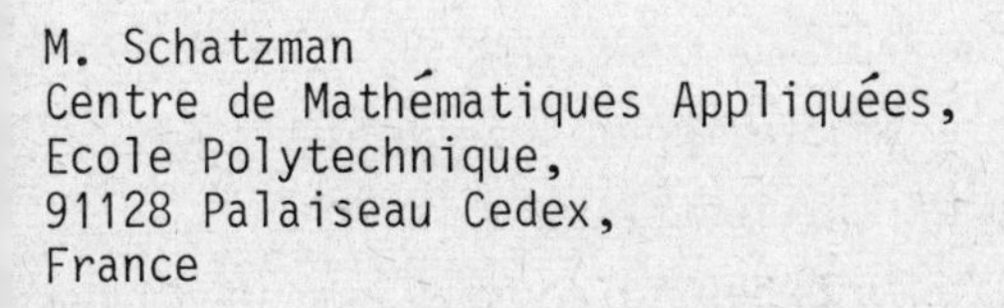

M. Schatzman
Centre de Mathématiques Appliquées,
Ecole Polytechnique,
91128 Palaiseau Cedex,
France

J M VEGAS

A Neumann elliptic problem with variable domain

1. INTRODUCTION

We consider the evolution equation

$$\left.\begin{aligned} \frac{\partial u}{\partial t} &= \Delta u + f(u) \quad \text{in} \quad \Omega \times (0,\infty) \\ \frac{\partial u}{\partial n} &= 0 \qquad\qquad\; \text{on } \partial\Omega \times (0,\infty) \end{aligned}\right\} \tag{1}$$

where Ω is a smooth domain in $\mathbb{R}^2$ and f is a smooth function, and we focus our attention on the possible bifurcation of stationary solutions of (1) which may arise when we vary the domain Ω. In order to do that, we introduce a 'continuous' family of domains Ω_ε which 'converges' as $\varepsilon \to 0$ to the union of two connected sets: $\Omega_0 = \text{'}\lim_{\varepsilon\to 0} \Omega_\varepsilon\text{'} = \Omega_0^L \cup \Omega_0^R$ (Fig. 1).

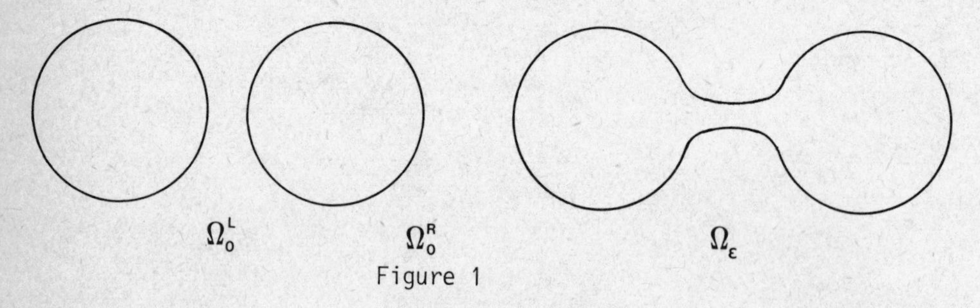

Figure 1

By the Liapunov-Schmidt method we reduce the problem of finding stationary solutions of (1) to a two-dimensional problem with a parameter ε, to which we apply the methods and concepts of singularity theory, obtaining the desired description of the bifurcations.

2. SETTING OF THE PROBLEM

In Hale and Vegas [5] we have shown that, if $\lambda^{(k)}(\Omega)$ represents the k-th eigenvalue of $-\Delta$ on Ω with Neumann boundary conditions, then

$$\lambda^{(2)}(\Omega_\varepsilon) \overset{\text{def}}{=} \lambda_\varepsilon \to 0 \text{ as } \varepsilon \to 0 \tag{2}$$

$\lambda^{(3)}(\Omega_\varepsilon)$ is bounded away from zero as $\varepsilon \to 0$ (3)

for a special family of domains Ω_ε with the shape shown in Fig. 1. These facts enable us to apply the Liapunov-Schmidt method by projecting on the suitable eigenspaces. More precisely, we select w_ε = normalized eigenfunction corresponding to λ_ε with the property:

$$\int_{\Omega_0^R} w_\varepsilon \, dx > 0,$$

and define $U_\varepsilon = \text{span } \{1, w_\varepsilon\}$, $\hat{L}_\varepsilon = U_\varepsilon^\perp$, $\hat{H}_\varepsilon = H^1(\Omega_\varepsilon) \cap \hat{L}_\varepsilon$ and $P_\varepsilon : L^2(\Omega_\varepsilon) \to U_\varepsilon$ as the orthogonal projection

$$P_\varepsilon f = |\Omega_\varepsilon|^{-1} \int_{\Omega_\varepsilon} f \, dx + w_\varepsilon \int_{\Omega_\varepsilon} f w_\varepsilon \, dx, \ |\cdot| = \text{Lebesgue measure} .$$

By writing $u = \alpha + \beta w_\varepsilon + \hat{u}$ $(\hat{u} \in \hat{L}_\varepsilon, \ \alpha, \beta \in \mathbb{R})$, $\Delta u + f(u) = 0$ is equivalent to

$$P_\varepsilon(\Delta(\alpha + \beta w_\varepsilon + \hat{u}) + f(\alpha + \beta w_\varepsilon + \hat{u})) = 0 \tag{4}$$

$$(I - P_\varepsilon)(\Delta(\alpha + \beta w_\varepsilon + \hat{u}) + f(\alpha + \beta w_\varepsilon + \hat{u})) = 0. \tag{5}$$

If f is smooth and all its derivatives are bounded in $\mathbb{R}$, then it defines a smooth map from $H^1(\Omega_\varepsilon)$ to $L^2(\Omega_\varepsilon)$. Moreover, if $f(0) = 0$ and Lip f is small (less than the lower bound in (3) above), property (3) enables us to apply the implicit function theorem, obtaining a unique solution of (5) $\hat{u} = \hat{u}(\alpha, \beta, \varepsilon)$ for $\varepsilon \geqslant 0$, $\alpha, \beta \in \mathbb{R}$. $\hat{u}$ is smooth in α, β for each $\varepsilon \geqslant 0$ fixed. Substituting in (4) and writing the complete analytic expression for P_ε, we obtain the <u>bifurcation equations</u>:

$$G(\alpha, \beta, \varepsilon) \overset{\text{def}}{=} \int_{\Omega_\varepsilon} f(\alpha + \beta w_\varepsilon + \hat{u}(\alpha, \beta, \varepsilon)) \, dx = 0 \tag{6}$$

$$H(\alpha, \beta, \varepsilon) \overset{\text{def}}{=} -\beta \lambda_\varepsilon + \int_{\Omega_\varepsilon} f(\alpha + \beta w_\varepsilon + \hat{u}(\alpha, \beta, \varepsilon)) w_\varepsilon \, dx = 0 . \tag{7}$$

Thus, u is a stationary solution of (1) if and only if $u = \alpha + \beta w_\varepsilon + \hat{u}$, where $\hat{u} = \hat{u}(\alpha, \beta, \varepsilon)$ and (α, β) satisfy (6) and (7).

In Vegas [6] we show that G and H, together with all their derivatives in α, β are continuous in ε, and, therefore, it makes sense to study (6), (7) as a perturbation of the corresponding problem for $\varepsilon = 0$, that is, equation (1) on the separate domains Ω_0^L and Ω_0^R. In particular, if all the zeros of f are simple, to every locally constant solution of (1) on Ω_0 there corresponds, for ε sufficiently small, a solution of (1) on Ω_ε which is 'almost constant' on Ω_0^L and Ω_0^R. See Hale and Vegas [5] for details.

If f has a multiple zero, say $f(0) = f'(0) = 0$, we introduce a small parameter λ in f, so that we may compare the effects of variations in the function with those perturbations of the domain. The bifurcation equations are thus (smoothly) dependent on λ, and (6), (7) becomes a bifurcation problem with two equations, two unknowns (α,β) and two parameters (λ,ε), which is completely degenerate for $\lambda = \varepsilon = 0$.

If we call $\lambda_\varepsilon = \nu$ in (7) and consider it as an independent parameter, we can apply the scaling techniques described in Chow et al. [2] (slightly modified in our case due to the presence of the parameter ε), and obtain a finite family of bifurcation curves in the (λ,ν)-plane depending on ε: $\lambda = \lambda_j(\nu,\varepsilon)$. Resetting $\nu = \lambda_\varepsilon$, we find a finite set of curves $\lambda = \lambda_j(\lambda_\varepsilon,\varepsilon) \overset{def}{=} \tilde{\lambda}_j(\varepsilon)$ which divide a neighbourhood of (0,0) in the (λ,ε)-plane in sectors in each of which the number of stationary solutions of (1) is constant.

THEOREM. (Vegas [6]): Let the domains Ω_ε be symmetric with respect to the y-axis, and let $f(\lambda,u) = \lambda u - g(u)$, where $g(u) = au^p + O(|u|^{p+1})$ as $|u| \to 0$.

Then, if $g(-u) = -g(u)$ for all $u \in \mathbb{R}$ and $a > 0$, there are exactly three bifurcation curves (Fig. 2),

$$\lambda = 0$$

$$\lambda = \lambda_\varepsilon$$

$$\lambda = \frac{p}{p-1}\lambda_\varepsilon + o(\lambda_\varepsilon)$$

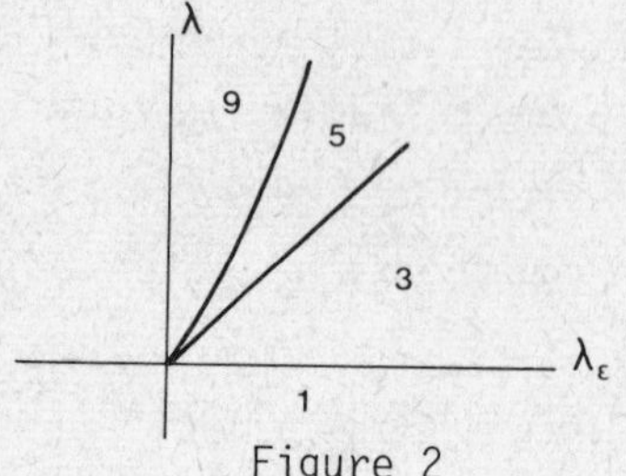

Figure 2

which divide a neighbourhood of $(\lambda,\varepsilon) = (0,0)$ into 4 regions, in each of which the number of solutions is as given in the figure. $\lambda = 0$ and $\lambda = \lambda_\varepsilon$ correspond to primary bifurcations at simple eigenvalues, while at $\lambda = \frac{p}{p-1}\lambda_\varepsilon + o(\lambda_\varepsilon)$ (or, equivalently, at $\lambda_\varepsilon = \frac{p-1}{p}\lambda + o(\lambda)$) two simultaneous secondary bifurcations occur.

Our purpose in this short communication is to show how, at least when $p = 3$, these results may be obtained in a different (and probably more intuitive) way, based on the methods of catastrophe theory, and, in particular, on Zeeman's work on the 'double-cusp' singularity.

2. APPLICATION OF THE METHOD OF UNFOLDINGS

The gradient (or variational) character of equation (1) implies that the bifurcation equations (6), (7) also have a gradient structure (Chow and Hale [1]). In our case, it is not difficult to see that, if F is a primitive of f, the 'potential'

$$V(\alpha,\beta,\varepsilon) = -\frac{1}{2}\int_{\Omega_\varepsilon} |\nabla(a + \beta w_\varepsilon + \hat{u}(\alpha,\beta,\varepsilon)|^2 \, dx$$

$$+ \int_{\Omega_\varepsilon} F(\alpha + \beta w_\varepsilon + \hat{u}(\alpha,\beta,\varepsilon)) \, dx$$

$$= -\frac{1}{2}\beta^2\lambda_\varepsilon - \frac{1}{2}\int_{\Omega_\varepsilon} |\nabla\hat{u}|^2 \, dx + \int_{\Omega_\varepsilon} F(\alpha+\beta w_\varepsilon+\hat{u}(\alpha,\beta,\varepsilon)) \, dx$$

satisfies

$$\frac{\partial V}{\partial \alpha}(\alpha,\beta,\varepsilon) = G(\alpha,\beta,\varepsilon); \; \frac{\partial V}{\partial \beta}(\alpha,\beta,\varepsilon) = H(\alpha,\beta,\varepsilon).$$

In particular, if $f(u) = \lambda u - g(u)$, where λ is a small parameter, then

$$V = V(\alpha,\beta,\lambda,\varepsilon) = \frac{1}{2}\lambda|\Omega_\varepsilon|\alpha^2 + \frac{1}{2}(\lambda-\lambda_\varepsilon)\beta^2 - \frac{1}{2}\int_{\Omega_\varepsilon} |\nabla\hat{u}(\alpha,\beta,\lambda,\varepsilon)|^2 \, dx$$

$$+ \int_{\Omega_\varepsilon} K(\alpha + \beta w_\varepsilon + \hat{u}(\alpha,\beta,\lambda,\varepsilon)) \, dx \qquad (8)$$

where $K'(u) = g(u)$.

If we now assume that $|\Omega_0| = 1$, so $|\Omega_0^L| = |\Omega_0^R| = \frac{1}{2}$ and $w_0 \equiv 1$ on Ω_0^R, $\equiv -1$ on Ω_0^L, then

$$V(\alpha,\beta,0,0) = \frac{1}{2}[K(\alpha+\beta) + K(\alpha-\beta)].$$

Let us consider the case $p = 3$, so that $g(u) = -u^3 + O(|u|^4)$. Then

$$V(\alpha,\beta,0,0) = \alpha^4 + \beta^4 + 6\alpha^2\beta^2 + O((|\alpha|+|\beta|)^5). \qquad (9)$$

By applying the standard techniques of singularity theory, one can show that $V(\alpha,\beta,0,0)$ has at $(0,0)$ a singularity of codimension 8 and that a universal unfolding for it is

$$W = \alpha^4 + \beta^4 + 6\alpha^2\beta^2 + A\alpha + B\beta + C\alpha^2 + D\beta^2 + E\alpha\beta + F\alpha^2 + G\beta^3 + H\alpha^2\beta^2 \tag{10}$$

where A,B,C,D,E,F,G and H are parameters. Let us recall the meaning of function W: if $U = U(\alpha,\beta)$ is a C^∞-small perturbation of $V(\alpha,\beta,0,0)$, then there exist a set of parameters $A_0, B_0,\ldots,H_0$ and a smooth change of variables $(\alpha,\beta) \to (\bar{\alpha},\bar{\beta})$ such that $U(\alpha,\beta) = W(\bar{\alpha},\bar{\beta},A_0,B_0,\ldots,H_0)$. This means that eight parameters are sufficient for the qualitative analysis of all possible perturbations of $V(\alpha,\beta,0,0)$. For a complete discussion of this concept, together with a special analysis of our function $V(\alpha,\beta,0,0)$, see Zeeman [8]; this singularity was called by him the 'double-cusp catastrophe'.

However, if we impose some symmetry hypotheses on our original problem, the number of parameters may be substantially reduced. In Vegas [6] we show that if the domains Ω_ε are symmetric with respect to the y-axis, then

$$G(\alpha,-\beta,\lambda,\varepsilon) = G(\alpha,\beta,\lambda,\varepsilon);\ H(\alpha,-\beta,\lambda,\varepsilon) = -H(\alpha,\beta,\lambda,\varepsilon),$$

and if, furthermore, we have $f(-u) = -f(u)$, then

$$G(-\alpha,\beta,\lambda,\varepsilon) = -G(\alpha,\beta,\lambda,\varepsilon);\ H(-\alpha,\beta,\lambda,\varepsilon) = H(\alpha,\beta,\lambda,\varepsilon).$$

In terms of the 'potential' V, this means that

$$V(-\alpha,\beta,\lambda,\varepsilon) = V(\alpha,\beta,\lambda,\varepsilon);\ V(\alpha,-\beta,\lambda,\varepsilon) = V(\alpha,\beta,\lambda,\varepsilon).$$

Therefore, $V(\alpha,\beta,\lambda,\varepsilon)$ has an extra property as a perturbation of $V(\alpha,\beta,0,0)$: like this function, it is even in both α and β. For this class of perturbations, we do not need the full 8-parameter set to construct the universal unfolding W; 3 parameters are enough: in fact, W is even in α and β if and only if $A = B = E = F = G = 0$. Hence

$$\tilde{W} = \alpha^4 + \beta^4 + 6\alpha^2\beta^2 + C\alpha^2 + D\beta^2 + H\alpha^2\beta^2 \tag{11}$$

is the universal unfolding for even perturbations of $V(\alpha,\beta,0,0)$.

Our objective is to analyse the solutions of the bifurcation equations (6),

(7), or, equivalently, the critical points of $V(\cdot,\cdot,\lambda,\varepsilon)$. This is equivalent to studying the critical points of W (or $\tilde{W}$). After some simplifications, the equations which these points must satisfy are

$$\alpha\left(\alpha^2 + \left(3 + \frac{H}{2}\right)\beta^2 + \frac{C}{2}\right) = 0 \tag{12}$$

$$\beta\left(\beta^2 + \left(3 + \frac{H}{2}\right)\alpha^2 + \frac{D}{2}\right) = 0. \tag{13}$$

The zero sets of (12) and (13) have the forms shown in Fig. 3.

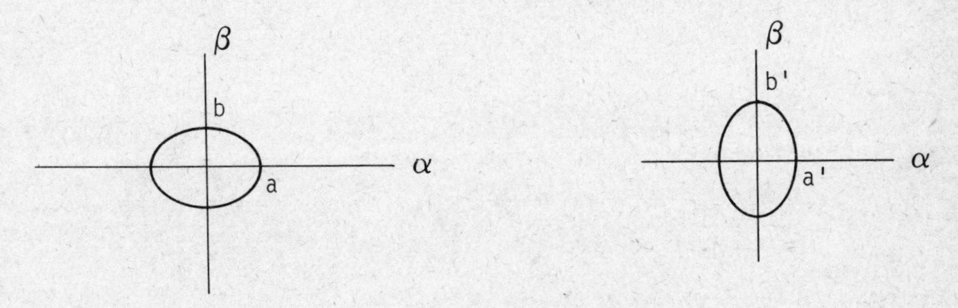

$$a = \left(-\frac{C}{2}\right)^{1/2};\ b = \left(\frac{-C/2}{3+\frac{H}{2}}\right)^{1/2}$$

(if $C < 0$)

$$a' = \left(\frac{-D/2}{3+\frac{H}{2}}\right)^{1/2};\ b' = \left(-\frac{D}{2}\right)^{1/2}$$

(if $D < 0$)

Figure 3

Therefore, the number of critical points of W can be 1, 3, 5 or 9, depending on the value of C, D, H as shown in Fig. 4.

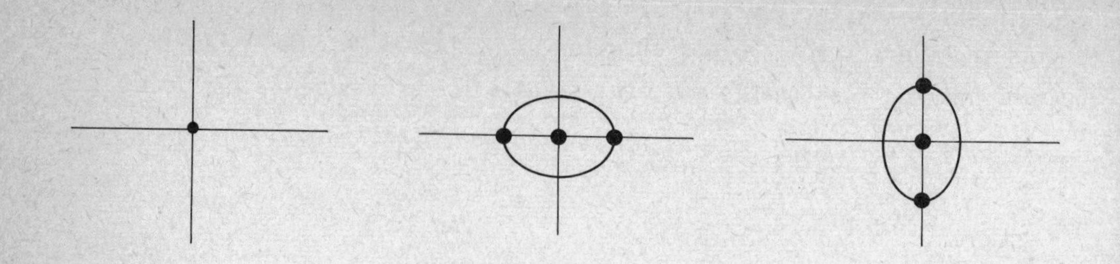

(1) $C \geqslant 0,\ D \geqslant 0$ (3) $C \geqslant 0,\ D < 0$ (3') $C < 0,\ D \geqslant 0$

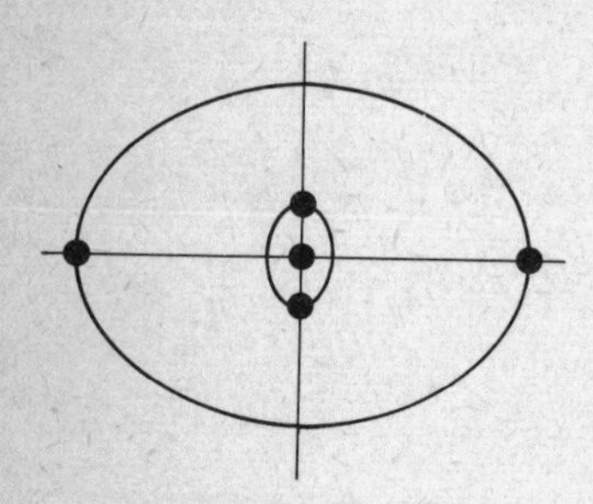

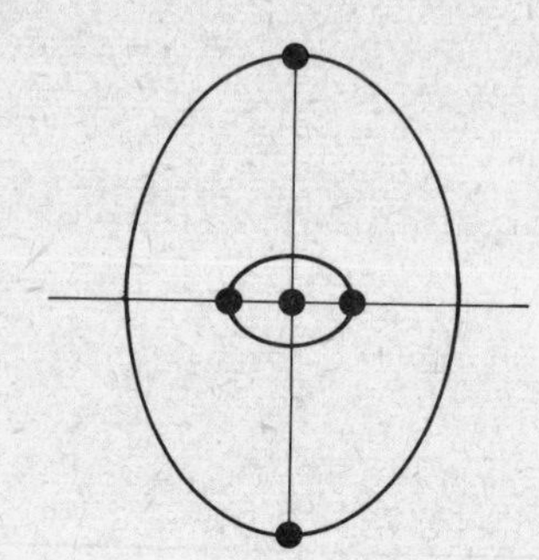

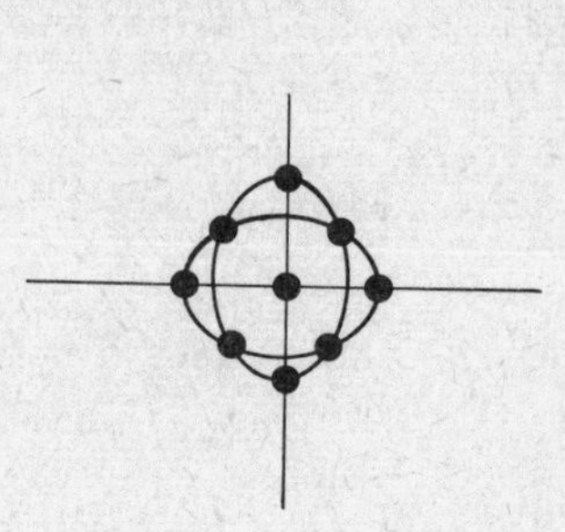

(5) $C < 0, D < 0, \frac{D}{C} < \frac{1}{3+\frac{H}{2}}$ (5') $C < 0, D < 0, \frac{D}{C} > 3+\frac{H}{2}$ (9) $C < 0,\ D < 0,\ \frac{1}{3+\frac{H}{2}} < \frac{D}{C} < 3 + \frac{H}{2}$

Figure 4

The bifurcations occur either when a new ellipse is 'created' by the change of sign of C or D, or when the ellipses are tangential (Fig. 5).

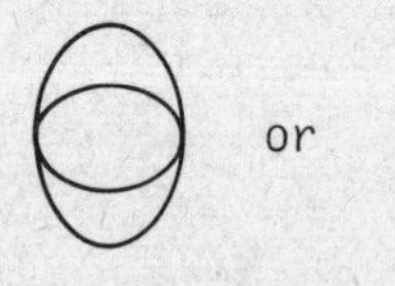

or

Figure 5

These situations happen when $\frac{D}{C} = \frac{1}{3 + \frac{H}{2}}$ and $\frac{D}{C} = 3 + \frac{H}{2}$, respectively. Thus, we see that parameter H, as long as it remains small, plays no fundamental role in the qualitative structure of these bifurcations, for it only appears in the term $3 + \frac{H}{2}$. The important parameters are C and D.

Now, what interpretation do C and D have with regard to our original P.D.E? If we look at the formula for $V(\alpha,\beta,\lambda,\varepsilon)$, and take into account that

$$||\hat{u}(\alpha,\beta,\lambda,\varepsilon)||_{H^1(\Omega_\varepsilon)} = O((|\alpha| + |\beta|)^3)$$

(this is shown in Vegas [6]), then

$$V(\alpha,\beta,\lambda,\varepsilon) = \frac{1}{2}\lambda|\Omega_\varepsilon|\alpha^2 + \frac{1}{2}(\lambda-\lambda_\varepsilon)\beta^2 + O((|\alpha| + |\beta|)^3).$$

Thus, C should be associated with $\frac{1}{2}\lambda|\Omega_\varepsilon|$, and D with $\frac{1}{2}(\lambda-\lambda_\varepsilon)$. Since $H = O(\varepsilon)$, the bifurcations occur when

$$1 - \frac{\lambda_\varepsilon}{\lambda|\Omega_\varepsilon|} = \frac{1}{3+O(\varepsilon)}\text{, that is, } \lambda = \frac{3}{2}\lambda_\varepsilon + o(\lambda_\varepsilon)$$

and when

$$1 - \frac{\lambda_\varepsilon}{\lambda|\Omega_\varepsilon|} = 3 + O(\varepsilon),$$

which is impossible if $\lambda > 0$.

Therefore, if λ remains fixed and ε increases so that λ_ε also increases, we start (when $\varepsilon = 0$) in the region (9) (since $\frac{D}{C} = 1 - \frac{\lambda_\varepsilon}{\lambda} + O(\varepsilon)$), we continue to region (5) when $\lambda = \frac{3}{2}\lambda_\varepsilon + o(\lambda_\varepsilon)$, to end up in (3) after crossing the value $\lambda_\varepsilon = \lambda$.

These computations completely confirm the results proved in Vegas [6] by the scaling method, and give a geometric explanation of the change in the number of solutions. Moreover, since one can show (Vegas [7]) that the stability of the stationary solutions of (1) can be deduced from their character of minimum or saddle points of the 'potential' V, and this function is equivalent (modulo change of variables) to the unfolding $\tilde{W}$, we can analyse these stability properties directly by computing the Jacobian of W at the critical points, obtaining the results shown in Fig. 6 (u means 'unstable', and s means 'stable'). In a more conventional way, we can draw the bifurcation diagram (Fig. 7).

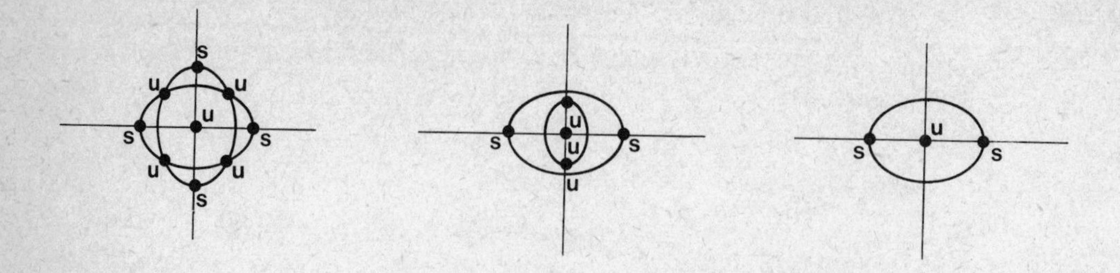

Figure 6

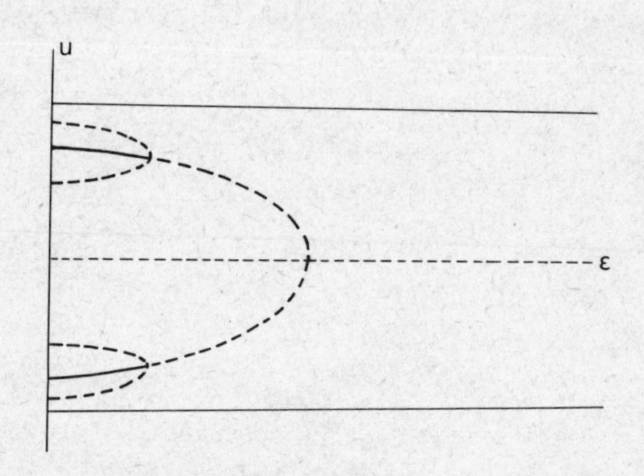

(—— = stable ; ------ = unstable)

Figure 7

3. FINAL REMARKS

1. A complete analysis of the double cusp with symmetry may be found in Golubitsky and Schaeffer [4]; their viewpoint is different from ours (and much more general) in that they do not use the gradient structure of the problem while assigning great importance to the bifurcation parameters. See also Golubitsky and Schaeffer [3].

2. The analysis presented here should be considered only as a heuristic explanation of the working mechanisms of some bifurcations whose existence has been proved elsewhere. Indeed, one of the major problems involved is

the continuous but nonsmooth dependence of the bifurcation functions G and H upon ε. Thus, application of the concepts and methods of singularity theory to nonsmooth unfoldings becomes risky, to say the least. However, in the spirit of Golubitsky and Schaeffer [3], we believe that any attempt to put a bifurcation problem within the framework of singularity theory and to find a universal unfolding for it is definitely worth the effort.

REFERENCES

[1] S.N. Chow and J.K. Hale, Methods of Bifurcation Theory, Grundlehren der Mathematischen Wissenschaften, vol. 251, Springer-Verlag, New York (1982).

[2] S.N. Chow, J.K. Hale and J. Mallet-Paret, Applications of generic bifurcation, II, Arch. Rat. Mech. Anal. 62 (3), 209-235 (1976).

[3] M. Golubitsky and D. Schaeffer, A theory for imperfect bifurcation via Singularity Theory. Comm. Pure Appl. Math. 32, 21-98 (1979).

[4] M. Golubitsky and D. Schaeffer, Imperfect bifurcation in the presence of symmetry, Commun. Math. Phys. 67, 205-232 (1979).

[5] J.K. Hale and J. Vegas, A nonlinear parabolic equation with varying domain, Arch. Rat. Mech. Anal. (to appear).

[6] J. Vegas, Bifurcations caused by perturbing the domain in an elliptic equation, J. Differential Equations (to appear).

[7] J. Vegas, Linear stability of bifurcating branches of equilibria (to appear).

[8] E.C. Zeeman, The Umbilic Bracelet and the Double-cusp Catastrophe, in Structural Stability, the Theory of Catastrophes and Applications in the Sciences, Springer Lecture Notes in Mathematics, vol. 525, 328-366 (1976).

J.M. Vegas
Departamento de Ecuaciones Funcionales,
Facultad de Ciencias Matemáticas,
Universidad Complutense de Madrid,
Madrid-3,
Spain

L VÉRON

Behaviour of solutions of nonlinear elliptic equations near a singularity of codimension 2

1. ISOTROPIC SINGULARITIES IN $\mathbb{R}^2$

Let Ω be an open subset of $\mathbb{R}^2$ containing 0, f a continuous function defined in Ω and g a continuous nondecreasing function defined in $\mathbb{R}$. The main problem that we consider in this section is the description near 0 of any $u \in C^1(\Omega-\{0\})$ satisfying

$$-\Delta u + g(u) = f \tag{1.1}$$

in the sense of distributions in $\Omega - \{0\}$. Recently many results concerning the behaviour of any singular solution in $\mathcal{D}'(\Omega-\{0\})$ of

$$-\Delta u + u\,|u|^{q-1} = 0 \tag{1.2}$$

have been given by Veron in [9]. In particular he gave a full description of the isolated singularities when $q \geqslant 3$.

THEOREM 1.1. Suppose $q \geqslant 3$ and $u \in C^2(\Omega-\{0\})$ satisfies (1.2) in $\mathcal{D}'(\Omega-\{0\})$; then

(i) either $|x|^{2/(q-1)}\,u(x)$ converges to a constant which can take only two values $\pm\,(2/(q-1))^{2/(q-1)}$, as x tends to 0,

(ii) or $u(x)/\text{Log}(1/|x|)$ converges to some constant c which can take any value, and u satisfies in $\mathcal{D}'(\Omega)$

$$-\Delta u + u\,|u|^{q-1} = 2\pi c\,\delta_0. \tag{1.3}$$

In particular u is nonsingular if c = 0.

Moreover there truly exist singularities of the two types described in Theorem 1.1. For example the function u_s defined in $\mathbb{R}^2-\{0\}$ by

$$u_s(x) = (2/(q-1))^{2/(q-1)}\,|x|^{-2/(q-1)} \tag{1.4}$$

is a singular solution of (1.2); and for any real c there exists a unique (in some suitable class) u satisfying (1.3) in $\mathcal{D}'(\mathbb{R}^2)$.

When $1 < q < 3$ the singularities of (1.2) are not necessarily isotropic: if we consider the following nonlinear eigenvalue problem on S^1

$$- \frac{\partial^2 v}{\partial \theta^2} + v \ |v|^{q-1} = \left(\frac{2}{q-1}\right)^2 v, \tag{1.5}$$

then there exist solutions of (1.5) changing sign. For such a solution v the function u defined in $\mathbb{R}^2 - \{0\}$ by

$$u(x) = |x|^{-2/(q-1)} \ v(x/|x|) \tag{1.6}$$

is a solution of (1.2) with a nonisotropic behaviour at 0. However, all the nonnegative singular solutions of (1.2) have an isotropic behaviour near 0 for $q > 1$ (see [9] Theorem 4.1):

(i) either $|x|^{2/(q-1)}$ u(x) converges to $(2/(q-1))^{2/(q-1)}$ as x tends to 0

(ii) or $u(x)/\mathrm{Log}(1/|x|)$ converges to some nonnegative constant c which can take any value and u satisfies (1.3) in $\mathcal{D}'(\Omega)$.

For a more general nonlinearity g we have the following result of isotropy obtained by Vazquez and Veron [7].

THEOREM 1.2. Suppose f is continuous in Ω, g is continuous and nondecreasing on $\mathbb{R}$ and $u \in C^1(\Omega - \{0\})$ is a solution of (1.1) in $\mathcal{D}'(\Omega - \{0\})$ satisfying $\lim_{x\to 0} x\, u(x) = 0$. Then $u(x)/\mathrm{Log}(1/|x|)$ admits a limit in $\mathbb{R} \cup \{+\infty, -\infty\}$; if the limit is zero u can be extended as a C^1 function in Ω satisfying (1.1) in $\mathcal{D}'(\Omega)$.

Remark 1.1. The condition $\lim_{x\to 0} x\, u(x) = 0$ is fulfilled as soon as g satisfies the growth condition ([10])

$$\lim_{|r|\to +\infty} g(r)/r^3 = +\infty. \tag{1.7}$$

Moreover we deduce from Theorem 1 of [4] that if $g(r)/r$ is bounded for $r > 1$ (resp. $r < -1$), then the limit of $u(x)/\mathrm{Log}(1/|x|)$ at 0 is $< +\infty$ (resp. $> -\infty$). As for the condition $\lim_{x\to 0} x\, u(x) = 0$, it is optimal when $g = 0$: the function

$$(x_1,x_2) \mapsto \frac{x_1}{x_1^2 + x_2^2}$$

is harmonic in $\mathbb{R}^2-\{0\}$ and nonisotropic at 0.

If we suppose that g satisfies the following condition:

$$\left.\begin{array}{l} \text{for any } a > 0,\ e^{-ar}\ g(r) \text{ and } e^{ar}\ g(-r) \text{ admit} \\ \text{a limit in } \mathbb{R} \cup \{+\infty, -\infty\} \text{ as } r \text{ tends to } +\infty, \end{array}\right\} \tag{1.8}$$

we can give an improvement of Theorem 1.2 in introducing the exponential orders of growth of g:

$$a_g^+ = \sup \{a \in \mathbb{R}^+ : \lim_{r \to +\infty} e^{-ar}\ g(r) = +\infty\}, \tag{1.9}$$

$$a_g^- = \sup \{a \in \mathbb{R}^+ : \lim_{r \to -\infty} e^{ar}\ g(r) = -\infty\}. \tag{1.10}$$

THEOREM 1.3. Under the hypotheses of Theorem 1.2 and (1.8), the limit c of $u(x)/\mathrm{Log}(1/|x|)$ at 0 satisfies

$$-2/a_g^- \leqslant c \leqslant 2/a_g^+. \tag{1.11}$$

If c is finite, $g(u) \in L^1_{loc}(\Omega)$ and u satisfies in $\mathcal{D}'(\Omega)$:

$$-\Delta u + g(u) = f + 2\pi c\ \delta_0. \tag{1.12}$$

Remark 1.2. If $a_g^+ = 0$ (resp. $a_g^- = 0$) we set $2/a_g^+ = +\infty$ (resp. $-2/a_g^- = -\infty$). As for the existence of solutions of (1.12), for any c finite satisfying (1.11), any $f \in L^1_{loc}(\Omega)$ with $df \in L^1(\Omega)$ (where $d(\cdot) = \mathrm{dist}\ (\cdot,\partial\Omega)$) and any $\phi \in L^\infty(\partial\Omega)$ there exists a unique u satisfying

$$\left.\begin{array}{l} -\Delta u + g(u) = 2\pi c\ \delta_0 + f, \\ u_{|\partial\Omega} = \phi, \end{array}\right\} \tag{1.13}$$

in the following weak sense: we say that u solves (1.13) if $u \in L^1(\Omega)$, $dg(u) \in L^1(\Omega)$ and

$$\left.\begin{aligned}&\int_\Omega (-u\Delta\zeta + g(u)\zeta)\ dx = 2\pi c\zeta(0) + \int_\Omega f\zeta\ dx - \int_{\partial\Omega} \phi \frac{\partial\zeta}{\partial\nu}\ d\sigma,\\ &\text{for any } \zeta \in W^{2,\infty}(\Omega) \cap W_0^{1,\infty}(\Omega).\end{aligned}\right\} \tag{1.14}$$

If $a_g^+ = a_g^- = +\infty$, c is necessarily zero and u is nonsingular in Ω. We shall see a generalization of that fact in the next section.

Idea of the proof of Theorem 1.2. Without loss of generality we suppose

$$\Omega \supset \{x \in \mathbb{R}^2 : \ |x| \leqslant 1\}.$$

Let (r,θ) be the polar coordinates in $\mathbb{R}^2$, $r \in [0,1]$, $\theta \in S^1$; we set

$$\bar{u}(r) = \frac{1}{2\pi}\int_{S^1} u(r,\theta)\ d\theta, \quad \bar{f}(r) = \frac{1}{2\pi}\int_{S^1} f(r,\theta)\ d\theta,$$

$v(t,\theta) = u(e^{-t},\theta)$, $\bar{v}(t) = \bar{u}(e^{-t})$, $h(t,\theta) = f(e^{-t},\theta)$ and $\bar{h}(t) = \bar{f}(e^{-t})$. As u is C^1 in $\Omega-\{0\}$, we deduce from regularity results on elliptic equations that it belongs to $W^{2,p}_{loc}(\Omega-\{0\})$ for any $p < +\infty$; hence equation (1.1) is satisfied almost everywhere and v satisfies

$$\frac{\partial^2 v}{\partial t^2} + \frac{\partial^2 v}{\partial\theta^2} = e^{-2t}(g(v)-h), \tag{1.15}$$

a.e. in $[0, +\infty) \times S^1$.

Step 1

$$\int_0^{+\infty} \|v(t,\cdot) - \bar{v}(t)\|_{L^\infty(S^1)}\ dt < +\infty.$$

From the monotonicity of g we have

$$\int_{S^1} (g(v) - \frac{1}{2\pi}\int_{S^1} g(v)\ d\theta)(v-\bar{v})\ d\sigma \geqslant 0.$$

Moreover

$$-\int_{S^1} \frac{\partial^2}{\partial\theta^2}(v-\bar{v})(v-\bar{v})\ d\theta \geqslant \int_{S^1} (v-\bar{v})^2\ d\theta,$$

so we deduce from (1.15)

$$\int_{S^1}(v-\bar{v})\,\frac{\partial^2}{\partial t^2}(v-\bar{v})\;d\theta - \int_{S^1}(v-\bar{v})^2\;d\theta \geqslant e^{-2t}\int_{S^1}(\bar{h}-h)(v-\bar{v})\;d\theta. \qquad (1.16)$$

If we set $X(t) = (\int_{S^1}(v-\bar{v})^2(t)\;d\theta)^{1/2}$ we get

$$\frac{d^2X}{dt^2} - X \geqslant e^{-2t}\;\|h-\bar{h}\|_{L^\infty((0,+\infty)\times S^1)}. \qquad (1.17)$$

From classical comparison results and the fact that $X(t) = o(e^t)$ we deduce that there exists $c > 0$ such that

$$X(t) \leqslant ce^{-t}, \qquad (1.18)$$

for $t > 0$. Set

$$H(t) = \frac{1}{2}\,X^2(t) + \int_0^t\int_0^s\int_{S^1} e^{-2\sigma}(h-\bar{h})(v-\bar{v})\;d\theta\;d\sigma\;ds.$$

The function H is convex on $[0,+\infty)$ and as $H(t)/t$ is bounded on $[0,+\infty)$ it is the same with $\frac{dH}{dt}(t)$. We deduce from (1.16) that

$$\frac{d^2H}{dt^2} + \int_{S^1}(v-\bar{v})^2\;d\theta \geqslant c\;\|v-\bar{v}\|^2_{L^\infty(S^1)}, \qquad (1.19)$$

for $t > 0$ and some $c > 0$. Integrating (1.19) yields the desired estimate.

<u>Step 2</u>

$$\lim_{t\to+\infty}\;\|v(t,\cdot) - \bar{v}(t)\|_{L^\infty(S^1)} = 0.$$

We construct a sequence $\{t_n\}$ such that

(i) $1 \leqslant t_{n+1} - t_n \leqslant 3$ for $n \geqslant 0$,

(ii) $\lim_{n\to+\infty}\;\|v(t_n,\cdot) - \bar{v}(t_n)\|_{L^\infty(S^1)} = 0.$

Set $\varepsilon > 0$; there exists an integer n_0 such that for $n \geqslant n_0$

(iii) $\|v(t_n,\cdot) - \bar{v}(t_n)\|_{L^\infty(S^1)} < \varepsilon$,

(iv) $\int_{t_n}^{t_{n+1}} \int_{t_n}^{s} e^{-2\sigma} \|h\|_{L^\infty(S^1)} \, d\sigma < \varepsilon$.

Let ψ_n be the solution of

$$\left.\begin{aligned} &\frac{\partial^2\psi_n}{\partial t^2} + \frac{\partial^2\psi_n}{\partial\theta^2} = e^{-2t}\,(g(\psi_n) - \|h\|_{L^\infty(S^1)}) \text{ in } (t_n,t_{n+1})\times S^1, \\ &\psi_n(t_n,\cdot) = \bar{v}(t_n) + \varepsilon;\quad \psi_n(t_{n+1},\cdot) = \bar{v}(t_{n+1}) + \varepsilon. \end{aligned}\right\} \tag{1.20}$$

Comparing ψ_n and v in $(t_n,t_{n+1}) \times S^1$, we get

$$0 \leqslant \psi_n - v \leqslant 3\varepsilon \quad \text{in } (t_n,t_{n+1}) \times S^1. \tag{1.21}$$

As ψ_n is independent of θ, (1.21) is also valid if we replace v by $\bar{v}$ and we get

$$-3\varepsilon \leqslant v\text{-}v \leqslant 3\varepsilon, \text{ in } (t_n,t_{n+1}) \times S^1, \tag{1.22}$$

which implies $\lim\limits_{t\to+\infty} \sup \|v(t,.) - \bar{v}(t)\|_{L^\infty(S^1)} \leqslant 3\varepsilon$.

<u>Step 3</u>. $u(x)/\text{Log}(1/|x|)$ is <u>converging</u>. First we note that if $\{\bar{u}(r)\}$ remains bounded it is the same with $\{u(x)\}$; hence $\lim\limits_{x\to 0} u(x)/\text{Log}(1/|x|) = 0$. So we suppose that $\{\bar{u}(r)\}$ is unbounded and that there exists a sequence $\{r_n\}$ going to 0 such that $\lim\limits_{n\to+\infty} \bar{u}(r_n) = +\infty$ (for example). Hence $\lim\limits_{n\to+\infty} u(r_n,\theta) = +\infty$ uniformly on S^1. From the maximum principle $\lim\limits_{x\to 0} u(x) = +\infty$. Set

$$\lambda = \lim_{t\to+\infty} \left(\frac{1}{2\pi}\int_{S^1} g(v(t,\theta)\, d\theta - \bar{h}(t))\right); \lambda \in \mathbb{R} \cup \{+\infty\};$$

and $\bar{v}$ satisfies

$$\frac{d^2\bar{v}}{dt^2} = e^{-2t}\left(\frac{1}{2\pi}\int g(v(t,\theta)\, d\theta - \bar{h}(t))\right), \tag{1.23}$$

in $(0,+\infty)$. In order to prove the convergence of $u(x)/\text{Log}(1/|x|)$, it is sufficient to prove the one of $\bar{v}(t)/t$ as t tends to $+\infty$.

If $\lambda > 0$ (or $\lambda = +\infty$) then $\bar{v}$ becomes convex as $t \geqslant T$, so $\bar{v}(t)/t$ admits a limit as t tends to $+\infty$.

If $\lambda \leqslant 0$, then λ is finite, so $\frac{d^2\bar{v}}{dt^2}$ is integrable on $[0, +\infty)$. Hence $\frac{d\bar{v}}{dt}(t)$ admits a finite limit as t tends to $+\infty$ and it is the same with $\bar{v}(t)/t$.

Step 4. End of the proof of Theorem 1.2. Suppose $\lim_{x\to 0} u(x)/\mathrm{Log}(|x|) = 0$ and set μ the solution of

$$\left.\begin{aligned} -\Delta\mu &= (h-g(0))^+ \text{ in } \{x : \ |x| < 1\} \\ \mu(x) &= u^+(x) \quad \text{for } |x| = 1. \end{aligned}\right\} \tag{1.24}$$

For any $\varepsilon > 0$, $u(x)$ is majorized in $\{x : 0 < |x| \leqslant 1\}$ by $\mu(x) + \varepsilon\,\mathrm{Log}(1/|x|)$. If we let $\varepsilon \to 0$ we get $u(x) \leqslant \mu(x)$. In the same way u is minorized in $\{x : |x| \leqslant 1\}$ by a continuous function. Following Theorem 1 of Serrin [4] we deduce that (1.1) holds in $\mathcal{D}'(\Omega)$ and $u \in C^1(\Omega)$.

2. REMOVABLE SINGULARITIES OF CODIMENSION 2

In this section Ω is an open subset of $\mathbb{R}^N$, $N \geqslant 1$, Σ a C^1 closed submanifold of Ω with codimension d and g a continuous function defined in $\Omega \times \mathbb{R}$. The problem is to know under what condition on g any u belonging to $C^1(\Omega-\Sigma)$ and satisfying

$$-\Delta u + g(\cdot,u) = 0, \tag{2.1}$$

in $\mathcal{D}'(\Omega-\Sigma)$ can be extended as a C^1 function in Ω satisfying (2.1) in $\mathcal{D}'(\Omega)$.

For the case in which g has a power-like growth an answer has been given by Brézis and Veron [1] for an isolated singularity and Veron [8] for a singularity of codimension $d > 2$. The most general result in this direction is [8].

THEOREM 2.1. Suppose we have

$$\left.\begin{aligned} &\liminf_{r\to+\infty} \ g(x,r)/r^{d/(d-2)} > 0, \\ &\limsup_{r\to-\infty} \ g(x,r)/|r|^{d/(d-2)} < 0, \end{aligned}\right\} \tag{2.2}$$

uniformly with respect to $x \in \Omega$ and

$$\{x \in \mathbb{R}^N \mid \text{dist}(x,\Sigma) < \eta\} \subset \Omega$$

for some $\eta > 0$; then any $u \in C^1(\Omega-\Sigma)$ satisfying (2.1) in $\mathcal{D}'(\Omega-\Sigma)$ can be extended as a C^1 function in Ω satisfying (2.1) in $\mathcal{D}'(\Omega)$.

In the case of compact submanifolds of codimension 2 Vazquez and Veron proved the following result [7].

THEOREM 2.2. Suppose Σ is a C^1 compact submanifold of Ω of codimension 2 and g satisfies

$$\left.\begin{aligned} &\liminf_{r\to+\infty} e^{-ar}\ g(x,r) = +\infty, \\ &\limsup_{r\to-\infty} e^{ar}\ g(x,r) = -\infty, \end{aligned}\right\} \tag{2.3}$$

uniformly with respect to $x \in \Omega$, for any $a > 0$; then any $u \in C^1(\Omega-\Sigma)$ satisfying (2.1) in $\mathcal{D}'(\Omega-\Sigma)$ can be extended as a C^1 function in Ω satisfying (2.1) in $\mathcal{D}'(\Omega)$.

Proof of Theorem 2.2. Without loss of generality we can suppose that Ω is relatively compact and that u is bounded in $\Omega - S$, where S is any neighbourhood of Σ in Ω.

Step 1. For any $\varepsilon > 0$ there exists $B_\varepsilon > 0$ such that

$$|u(x)| \leqslant \varepsilon \text{ Log}(1/\text{dist}(x,\Sigma)) + B_\varepsilon, \tag{2.4}$$

for any $x \in \Omega-\Sigma$.

We set $d(\cdot) = \text{dist}(\cdot,\Sigma)$ and we choose a neighbourhood S of Σ such that for any $y \in S$ the ball $\{x \in \mathbb{R}^N : |x-y| \leqslant d(y)\}$ is included in Ω. From (2.3), for any $\varepsilon > 0$ there exist a and $C > 0$ such that

$$g(x,s) \geqslant a\ e^{2s/\varepsilon} - C, \tag{2.5}$$

for any $x \in \Omega$ and $s \geqslant 0$. We fix $x_0 \in S - \Sigma$ and set

$$\psi(x) = \lambda \text{ Log}(1/(R^2 - |x-x_0|^2)) + \mu \text{ for } 0 \leqslant |x-x_0| < R,$$

where $R < d(x_0)$ and λ and μ are to be chosen such that $-\Delta\psi + a\, e^{2\psi/\varepsilon} \geq C$ in $\{x \in \mathbb{R}^N : |x-x_0| < R\}$. For the sake of simplicity we set

$$\psi(r) = \lambda \operatorname{Log}(1/(R^2-r^2)) + \mu$$

and we get

$$-\Delta\psi + a\, e^{2\psi/\varepsilon} = -\frac{4\lambda R^2}{(R^2-r^2)^2} + a\, \frac{e^{2\mu/\varepsilon}}{(R^2-r^2)^{2\lambda/\varepsilon}}. \tag{2.6}$$

If we take $\lambda = \varepsilon$ and $\mu = \frac{\varepsilon}{2} \operatorname{Log} \{(C+4\varepsilon)R^2/a\}$, we get

$$-\Delta\psi + a\, e^{2\psi/\varepsilon} \geq C, \tag{2.7}$$

in $\{x \in \mathbb{R}^N : |x-x_0| < R\}$. Moreover we have

$$-\Delta u + a\, e^{2u/\varepsilon} \leq C, \tag{2.8}$$

a.e. in $\{x \in \mathbb{R}^N : u(x) \geq 0\}$. As $\lim_{r\to R} \psi(r) = +\infty$, we deduce from Kato's inequality and the maximum principle that $u(x) \leq \psi(x)$ for $0 \leq |x-x_0| < R$. In particular

$$u(x_0) \leq \psi(x_0) = \varepsilon \operatorname{Log}(1/R) + \frac{\varepsilon}{2} \operatorname{Log} \{(C+4\varepsilon)/a\}. \tag{2.9}$$

If we let $R \to d(x_0)$, we deduce (2.4) for u^+. In the same way we get a minorization for u.

<u>Step 2</u>. <u>End of the proof</u>. We suppose $N > 2$ and we define the following harmonic function in $\Omega - \Sigma$:

$$\mu(x) = \int_\Sigma \frac{d\sigma}{|x-\sigma|^{N-2}}. \tag{2.10}$$

Using local coordinates on Σ it is not difficult to see that there exist two positive constants such that

$$\mu(x) \geq C \operatorname{Log}(1/d(x)) - D, \tag{2.11}$$

for $x \in \Omega - \Sigma$ (see [7], Lemma 2.3). As we have

$$-\Delta u + ae^{u} \leqslant C, \tag{2.12}$$

a.e. on $\{x \in \mathbb{R}^N : u(x) \geqslant 0\}$, for some a and $C > 0$ we fix $\eta > 0$ and k such that $ae^k \geqslant C$ and we set

$$v_\eta(x) = \eta\mu(x) + k + \|u^+\|_{L^\infty(\partial\Omega)}. \tag{2.13}$$

The function v_η satisfies

$$-\Delta v_\eta + ae^{v_\eta} \geqslant C. \tag{2.14}$$

As $(u-v_\eta)^+$ vanishes in some neighbourhood of $\partial\Omega$, we deduce from Kato's inequality as in [1] that

$$u(x) \leqslant v_\eta(x) \tag{2.15}$$

in $\Omega - \Sigma$. If we let $\eta \to 0$ we obtain a uniform upper bound for u in Ω. In the same way u is uniformly minorized in Ω and we deduce from Theorem 1.1 of [3] that u can be extended in Ω as a C^1 solution of (2.1) in $\mathcal{D}'(\Omega)$.

The proof for N = 2 is left to the reader.

REFERENCES

[1] H. Brézis and L. Veron, Removable singularities of some nonlinear elliptic equations, Archs. ration. Mech. Anal. 75, 1-6 (1980).

[2] J. Nitsche, Über die isolierten Singularitäten der Lösungen von $\Delta u = e^u$, Math. Z. 69, 316-324 (1957).

[3] J. Serrin, Local behaviour of solutions of quasilinear equations, Acta Math. 111, 247-302 (1964).

[4] J. Serrin, Isolated singularities of solutions of quasilinear equations, Acta Math. 113, 219-240 (1965).

[5] J.L. Vazquez, On a semilinear equation in $\mathbb{R}^N$ involving bounded measures, Proc. Royal Soc. Edinburgh (to appear).

[6] J.L. Vazquez and L. Veron, Singularities of elliptic equations with an exponential nonlinearity (to appear).

[7] J.L. Vazquez and L. Veron, Isolated singularities of some semilinear elliptic equations (to appear).

[8] L. Veron, Singularités éliminables d'équations elliptiques nonlinéaires, J. Diff. Equ. 41, 87-95 (1981).

[9] L. Veron, Singular solutions of some nonlinear elliptic equations, Nonlinear Anal. 5, 225-242 (1981).

[10] L. Veron, Global behaviour and symmetry properties of singular solutions of nonlinear elliptic equations (to appear).

L. Veron
Département de Mathématiques,
Faculté des Sciences,
Parc de Grandmont,
37200 Tours,
France